Animal Social Networks

Animal Social Networks

EDITED BY

Jens Krause

Professor of Fish Biology and Ecology,
Humboldt University, Berlin, Germany and
Professor of Biology and Ecology of Fishes,
Leibniz-Institute of Freshwater Ecology and Inland Fisheries, Berlin, Germany

Richard James

Senior Lecturer,
University of Bath, UK

Daniel W. Franks

Reader in Complex Systems,
University of York, UK

Darren P. Croft

Associate Professor of Animal Behaviour,
University of Exeter, UK

OXFORD
UNIVERSITY PRESS

Great Clarendon Street, Oxford, OX2 6DP,
United Kingdom

Oxford University Press is a department of the University of Oxford.
It furthers the University's objective of excellence in research, scholarship,
and education by publishing worldwide. Oxford is a registered trade mark of
Oxford University Press in the UK and in certain other countries

Published in the United States of America by Oxford University Press
198 Madison Avenue, New York, NY 10016, United States of America

British Library Cataloguing in Publication Data
Data available

Library of Congress Control Number: 2014942172

ISBN 978–0–19–967905–8

Foreword

Thirty five years ago, David Attenborough concluded his landmark television series *"Life on Earth"* with a programme on humans. For its title he chose *"The Compulsive Communicators"*, suggesting that our passion to communicate is as much a key to our success as fins for fish or feathers for birds. This compulsion is especially evident in our modern society, not just in gossiping groups but also lone individuals, busily networking even as they walk or drive around town.

This inspiring collection of chapters reveals that although fish, birds and other animals might not share such compulsion to communicate, nevertheless their associations and interactions create intricate social networks which play a central role in their lives too. The book is a marvellous introduction to a newly discovered world. It shows how "network thinking" - involving nodes, edges, closeness and clustering – can help illuminate many aspects of social life, from the collective behaviour of groups, as shoals and flocks form and move in extraordinary synchrony, to the evolution of cooperation, and to how ideas and diseases are transmitted through populations.

The discovery of these social networks is partly a result of new techniques. Individuals can now easily be tracked with loggers and their positions and associations mapped in space and time. This often results in millions of data points, which would have been overwhelming just a few years ago, but these can now be readily analysed, thanks to advances in computational power and methodology. The resulting network diagrams and metrics reveal how individuals are connected to others not only by their direct interactions with mates, family and immediate neighbours, but via linked chains to distant individuals who they will never meet, links which nevertheless have important implications both for their own welfare and for population biology.

However, the importance of networks for social behaviour has emerged from a change in thinking, too. The reductionist, bottom up, approach of behavioural ecology emphasised how the behaviour of individuals, selected to maximise their inclusive fitness, led to various outcomes for social organisation. But as Robert Hinde pointed out, forty years ago, we need to consider the two-way interaction between individuals and social systems. Social systems are just as much part of the stage on which individuals play their behaviour as ecological factors, so they are an essential part of the selective landscape which determines how decisions are moulded by natural selection. Individual decision making influences social organisation, but the arrow also goes the other way, feeding back from social systems to influence how individual might best behave.

The book serves as an excellent primer for those who want to master the new terminology and understand how networks are measured, analysed and interpreted. As many of the authors point out, a static depiction of a network tells only part of the story; interactions change through time, dynamics which are often key to understanding how disease and information spreads. Spread might take place through direct interaction (grooming or fighting, for example) or merely by spatial association (individuals using the same sites). The nodes in a network might change behaviour too, as individuals are manipulated by parasites or react to neighbours. Following such changes through space and time will be a challenge, especially if keystone individuals in a network (so called "super-spreaders" of traits) are hard to find.

There are wonderful examples, too, of network analysis providing novel insights: how competitors jostle for position to enhance their relative attractiveness to prospective mates, predicting who will first discover a novel food source, and how social learning spreads innovations and establishes cultural traditions in the wild.

The chapters also point to fertile fields for future research. How do differences in personality and developmental history influence individual positions in a network and their influence on the spread of traits? What mechanisms cause co-operators to cluster and hence promote the evolution of cooperative behaviour? How can we manipulate networks to minimise the spread of disease, enhance the spread of information or maximise welfare? How can we avoid destabilising social groups when individuals are added or removed for conservation? So the book is not only a marvellous introduction to this exciting new field, it will also help to inspire the next generation of studies.

Nicholas B. Davies

Acknowledgements

Publishing an edited book is always a huge undertaking and we would like to thank all those colleagues who reviewed chapters in this book or who contributed in other ways: Lucy Aplin, Natasha Boyland, Lauren Brent, Julien Cote, Torben Dabelsteen, Daniel Charbonneau, Mathew Edenbrow, Mathias Franz, Stephanie Godfrey, Charlotte Hemelrijk, Vincent Janik, Robert Jeanne, Raphael Jeanson, Susanne Joop, Stefan Krause, Ralf Kurvers, Shuyan Liu, Christof Neumann, Paul Rose, Bernhard Voelkl, Ashley Ward, Mike Webster, and Max Wolf.

The Zentrum für interdisziplinäre Forschung, Bielefeld University, provided us with generous funding for an international workshop which helped kick-start this book. Their organizational and financial support is gratefully acknowledged. In addition, we would like to thank Ian Sherman from Oxford University Press for his help with the book proposal when the project was at an early stage and Lucy Nash, also from Oxford University Press, for her help, support, and patience throughout the entire publishing process. Funding was provided to JK by the Leibniz Competition (SAW-2013-IGB-2) and the Leibniz-Institute of Freshwater Ecology and Inland Fisheries, and to DPC by the Leverhulme Trust and the Natural Environment Research Council.

Contents

List of contributors xvi

Section 1 Introduction to Animal Social Networks 1

1 General introduction **3**
Jens Krause, Richard James, Daniel W. Franks, and Darren P. Croft

What is a social network and why is it important? 3
Book structure and content 3

2 A networks primer **5**
Richard James

Basics 5
Measuring network structure 6
Node-based measures of structure 6
Network-level measures of structure 7
Clusters or communities 7
Model networks and network models 8

Section 2 Patterns and Processes in Animal Social Networks 9
Darren P. Croft

3 Assortment in social networks and the evolution of cooperation **13**
Darren P. Croft, Mathew Edenbrow, and Safi K. Darden

Introduction to cooperation 13
Theoretical work on the evolution of cooperation in structured populations 14
Pathways to assortment by cooperation in social networks 17
Non-random distribution of individuals in space and time 17
Social structuring in the absence of spatial segregation 18
Spatial and temporal assortment by simple behavioural rules 18
Conditional cooperation 20
Cooperation in social networks: conclusions and future directions 22
Acknowledgements 23

4 Mating behaviour: sexual networks and sexual selection **24**
Grant C. McDonald and Tommaso Pizzari

Introduction 24
Sexual selection 24
Sexual selection in structured populations 26
The logic of sexual networks 27
Intrasexual interactions and the measurement of sexual selection 28
Precopulatory selection 30
Postcopulatory selection 30
Intersexual interactions, mating patterns, and the operation of selection 32
Mating system ecology 35
Individual network measures 36
Sexual selection and sexual networks: conclusions 37
Acknowledgements 37

5 Quantifying diffusion in social networks: a Bayesian approach **38**
Glenna Nightingale, Neeltje J. Boogert, Kevin N. Laland, and Will Hoppitt

Introduction to social transmission in groups of animals 38
Network-based diffusion analysis 39
Why do we need Bayesian network-based diffusion analysis? 40
Simulated diffusion data 41
Previous formulation of time of acquisition diffusion analysis 41
Bayesian formulation of time of acquisition diffusion analysis 43
Likelihood function for time of acquisition diffusion analysis 44
Prior specification 44
Generating posteriors using updating methods 45
Model discrimination 45
Results 46
Posterior parameter estimates 46
Markov chain Monte Carlo replication 49
Model discrimination 49
A Bayesian approach to quantifying diffusion on social networks: conclusions and future directions 49
Acknowledgements 52

6 Personality and social network analysis in animals **53**
Alexander D. M. Wilson and Jens Krause

Introduction to personality and social network analysis in animals 53
Network consistency and 'keystone' individuals 54
Fitness consequences of network positions 55
Networks and behavioural types 56
Networks and personality from a developmental perspective 58
Personality and social network analysis in animals: conclusions and future directions 59
Acknowledgements 60

7 Temporal changes in dominance networks and other behaviour sequences **61**
David B. McDonald and Michael E. Dillon

Introduction to the analysis of temporal changes in networks 61
Network formulation and triad census approach 63
Ranking algorithms 63
R scripts for analysing dominance data 65
Differences among ranking algorithms 65
Effect of contest order on Elo ranking 65
Comparing contest and outcome adjacency matrices over time 68
Analysing the contest matrix by quartiles 68
Analysing the outcome matrix by quartiles 70
Experimental and modelling approaches: conclusions and future directions 71
Acknowledgements 72

8 Group movement and animal social networks **73**
Nikolai W. F. Bode, A. Jamie Wood, and Daniel W. Franks

Introduction to group movement and animal social networks 73
Population level 75
Group level 77
Individual level 80
Group movement and animal social networks: conclusions and future directions 81
Acknowledgements 83

9 Communication and social networks **84**
Peter K. McGregor and Andrew G. Horn

Introduction to communication and social networks 84
Communication and network approaches 84
Signals, information, and communication 85
Information exchange and communication networks 85
Receiver diversity and communication in networks 86
Empirical successes of the communication network approach 87
Eavesdropping 87
Audience effects 89
Alarm call spread 90
Linking communication networks to social networks 91
Signals as methodological tools for studying social networks 91
Mapping communication networks as social networks 92
Communication networks and information flow 92
Communication and social networks: conclusions and future directions 94

10 Disease transmission in animal social networks **95**
Julian A. Drewe and Sarah E. Perkins

Introduction to disease transmission networks 95
The use of animal social networks to study infectious disease transmission 96
Networks and disease management 97

Collecting social network data to study disease transmission 97
How many host–parasite associations should be included in a transmission network? 98
Sampling considerations and the boundary effect 100
Weighted or unweighted networks: capturing transmission processes? 100
Choice of time interval for constructing parasite transmission networks 101
Data analysis: which network measures are relevant to disease transmission? 105
Network centrality and disease transmission 106
Relationships between network measures and host attributes 107
Disease transmission network dynamics 107
Host and parasite-driven parameters in transmission networks 107
Effects of infection on networks 108
Disease transmission in animal social networks: conclusions and future directions 109

11 Social networks and animal welfare **111**
Brianne A. Beisner and Brenda McCowan

Introduction to the use of social network analysis in animal welfare 111
Physical health in animal social networks 113
Disease transmission in animal networks 113
Psychological and social health in animal networks 116
Social aggression in animal social networks 116
Social stress and health in animal social networks 119
Social network analysis in animal welfare: conclusions and future directions 121
Acknowledgements 121

Section 3 Taxonomic Overviews of Animal Social Networks **123**
Jens Krause

12 Primate social networks **125**
Sally Macdonald and Bernhard Voelkl

Introduction to social network analysis in primatology 125
Why is social network analysis useful for primatologists? 125
A brief history of social network analysis in primatology 126
Levels of primate social network analysis 129
Primate social network analysis at the individual level 129
Primate social network analysis at the subgroup level 130
Primate social network analysis at the group level 131
Potential pitfalls and limitations in primate social networks 132
Group size in primate networks 132
Observation frequency in primate social networks 133
Specificity in primate social networks 134
Intraspecific variability in primate social networks 134
Social network analysis in primatology: conclusions and future directions 136
Acknowledgements 138

13 Oceanic societies: studying cetaceans with a social networks approach **139**
Shane Gero and Luke Rendell

Introduction to network analysis of cetacean societies 139
Oceanic social networks 139
Studying cetaceans using a social network approach 142
Identifying individuals in cetacean social networks 143
Interactions between individuals in cetacean social networks 143
Contributions from studies on cetaceans 143
Methodological advances in studying cetacean social networks 143
Cetacean networks and management: resilience and survival 144
Social roles and decision making in cetacean social networks 145
Drivers of network structure in cetacean societies 145
Cultural transmission and cetacean social networks 146
Social networks and communication networks among cetaceans 147
Current challenges and avenues for inquiry concerning cetacean social networks 147
Linking cetacean social networks with vocal complexity 147
Cetacean social network analysis: beyond associations 148
Collective motion and decision making in cetacean societies 148
Network analysis of cetacean societies: conclusions and future directions 149
Acknowledgements 149

14 The network approach in teleost fishes and elasmobranchs **150**
Jens Krause, Darren P. Croft, and Alexander D. M. Wilson

Introduction to networks in teleost fishes and elasmobranchs 150
Population structure of teleost fishes and elasmobranchs 151
Techniques for identifying individuals in teleost fishes and elasmobranchs 151
Guppies and sticklebacks—a case study 153
Outlook for population applications of social network studies 153
Familiarity and site fidelity in teleost fish and elasmobranchs 154
Cooperation in teleost fishes 155
Fish cognition and social learning 155
Collective behaviour and social networks in teleost fishes 156
Application of social network analysis to welfare in teleost fishes 156
Network analysis of teleost fishes and elasmobranchs: conclusions and future directions 158
Acknowledgements 159

15 Social networks in insect colonies **160**
Dhruba Naug

Introduction 160
Social interactions and their proximate basis 161
Structure of the colony interaction network 162
Function of the colony interaction network 164
Information collection and transfer 164
Colony work organization 165
Material transport 167
Conclusions and future directions 169

16 Perspectives on social network analyses of bird populations **171**
Colin J. Garroway, Reinder Radersma, and Camilla A. Hinde

Introduction to social network analysis in birds 171
Building avian social networks 173
Ringing and observation in avian social networks 173
Passive integrated transponder tags in the study of avian social networks 175
Telemetry tracking in the study of avian social networks 176
Choosing the best method for studying avian social networks 177
Exploring avian social networks 177
Social network analysis of avian societies 179
Eco-evolutionary processes affected by the structure of avian societies 179
Social phenotypes and gene × environment interactions in avian societies 179
The social characteristics of individuals, and emergent network structure in avian societies 180
The role of social context in natural and sexual selection in avian societies 180
Underlying social structure and collective behaviour in avian societies 180
Individual interactions and social structure in avian societies 181
The conservation and management of social units in avian societies 181
Genetic determinants of variation in social phenotypes in avian societies 182
The interaction of genetic and social structures in avian societies 182
Social network analysis in birds: conclusions and future directions 183
Acknowledgements 183

17 Networks of terrestrial ungulates: linking form and function **184**
Daniel I. Rubenstein

Introduction to terrestrial ungulate social networks 184
The play and the players in terrestrial ungulate social systems 184
The script, the data, and methodologies for ungulate social systems 185
Equids as model organisms 187
Similar social structures but different social networks among equids 188
Fission–fusion functionality in hierarchical equid societies 188
Identifying communities among equid societies 189
Structure shapes function in equid societies 191
Generalizing to other ungulates 192
Wild and free-ranging ungulate species 192
Domestic and captive animals 194
Terrestrial ungulate social networks: conclusions and future directions 194
Acknowledgements 196

18 Linking lizards: social networks in reptiles **197**
Stephanie S. Godfrey

Introduction to social networks in reptiles 197
The application of social networks to understanding reptile behavioural ecology 197
Detecting and describing social organization in reptiles 198
What shapes social organization in reptiles? 199
Thermal and ecological requirements for reptiles 199

Genetic structure and kin-based sociality in reptiles 200
Proximate mechanisms of social organization in reptiles 200
Visual and olfactory signalling and communication pathways in reptiles 200
Individual differences and network position in reptile societies 201
Consequences of social networks: parasite transmission among reptiles 201
Advantages of using networks in understanding reptile systems 202
Challenges of applying a network approach to reptiles 203
Marking, identifying, and monitoring reptiles 204
Identification of reptiles upon capture 204
Visual identification of reptiles 204
Remote monitoring of associations among reptiles 205
Social networks in reptiles: conclusions and future directions 207
Acknowledgements 207

Section 4 Animal Social Networks: Conclusions **209**

19 Animal social networks: general conclusions **211**
Jens Krause, Richard James, Daniel W. Franks, and Darren P. Croft

The changing nature of animal social network data 212
The changing nature of animal social network analysis 213

References 215
Index 255

List of Contributors

Brianne A. Beisner Animal Behavior Laboratory for Welfare & Conservation, School of Veterinary Medicine, University of California, Davis, USA

Nikolai W. F. Bode Department of Mathematical Sciences, University of Essex, UK

Neeltje J. Boogert Bute Building, University of St Andrews, UK

Darren P. Croft Centre for Research in Animal Behaviour, College of Life and Environmental Sciences, University of Exeter, UK

Safi K. Darden Centre for Research in Animal Behaviour, College of Life and Environmental Sciences, University of Exeter, UK

Nicholas B. Davies Department of Zoology, University of Cambridge, UK

Michael E. Dillon Department of Zoology and Physiology, University of Wyoming, USA

Julian A. Drewe The Royal Veterinary College, London, UK

Mathew Edenbrow Centre for Research in Animal Behaviour, College of Life and Environmental Sciences, University of Exeter, UK

Daniel W. Franks York Centre for Complex Systems Analysis, Departments of Biology and Computer Science, University of York, UK

Colin J. Garroway Edward Grey Institute, Department of Zoology, University of Oxford, UK

Shane Gero Sea Mammal Research Unit and Centre for Social Learning and Cognitive Evolution, School of Biology, University of St. Andrews, UK, and Department of Zoophysiology, Institute for Bioscience, Aarhus University, Denmark

Stephanie S. Godfrey School of Veterinary and Life Sciences, Murdoch University, Western Australia

Camilla A. Hinde Behavioural Ecology Group, Wageningen University, The Netherlands

Andrew G. Horn Department of Biology, Life Science Centre, Dalhousie University, Halifax, Nova Scotia, Canada

Will Hoppitt Department of Life Sciences, Anglia Ruskin University, Cambridge, UK

Richard James Department of Physics, University of Bath, UK

Jens Krause Faculty of Life Sciences, Humboldt University, Berlin, Germany, and Leibniz-Institute of Freshwater Ecology and Inland Fisheries, Berlin, Germany

Kevin N. Laland Centre for Social Learning and Cognitive Evolution, School of Biology, University of St. Andrews, UK

Sally Macdonald Courant Research Centre Evolution of Social Behaviour, Georg-August-Universität, Goettingen, Germany

Brenda McCowan Animal Behavior Laboratory for Welfare & Conservation, School of Veterinary Medicine, University of California, Davis, USA

David B. McDonald Department of Zoology and Physiology, University of Wyoming, USA

Grant C. McDonald Edward Grey Institute, Department of Zoology, University of Oxford, UK

Peter K. McGregor Centre for Applied Zoology, Cornwall College, UK

Dhruba Naug Department of Biology, Colorado State University, USA

Glenna Nightingale Centre for Social Learning and Cognitive Evolution, School of Biology, University of St. Andrews, UK

Sarah E. Perkins Cardiff School of Biosciences, UK

Tommaso Pizzari Edward Grey Institute, Department of Zoology, University of Oxford, UK

Reinder Radersma Edward Grey Institute, Department of Zoology, University of Oxford, UK

Luke Rendell Sea Mammal Research Unit and Centre for Social Learning and Cognitive Evolution, School of Biology, University of St. Andrews, UK

Daniel I. Rubenstein Department of Ecology and Evolutionary Biology, Princeton University, USA

Bernhard Voelkl Edward Grey Institute, Department of Zoology, University of Oxford, UK

Alexander D. M. Wilson Department of Biology, Carleton University, Canada

A. Jamie Wood Department of Biology, University of York, UK

SECTION 1

Introduction to Animal Social Networks

CHAPTER 1

General introduction

Jens Krause, Richard James, Daniel W. Franks, and Darren P. Croft

The network approach has taken hold in many different disciplines in mathematics, physics, sociology, and various areas of biology. A huge range of systems—regardless of whether they are of technological or biological origin—can be described in terms of discrete components that interact and are thus amenable to network analysis. The universal nature of the network approach has greatly contributed to its success because every innovation, be it conceptual or methodological, becomes almost instantly accessible in all fields of activity, thereby pushing the discipline forwards at great speed.

When the first reviews of the network approach for behavioural biologists appeared (Krause et al. 2007; Croft et al. 2008; Wey et al. 2008) this field of work was still at an exploratory stage. Since then, network analysis has become firmly embedded in behavioural biology, with a large, active, and rapidly growing community of practitioners. This development is based on the facts that (1) social networks provide an ideal framework within which to study social structure and (2) social structure is a fundamental piece of biological information that informs ecological and evolutionary processes at every level of organization (Kurvers, Krause, et al. 2014).

What is a social network and why is it important?

A social network can be defined as any number of individuals interconnected via social ties (sexual, cooperative, etc.) between them. Understanding social network structure is of great importance because the structural characteristics of a group or population will affect the biology of its constituent members—including finding and choosing a sexual partner, developing and maintaining cooperative relationships, and engaging in foraging and antipredator behaviour. An understanding of the structure of social networks can provide new insights into processes that occur on the network—insights that would not be possible from a consideration of dyadic interactions in isolation. For example, at the population level, the structure of a network will affect the speed at which information travels through a population. Furthermore, at the level of the individual, information on who is connected to whom in the network can make it possible to predict who will learn from whom, or who will infect whom with a disease.

Book structure and content

In this book, we bring together contributions from an array of researchers, with the aim of providing both an overview of the power of the network approach for understanding patterns and process in animal populations, and outlining how current methodological constraints and challenges can be overcome. This book provides an overview of the insights that network analysis has provided into major biological issues such as cooperation, mating, and communication, and how network analysis has enhanced our understanding of the social organization of several important taxonomic groups (e.g. primates, fishes, and birds). Each chapter provides

Animal Social Networks. Edited by Jens Krause, Richard James, Daniel W. Franks and Darren P. Croft.

an in-depth treatment of an area in which the network approach has played a particular role. Some readers may be more interested in the conceptually oriented chapters on patterns and processes (Section 2). Others might prefer to study the taxonomic chapters to read about species of immediate interest to them (Section 3). The final chapter gives a brief overview of the book as a whole and identifies both technological advances that have the potential to propel data collection forwards and methodological topics that are central to this field of research.

CHAPTER 2

A networks primer

Richard James

This short chapter can safely be skipped by anyone already familiar with the common terms (and their synonyms) used to describe networks and their structure. It is not intended as an exhaustive, or even a representative, introduction to networkology; such things can be found elsewhere (see for example Wasserman and Faust 1994 or Hanneman and Riddle 2005 for a social sciences perspective, or Newman 2010 for a physical sciences approach). Instead it is designed as a largely non-technical introduction to some of the terms used in later chapters to quantify the structure and dynamics of animal social networks.

Basics

A **network** (or, equivalently, a **graph**) consists of a set of n **nodes** (or **vertices**) and a set of E **edges** (or **links**, **connections,** or **arcs**). The simple toy network depicted in Figure 2.1 has $n = 5$ nodes and $E = 6$ edges. The layout of the nodes in such a visualization is often arbitrary (as here), and no meaning should be attached to the length or shape of the edges, nor to the fact that some of them (between nodes 1 and 4 and nodes 2 and 3 in this case) may cross. Figure 2.1b shows one of the ways to represent a network mathematically: the **adjacency matrix** $\underline{A}$. The edges in this particular network are **directed** and **unweighted**. The latter means that an edge is either present (giving a 1 in the adjacency matrix) or absent (a 0). For this reason, networks with unweighted edges are often referred to as **binary**. The edges are directed from an **actor** to a **receiver**. So, for example, the presence of the directed edge from node 4 to node 5 in Figure 2.1a is represented by the 1 in the 4th row, 5th column of the adjacency matrix in Figure 2.1b.

In nearly all animal social networks, each node represents a single, identifiable individual. Properties of each individual are usually referred to as **node attributes** and are often distinguished in network visualizations by choice of node size, colour, or shape. Pairs of nodes form a **dyad** when there is an edge between them. Many types of relationship may be represented by an edge between individuals A and B: 'A groomed B', 'A sat near B', 'A and B used the same roost', 'A and B mated', and so on.

Usually all edges in a network represent the same type of relation, but authors of several chapters in this book are interested in **multiplex** (or **multilevel**) networks, where more than one relation between dyads in the same set of animals is recorded and analysed. The relations may be of several behaviour types, or there may be a longitudinal set that is of the same relation but is recorded at different times. As more social data are collected electronically (Krause et al. 2013; Kühn and Burghardt 2013), continuous-time networks, often dubbed **temporal networks,** are becoming more common (see e.g. Blonder et al. 2012).

For the most part, though, the animal social network analysis in this book is concerned with the characterization of the **structure** (or **topology**) of single-relation **static networks** of the type depicted in Figure 2.1. Directed edges represent relations such as '1 groomed 2' or '4 attacked 5'. If we have some measure of the frequency and/or strength of the dyadic behaviour, we may wish to give **weight** to each edge. Figure 2.2 is a weighted version of the directed network of Figure 2.1.

Many types of dyadic relation—'A and B used the same roost', for example—are represented by

Animal Social Networks. Edited by Jens Krause, Richard James, Daniel W. Franks and Darren P. Croft.

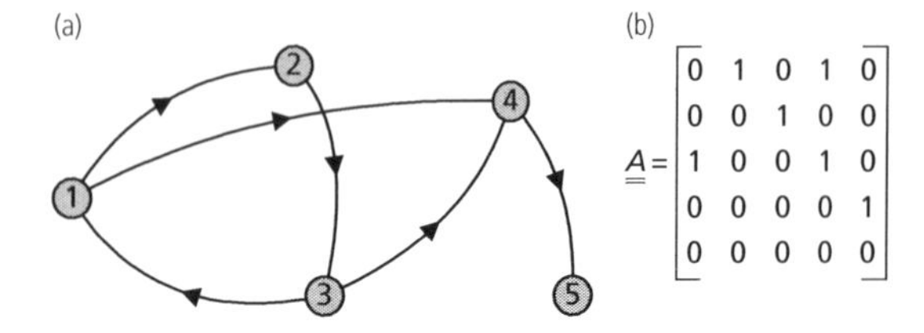

Figure 2.1 (a) A simple network with five nodes and six directed, unweighted edges, and (b) its adjacency matrix. A depiction of a network using symbols for nodes and lines for edges is often referred to as a **sociogram**.

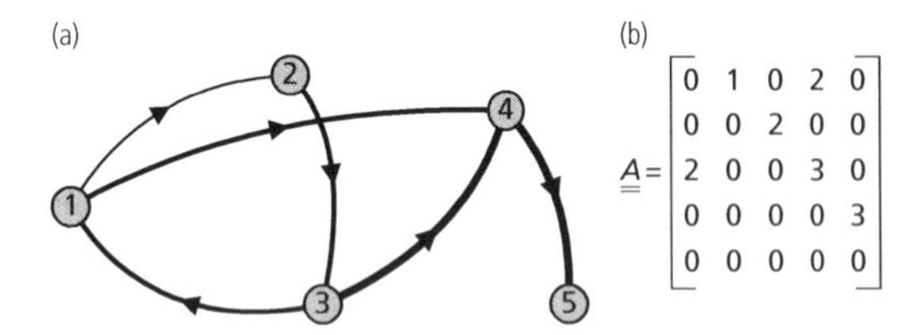

Figure 2.2 (a) A version of our toy network with weighted edges, and (b) its adjacency matrix.

undirected edges (Figure 2.3), which again may be weighted or unweighted. Note that each edge appears twice in the adjacency matrix of an undirected network. Networks derived from **association data** (as distinct from **interaction data**) are usually undirected. The associations may be truly dyadic (exactly two animals were in proximity) or derived from, for example, group membership. In

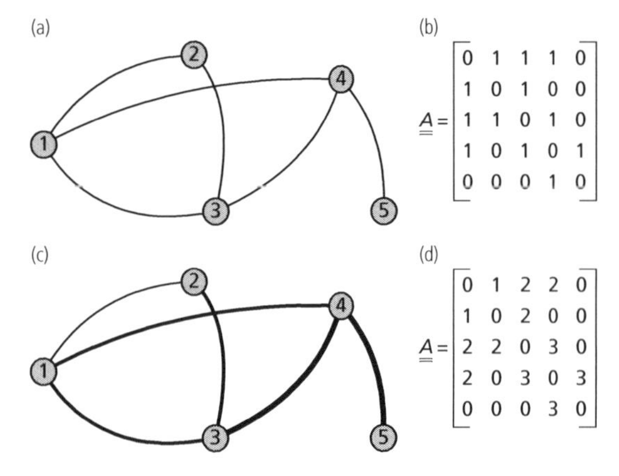

Figure 2.3 (a) An undirected, unweighted network with five nodes and six edges, and (b) its adjacency matrix. (c) An undirected, weighted network and (d) its adjacency matrix.

such associative networks, an **association index** is often used to quantify the edge weights. In principle there are other edge types (signed, for example, so that some edges are positive and some negative) but the '2 by 2' structure of weighted/unweighted and directed/undirected covers the examples in this book.

Measuring network structure

Having constructed an animal social network, the next job is to characterize its structure and then to analyse that structure in the light of what is known, or hypothesized, about the biology of the study population. Here we introduce some of the measures used to characterize network structure; most of them are **node-based measures** or are derived from them. It is important to realize at the outset that not all measures of network structure (or indeed the methods used to analyse them) are appropriate, or even have meaning, for all edge types or network structures.

Before we start, we need some jargon. A **path** is a sequence of nodes with an edge from each node to the next, and in which no node or edge is used twice. The number of edges defines the **path length**. Many measures make use of the shortest (**geodesic**) path between two nodes and call that quantity the path length. A network is **connected** if there is a path between every pair of nodes. If so, the network contains a single **component**; otherwise it consists of a set of disconnected components. It is sometimes useful to distinguish a path between two nodes from a **trail**, in which nodes (but not edges) can be revisited, and a **walk**, in which both nodes and edges can be re-used.

Node-based measures of structure

A lot of the analyses in this book use node-based measures of structure. The overwhelming majority of these are measures of **centrality**, used to differentiate the social importance or influence of members of the population. There are lots of node centrality measures to choose from; Borgatti (2005) and Wasserman and Faust (1994) explain many of them and offer very good advice on which to use for a given problem. There are two main families of centrality measures, derived either from tracing paths through the network, or more directly from the adjacency matrix $\underline{A}$ using linear algebra.

Among the path-following measures, by far the most commonly used is the **degree centrality** (usually just called **degree**). For an unweighted undirected network (Figure 2.3a), this is just the number of edges a node is connected to. For unweighted directed networks (Figure 2.1a), we need **out-degree** and **in-degree**, the number of edges leaving and arriving at a node, respectively. **Weighted degree**—the sum of the weights of the edges connected to a node—is often called **node strength**. The notion with all degree measures is that the most connected nodes are the most central. A related set of measures is the node (q) **reach**, which is the number of nodes that are a distance q away.

Other path-following node centrality measures include **betweenness, closeness**, **flow,** and **information centrality**. These—loosely speaking—count paths between pairs of nodes that pass through the node of interest. Among the matrix-derived measures are **Katz centrality** and **eigenvector centrality**. All of these were developed for use in human social sciences, mostly for binary networks. Other centrality measures discussed in this book (see Chapter 7) include **hub score, authority score,** and **power**.

A node's **clustering coefficient** is used in some chapters to measure 'cliquishness': the extent to which two of one's neighbours are themselves neighbours (forming a triangle of edges). This is a well-defined measure for undirected networks, although care is needed if edges are weighted. It is not defined for directed networks, where something like a **triad census** is used to count the various possible directed triangles.

Network-level measures of structure

The approach most commonly used to characterize an entire social network is to compute the average of a chosen node-based measure. A simple example is the **mean degree** $\bar{k}$. For an unweighted, undirected network (Figure 2.3a) this is just the average of all the row (or column) sums of $\underline{A}$. The entries in $\underline{A}$ add to twice the number of edges in this case, so $\bar{k} = 2E/n$. Often the mean node value will be standardized to aid (although sometimes only slightly) comparison between populations, species, or behavioural contexts. In the case of degree, the maximum possible node value is n–1, so a sensible standardized mean degree would be $\rho = \frac{2E}{n(n-1)}$. This particular measure is usually called the network **density**, as it is the fraction of all possible edges that are present in the network.

There are plenty of alternatives to mean node measures to describe overall network patterns, including correlations and distributions of various node-based measures. A qualitatively different approach is to enumerate counts of small **motif** structures in the network. The simplest motifs are dyads; a **dyad census** counts the three possible types of dyad between each possible pair of nodes in a directed network, namely **Mutual** (an edge both ways), **Asymmetric** (an edge one way but not the other) and **Null** (no edge), forming the **'MAN'** distribution of dyads. A triad census is an enumeration of the 16 possible combinations of MAN edges among 3 nodes, which include the **transitive triad** and the **cycle.** There are seven possible triads if mutual dyads are neglected, and four if the edges are not directed.

Surprisingly little use is made in network science of the **distribution** of node measures. The most notable exception is the **degree distribution** $P(k)$, which measures the probability that a node has degree k; if a simple count of nodes with a particular degree is given, this is more correctly denoted a **degree sequence**. A fat-tailed degree distribution might indicate the presence of **superspreaders**, or **hubs**, in the population, that is, individuals that may have a disproportionate influence on the spread of disease or information, for example. Correlational analyses are relatively common in animal social network analysis. It is possible to test for **homophily**, or **assortativity**, by seeing whether a particular node attribute (size or sex, for example) of a node is correlated with that of its neighbours. The attribute in question could itself be a network-derived node measure, as in the case of **degree assortativity**.

Clusters or communities

Many networks contain groups of nodes better connected among themselves than they are to the rest of the network (Figure 2.4). These clusters of well-connected nodes are usually referred to as **communities** in network science. Some animal social networks have been found to have well-defined communities, indicating a level of social structure between that of the dyad and the population, and often with a hierarchical structure. There is now a multitude of different schemes to look for network communities (Fortunato 2010), many of which involve the maximization of a quality factor, the network **modularity**.

If a network is found to have a well-defined community structure, analysis usually proceeds with a search for variables that explain their occurrence. In this sense, community structure and network assortativity are addressing similar properties, except that in the former, rather than choosing likely sources of node assortment a priori, we allow the observed network topology to determine cluster membership. It should be mentioned in passing that averages of node measures may be meaningless in a network with communities, and that some node measures are (overly) sensitive to the presence of communities.

Model networks and network models

A key component of animal social network analysis (and more so than in most other branches of network science) is to try to test the observed structure for statistical significance. The most common approach to date has been null hypothesis significance testing (also known as NHST—see Croft, Madden, et al. 2011), using a statistical null model informed by the biology of interest and the perceived bias in the behavioural observations that generated the network. Often the testing is achieved by comparing the observed network with an ensemble of random networks. These may be constructed from scratch or chosen from a wide range of model networks developed in other disciplines.

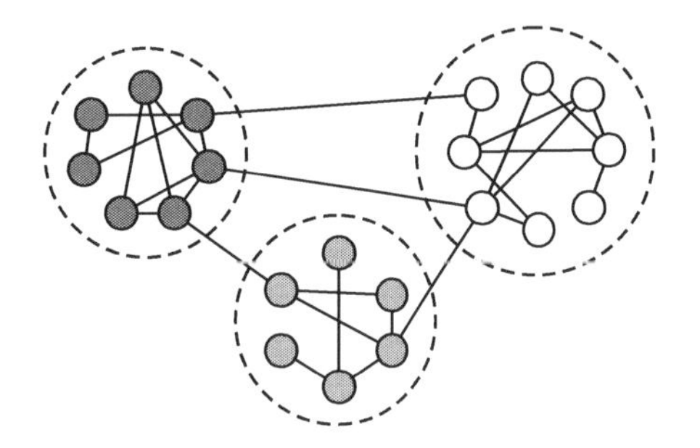

Figure 2.4 A network with a fairly clear partition of nodes into three communities.

Null models all have a random element, with or without constraints (when they become **conditionally uniform models**). Perhaps the simplest and best understood model network is the unconstrained **Erdős-Rényi random graph** (or **Poisson random graph**). Other 'off the peg' options (see Newman 2010) include **configuration models**, **small world networks**, and **scale-free networks**, each of which offers different sets of constraints. In the social sciences, the **U|MAN distribution** (Wasserman and Faust 1994) is a conditionally uniform model based on the distribution of MAN dyads outlined above. Among the bespoke null models for animal social networks are those that deal with group-derived association data (see Whitehead 2008b or Croft et al. 2008).

An alternative to generating ensembles of random networks is to generate ensembles of randomized node labels. **Node label permutation** is at the heart of the oft-used **Mantel test** for comparing two matrices and its variants, including the **quadratic assignment procedure** (QAP). Multivariate versions of these tests (**MR-QAP**) comprise one set of a number of potentially attractive statistical models developed for the analysis of human social networks (Hanneman and Riddle 2005; Snijders 2011). Other models that may be of use for animal social network analysis include **exponential random graph models** and **hidden Markov field** models. These tools of model-based inference are not much used in this volume, although they do offer hope for more robust statistical analysis of animal social networks in the future (Pinter-Wollman et al. 2014).

SECTION 2

Patterns and Processes in Animal Social Networks

Darren P. Croft

Social network analysis was initially deployed in the study of animal behaviour as a tool for describing the social structure of populations (e.g. Lusseau 2003; Croft et al. 2004). This approach allowed researchers to consider the social structure of animal populations at different social scales and provided a new tool kit for quantifying both local and global patterns of social structure (Croft et al. 2008). This work highlighted the heterogeneity in the social structure of both wild and captive animal populations (see Section 3). It is now clear that the vast majority of animal populations are not well mixed, and social interactions are influenced by space, time, and social factors. This section of the book is devoted to understanding the consequences of this social heterogeneity for the behaviour and welfare of individuals and processes that occur at the population level (e.g. disease and information transmission). The chapters in this section are structured by topic, each focussing on a key area where an understanding of the social network structure can provide insight into both the mechanisms and implications of population social structure.

This section of the book starts with a chapter by Croft et al. on the evolution of cooperation. Why should one individual perform a costly act so that another can receive a benefit? The key to answering this question is in understanding the patterns of social mixing in populations. All mechanisms proposed to maintain cooperation have one thing in common: they can all be considered as mechanisms that drive social assortment by cooperation in social networks (Aktipis 2008; Fletcher and Doebeli 2009; Nowak et al. 2010). In this chapter, Croft et al. review the theoretical work that has examined the role of social network structure in maintaining cooperation and discuss the biological mechanisms that can generate social assortment by cooperation in real-world social networks.

In Chapter 4, McDonald and Pizzari discuss how representing mating behaviour as a sexual network can provide new insight into sexual selection. Considering sexual interactions as a network is a relatively new area in non-human animals (McDonald et al. 2013). Traditionally, quantitative frameworks used to measure sexual selection assumed populations were panmictic (McDonald et al. 2013). As outlined above, however, this is rarely the case, and recent empirical and theoretical work clearly demonstrates that the structuring of mating networks can have a major effect on the strength of selection on sexual traits (Oh and Badyaev 2010; McDonald et al. 2013). In this chapter, the authors review the recent theoretical and empirical advances that have been made by considering mating behaviour as a sexual network. The authors illustrate how social network analysis can help biologists to unravel the complex dynamics of mating systems in natural populations, and the consequences of this structure for evolution.

In Chapter 5, Nightingale et al. discuss methods for considering how novel behaviours diffuse (spread) through a population. Of particular interest are behaviours that are socially transmitted whereby there is a causal influence by one individual on the rate at which another individual acquires and/or performs a behavioural trait (Hoppitt and Laland 2013). The challenge for studies of social transmission is in distinguishing social transmission from simple diffusion (where the observed trait may spread through a population without social transmission). Nightingale et al. outline how network-based diffusion analysis can be used to infer social transmission when the acquisition of the trait of interest follows the social network structure of a group of animals.

In Chapter 6, Wilson and Krause discuss the interplay between animal personality and social networks. There is great interest in the evolution of animal personality and its importance for ecology and evolution (Sih et al. 2012; M. Wolf and Weissing 2012). Currently, however, relatively little is known regarding the role of animal personality in complex social dynamics (Krause et al. 2010). When measuring animal personality, few studies have looked beyond dyadic relationships and placed the personality of individuals in the context of a social network. As outlined in Chapter 5, social network analysis provides an opportunity to gain an understanding of the role that different personalities play in groups, communities, and populations regarding information or disease transmission, or in terms of cooperation.

Methods for studying social dominance clearly recognize that there is a network of interactions (i.e. a dominance hierarchy); however, surprisingly little use has been made of more advanced network methods for the study of dominance structures in animal populations (Shizuka and McDonald 2012). In Chapter 7, McDonald and Dillon discusses how social network analysis can be used to quantify the structure of dominance interactions. This approach allows for the dominance trajectories of individuals to be tracked, that is, determining whether individuals ascend or descend in social rank.

There has been considerable interest over the last decade in the mechanisms that underpin the collective behaviour of social groups (Sumpter 2010). Across taxonomic groups, animals coordinate movement, fish school, birds flock, and ungulates herd (Krause and Ruxton 2002). Traditionally, models of collective movement assumed that all individuals were equally likely to interact (Couzin and Krause 2003); consequently, these models did not capture the social heterogeneity that exists in real-world social systems. In Chapter 8, Bode and Franks provide an overview of the consequences of non-random social preferences (for example, based on familiarity) for the movement dynamics of social groups.

One area where network thinking has been prominent for some time is in the study of animal communication, a topic covered in Chapter 9 by McGregor and Horn. Here, a communication network is defined as several individuals within signalling and receiving range of one another (McGregor 2005). Because of the physics of signal transmission and the ephemeral nature of signals, communication networks tend to be small and relatively simple in their structure (McGregor 2005). However, this research area is expanding to consider neighbourhood effects where signals are propagated through social networks. As this research area develops further, it is clear that social network analysis will prove useful. In the study of animal social networks, communication is a key component used to define the social associations (Whitehead 2008a). While many studies have used the potential for communication as the definition of a social association, few studies have defined social interactions based on observed communication events. As outlined in this chapter, much progress could be made by bringing together work on animal communication networks with other aspects of social network analysis in animal societies.

One of the earliest applications of network theory to biological processes was in the context of disease transmission. In Chapter 10, Drewe and Perkins provide an overview of how social network analysis can be used to understand disease epidemiology in animal populations. Early theoretical work on disease transmission assumed random mixing of individuals in a population. As discussed above, contact patterns in populations are heterogeneous, and this heterogeneity will have implications for infectious dynamics. Drewe and Perkins review the theoretical work on disease transmission in social networks and provide an overview of the empirical studies that have tested the assumptions and predictions from these models.

Advancing our knowledge of social behaviour is of fundamental importance for maximizing the productivity and welfare of animals held in captivity. Previous work on the link between animal welfare and social behaviour is generally restricted to studies focussed on simple group attributes such as group size, density, and composition, and previous work has generally ignored the implications of social heterogeneity for animal welfare. In Chapter 11, Beisner and McCowan outline how social network analysis can be used as a management tool to improve animal welfare. The chapter outlines recent work that has used social network analysis in the context of animal welfare and provides an overview of where future applications are likely to have the largest impact in improving the physiological, psychological, and social well-being of animals held in captivity.

Contained within these chapters are examples of how network thinking can provide new insight into key biological questions and problems. Many of the research areas presented are still in their infancy, with some only having a handful of empirical studies. However, the potential for the future is clear. As outlined in these chapters, theoretical work makes clear predictions for the consequences of structured populations for individuals and populations. The challenge for future research on animal social networks is to move from studies that simply describe the patterns of social structure to studies that quantity the consequences of this structure for individuals, populations, and ecological and evolutionary processes. Traditionally, the approach to studying animal social networks has been to aggregate data over a sampling period and to consider the social network as a static representation (Croft et al. 2008). This approach, however, is very restrictive as the social structure of real-world populations is dynamic, and processes on social networks are unlikely to be in a steady state (Blonder et al. 2012). As highlighted in a number of chapters, to understand the interaction between patterns and processes on social networks, it is necessary to incorporate these dynamics. Understanding temporal dynamics is also essential to capture the feedback loops between process and social structure. For example, the time ordering of social interactions will have implications for the transmission of disease in a population (Blonder et al. 2012). However, disease is also likely to influence population social network structure, as individuals adjust their social ties in response to disease (Croft, Edenbrow, et al. 2011). Capturing the temporal aspects of social networks present new challenges; however, recent work is making excellent progress in this regard (see Chapter 5 for an example). The temporal dynamics of networks have been of great interest in other disciplines, including in the study of human social networks, and a number of methods and approaches exist that may prove useful in the study of the temporal dynamics of animal social networks (Blonder, et al. 2012; Pinter-Wollman et al. 2013). We are still to unlock the full potential of social network analysis for understanding the mechanisms and functions underpinning animal populations; however, the ideas and methods outlined in these chapters provide an excellent guide to how we may achieve this.

CHAPTER 3

Assortment in social networks and the evolution of cooperation

Darren P. Croft, Mathew Edenbrow, and Safi K. Darden

Introduction to cooperation

One of the longest standing interdisciplinary challenges lies in unravelling the mechanisms underpinning the evolution of cooperation (Darwin 1859; W. Hamilton 1963; Pennisi 2009). A cooperative interaction involves one individual performing an act that is beneficial to one or more other individuals. However, the question of why one individual should pay a cost so others can receive a benefit appears to generate an evolutionary paradox.

Fundamental to understanding the evolution of cooperation is understanding the costs paid by actors and the benefits gained by receiving individuals. In some contexts, the cooperative act may have immediate benefit for both the actor and receiver in the form of a by-product mutualism (J. Brown 1983). In contrast, there are many examples of cooperation where the act of cooperation by the actor results in a directly incurred cost with no immediate benefit to that individual (see Sachs et al. 2004 for a review). For the frequency of a cooperative phenotype to increase in a population, the carrier of the cooperative trait must end up with a higher fitness than the average population members. Thus, individuals that put themselves at a disadvantage so others can receive benefits will only increase in frequency if the benefits received from others outweigh the costs of the disadvantage.

W. Hamilton (1964a) proposed an elegant solution to the evolution of cooperation when the actor and receiver of the cooperative act are related. More specifically, Hamilton's rule (W. Hamilton 1964a) states that the coefficient of relatedness r must be larger than the cost–benefit ratio of the cooperative act. In this way, individuals gain indirect fitness benefits by increasing the reproductive success of their relatives. Indeed, much of the work over the last four decades examining cooperation in animal populations has focussed on species that live in discrete family groups where kin selection is a major driving force behind cooperation (Bourke and Franks 1995; Solomon and French 1996; Cockburn 1998; Clutton-Brock 2002).

However, while kin selection appears to be widespread (see Hatchwell 2010 for a review) and provides a satisfactory explanation for cooperation, many social species live in dynamic societies where individuals regularly interact and cooperate with non-kin (Krause and Ruxton 2002). Recently, a number of studies have begun to integrate molecular approaches with social network studies to investigate the role of kinship in structuring dynamic animal societies. This work has shown that, even in cooperatively breeding groups such as meerkats (*Suricata suricatta*), individuals may interact and cooperate with non-kin (Madden et al. 2012). Moreover, recent work on a wild population of guppies (*Poecilia reticulata*), which are known to cooperate during predator inspection (Croft et al. 2006), found that the social network for a population living under high predation risk was well mixed, with no assortment by kinship (Croft et al. 2012). How then can cooperation be maintained among non-kin?

The question of what maintains cooperation among non-kin has received considerable game-theoretic attention over the last three decades (Nowak 2006). The majority of the attention has

Animal Social Networks. Edited by Jens Krause, Richard James, Daniel W. Franks and Darren P. Croft.

been founded upon the prisoner's dilemma (Axelrod and Hamilton 1981) and to a lesser extent the snowdrift game (also known as the hawk–dove or chicken game) (Maynard Smith 1982). These theoretical approaches have generated numerous models, including direct reciprocity (Trivers 1971), indirect reciprocity (Nowak and Sigmund 1998), generalized reciprocity (Pfeiffer et al. 2005; Nowak and Roch 2007; Rutte and Taborsky 2007), network reciprocity (Ohtsuki and Nowak 2007) (a generalization of spatial reciprocity (Nowak and May 1992) to evolutionary graph theory (Lieberman et al. 2005)), group selection (D. Wilson 1975) and by-product benefits (J. Brown 1983), to name but a few. These competing theories have in turn sparked much debate in the literature (E. Wilson 2005; K. Foster et al. 2006; Nowak 2006; West et al. 2006; D. Wilson 2008).

From a theoretical perspective, discussions have ranged from the necessity of kin selection as a mechanism for the evolution of cooperation (K. Foster et al. 2006; West et al. 2006) to the validity of multilevel (or group) selection as an explanation for cooperation (E. Wilson 2005; D. Wilson 2008; Leigh 2010). From an empirical perspective, discussions have focussed on the extent to which non-human animals are capable of the complex and cognitively demanding bookkeeping strategies that are required for some models such as direct reciprocity (Milinski and Wedekind 1998; J. Stevens et al. 2005). This chapter is not intended to be a review of the many mechanisms that can lead to the evolution of cooperation, for which there have been many reviews written recently (Sachs et al. 2004; Nowak 2006; West et al. 2007; Clutton-Brock 2009). In contrast, the aim of this chapter is to focus on the mechanisms that underpin the structure of social networks for cooperation and to illustrate how understanding these mechanisms can provide new insights into the evolution and maintenance of cooperation in animal populations.

Cooperation by definition involves social interactions; in this chapter, we ask what general features of the population social structure are central to the different models proposed to explain cooperation. A number of authors have recently suggested that assortment between cooperators is the most fundamental requirement for the evolution of cooperation (Aktipis 2008; Fletcher and Doebeli 2009; Nowak et al. 2010). Cooperators can prevail because they interact with one another; and through this assortment, clusters of cooperators gain higher fitness payoffs than defectors in the population (Aktipis 2008; Fletcher and Doebeli 2009; Nowak et al. 2010).

In this chapter, we focus on the proximate mechanisms that can drive assortment by cooperation in social networks. We focus our attention on the evolution of cooperation among non-kin and the mechanisms that can underpin such cooperation. We review the theoretical literature that has investigated the interaction between social network structure and cooperation, and draw general conclusions from this body of work. In particular, we highlight the structural properties of social networks that appear to be fundamental for the maintenance of cooperation. We discuss empirical case studies that have examined patterns of cooperation in real-world social networks and relate these back to the predictions arising from theory. We then review the biological mechanisms that may drive the structural properties of social networks thought to be fundamental for the evolution and maintenance of cooperation. Finally, we propose future directions for research that are likely to yield further insights into the evolution of cooperation in real-world populations.

Theoretical work on the evolution of cooperation in structured populations

Elementary theory suggests that in a well-mixed population (in which all individuals are equally likely to interact), average fitness is maximized when all members of the population are cooperative (Maynard Smith 1982). However, cooperators are open to exploitation by invading defectors, and cooperation is thus evolutionary unstable (Maynard Smith 1982). It has long been recognized that social interactions in real animal populations are far removed from this traditional mean field assumption of a well-mixed population (Nowak and May 1992). For example, individuals tend to use space in a non-random way, and the spatial distributions of populations tend to mean that neighbours will interact more often than non-neighbours.

In fact, early theoretical work modelling the evolution of cooperation in structured populations focussed on spatial effects on evolutionary game dynamics (Nowak and May 1992; Nowak and Sigmund 1993; Nowak et al. 1994). Traditionally, in these games individuals are arranged on a regular lattice, and interactions occur among nearest neighbours in the lattice (Nowak and May 1992) (Figure 3.1a). It was clear from this work on static lattices that when social interactions are non-random and cooperators cluster in space, cooperation can evolve and be maintained (Nowak and May 1992).

More recently the role of population social structure in driving cooperation has been explored using evolutionary graph theory, which studies evolution on graphs (networks) (Figure 3.1b) (Lieberman et al. 2005; Ohtsuki et al. 2006). Here, individuals occupy nodes of a graph, and the edges denote who interacts with whom. Early work using this approach demonstrated that when the graph is regular (all individuals have the same network degree k (number of neighbours)) and the population size is large ($N >> k$), then a simple rule can explain the evolution of cooperation: $b/c > k$, where b represents benefit and c represents cost. This simple rule states that the benefit–cost ratio must exceed the average number of neighbours. Here, cooperators can prevail by forming clusters of cooperators in the network where cooperation is reciprocated, a process known as network reciprocity.

The application of evolutionary graph theory is significant because it has highlighted the importance of structure, particularly heterogeneous structure, in determining the outcome of evolutionary games, and also provided powerful analysis tools for theoretical and empirical work. However, there are limitations to this approach as it has primarily been applied so far. Although considering the social structure as a network of interacting agents increases the biological realism of the model, there are many aspects of evolutionary graph theory that are somewhat abstract from real social dynamics.

For example, evolutionary dynamics on graphs are dependent on an update rule in which the structure of the graphs is usually fixed on an evolutionary timescale. Individuals that occupy nodes in the graph are often replaced at time steps by a 'death–birth rule', or individual strategies by an 'imitation rule', with the patterning of nodes and edges maintained (static network structure) (Laws et al. 1975). With this type of approach, individuals are chosen at random to die in the graph and are then replaced by one of their neighbours, who compete for this empty site proportional to their fitness; alternatively, individuals imitate neighbours at random or based on a simple rule such as, for example, one based on the success of a neighbour. Clearly, such scenarios do not capture the dynamics of real-world social networks, where social interactions and relationships are dynamic, and individuals can update their social partners and adjust their behavioural strategy.

This limitation has been addressed in part by models of dynamic graphs, where the actions of individuals (nodes) induce changes to the network, and vice versa, so that both the properties of the nodes and connections between nodes are dynamic (Sukumar 1992; Skyrms and Pemantle 2000; Pacheco et al. 2006; Santos, Pacheco, et al. 2006a,b). Dynamics have also been modelled by introducing

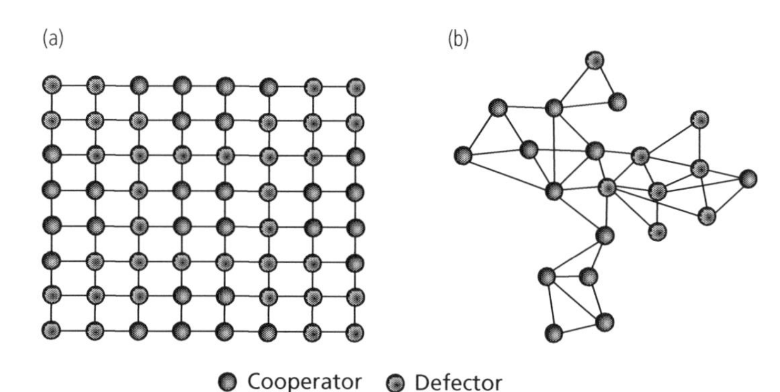

Figure 3.1 (a) An example of a regular spatial lattice showing spatial assortment by cooperation. (b) An example of a network showing assortment by cooperation.

features such as preferential attachment, alterations of connectivity, loss and gain of nodes, and individual decisions on the establishment and removal of ties to others (Pacheco et al. 2006). Work modelling the effect of a dynamical network structure driven by 'active linking' (link dynamics generated by the formation of new links and loss of existing links) demonstrates that if the timescale upon which ties between nodes change is less than the timescale upon which strategies (i.e. cooperation versus defection) change ('fast active linking'), then the pay-off matrix is altered (Pacheco et al. 2006). In the altered matrix, the ratio of the advantage of interactions that are assortative to those that are dissasortative will increase. In other words, dynamic structuring of social ties in the network can increase benefits accrued from cooperative behaviour, and cooperation will flourish (Santos, Pacheco, et al. 2006a). Work has further shown that the 'reorganization' of social ties increases average connectivity in the network while maintaining heterogeneity and that, under these conditions, network topology must continue to change in order for cooperation to be maintained (Santos, Pacheco, et al. 2006a).

Another approach, which has been used to model evolutionary dynamics in human populations, is to classify the social interactions an individual has by that individual's membership to well-defined sets (Tarnita et al. 2009). For example, individuals may work for a particular company, play sports at a particular sports club, and attend local community groups. Sets constitute the contextual entity in which individuals meet at a certain frequency and, as such, define population social structure. Individual behaviour and set memberships can change over time, making this a dynamical approach. In essence, dynamics are induced by individuals inheriting strategies and set membership with a certain probability of mutation, and by imitating, again with a certain probability, the strategies and set memberships of others who are successful.

Based on these principles, evolutionary set theory has been proposed as a framework for modelling evolution on structured populations, including the evolution of the structure itself (Tarnita et al. 2009). In this approach, the population is not formally modelled as a network; however, it can be represented as a network at a given point in time and constitutes a dynamical graph theoretical approach. To the best of our knowledge, evolutionary set theory has not been applied to non-human animals, but there clearly are systems in which social interactions occur in sets. For example, interindividual interactions can be defined based on use of specific localities (home ranges, foraging patches, roosting sites, etc.) and the timing of their use (temporal patterns of resource use).

How do theoretical predictions from evolutionary graph theory match what we see in empirical networks? Just to summarize, evolutionary graph theory suggests that cooperators are successful in heterogeneous networks (i.e. variance in degree distribution and patterns of connectivity) (Santos and Pacheco 2006) and in networks with low average connectivity (high social viscosity) (Ohtsuki et al. 2006), both of which increase the likelihood and benefits of assortment by cooperative strategy (Santos and Pacheco 2006). In addition, theoretical work on dynamical graphs has highlighted the central role of dynamic network structure in the determining the outcome of evolutionary games on graphs (Sukumar 1992; Skyrms and Pemantle 2000; Santos, Pacheco, et al. 2006a,b).

Three recent studies have explored how empirical social networks are structured by cooperation and found evidence of assortment by cooperative behaviour in real-world social networks (Croft et al. 2009; Apicella et al. 2012; Daura-Jorge et al. 2012). Croft and colleagues (2009) found that, in a wild Trinidadian guppy population, strong social network ties were assorted by the propensity of individuals to perform predator inspection, a cooperative behaviour in which individuals approach and inspect a predator to gain information on the threat posed by the predator (Pitcher et al. 1986). In a free-ranging dolphin (*Tursiops truncatus*) population, Daura-Jorge and colleagues (2012) found that the strongest ties in the network were among cooperative individuals and that clusters in the networks contained almost exclusively either cooperators or non-cooperators. Finally a study by Apicella et al. (2012) on a human hunter-gatherer population in Tanzania found similar results, namely, that ties in the network were more likely to exist between individuals that exhibited similar levels of cooperative behaviour and that individuals cluster in the network by their cooperativeness.

Some researchers have taken the work on empirical networks a step further and explored the evolution and maintenance of cooperation experimentally. In a large experiment ($n = 1229$) performed by Gracia-Lázaro and colleagues (2012), the heterogeneity of a static social network did not play a role in the maintenance of cooperative behaviour. Indeed, other experimental work in humans applying graph theory has shown that cooperation does not prevail unless graphs are dynamic (Rand et al. 2011). Rand and co-workers (2011) found cooperation was favoured when rapid rewiring of network ties was permitted, and inhibited when rewiring was slow. Moreover, in networks where rapid rewiring was permitted, degree heterogeneity was observed, resulting in interactions between cooperators occurring at a greater frequency than cooperator–defector or defector–defector interactions. Cooperator–cooperator interactions were also found to be longer in duration compared to interactions including a defector.

Finally, a number of studies have begun to model the evolution and maintenance of cooperation using real-world social network data. The findings of these recent studies fit closely with theoretical predictions. Voekl and Kasper (2009) incorporated real-world data from 70 primate groups to investigate the probability of fixation of cooperation in the observed social structures of these groups compared to structures derived from theoretical models. In their analysis, they found strong evidence for network structures that support the evolution and maintenance of cooperative behaviour. Furthermore, they found that network heterogeneity generated from both graph topology and the variation in strengths of connections between individuals contributed to the fixation probability of cooperation. In addition, a study on the evolution of generalized reciprocity by van Doorn and Taborsky (2012) modelled the occurrence of altruistic behaviour on graphs of real-world social networks and found that cooperation could be maintained via generalized reciprocity in these real-world networks from various fish, bird, and mammal populations.

Pathways to assortment by cooperation in social networks

All mechanisms proposed to underpin cooperation lead to assortment in either physical space, social space (i.e. on social networks or in sets), or phenotypic space. The mechanisms for the evolution of cooperation can thus be seen as different ways of generating assortment (Fletcher and Doebeli 2009; Nowak et al. 2010). Thus, the key to unlocking the paradox of cooperation among non-kin is in identifying the proximate mechanisms that generate assortment among cooperators. There are many mechanisms that have been proposed to drive assortment by cooperation, and they may not be mutually exclusive; that is, cooperation can be underpinned by more than one mechanism.

Non-random distribution of individuals in space and time

The most basic way that assortment can occur is via population viscosity and spatial structure, which will mean that individuals with similar strategies are more likely to interact with one another (W. Hamilton 1971; Nowak and May 1992; West et al. 2002; Le Galliard et al. 2003; P. Taylor et al. 2007; Lion and van Baalen 2008). The degree of spatial structure and population viscosity will in part depend on patterns of demography and local competition. For example, limited dispersal will increase interactions between related individuals, which may also lead to increased competition between kin, and this competition can balance out the benefits of social viscosity (West et al. 2002).

It is well documented that there is often a non-random spatial distribution of individuals based on phenotypic traits such as size, sex, and age (Krause and Ruxton 2002; Ruckstuhl and Neuhaus 2005). If we take sexual segregation, for example, males and females of many species are known to occupy different habitats (Wearmouth and Sims 2008). Furthermore, spatial segregation by phenotypic traits may be an important mechanism in driving the assortment of social groups (Croft et al. 2003). For example, it is well documented that fish of different size may occupy habitats with different water depths, and this phenomenon may be an important mechanism driving size assortative shoaling by body size in fish (Croft et al. 2003).

However, the extent to which populations spatially segregate by cooperative tendency remains unknown. There is considerable evidence that individuals differ repeatedly in their cooperative behaviour in such a way that the propensity to cooperate may be considered a personality trait (K. Arnold et al. 2005; Bergmüller and Taborsky 2007; Charmantier et al. 2007; Schürch and Heg 2010a,b; see also Bergmüller et al. 2010 for a general discussion). For example, guppies show consistent individual differences in their cooperative behaviour during predator inspection (see Figure 3.2). Where cooperative behaviour is correlated with other behavioural traits (e.g. boldness, as in the context of predator inspection), one could imagine scenarios that could drive spatial segregation by cooperative phenotypes (for example, due to habitat differences in predation risk).

Social structuring in the absence of spatial segregation

As outlined in 'Non-random distribution of individuals in space and time', real-world populations are often socially segregated by phenotypic traits such as size and sex (Krause and Ruxton 2002; Ruckstuhl and Neuhaus 2005). If levels of cooperative behaviour are correlated with other phenotypic traits, then assortment of the population based on these phenotypic traits may lead to assortment of cooperators. The potential for such effects have been modelled as games in phenotypic space (Antal et al. 2009).

Antal et al. (2009) modelled conditional cooperators that cooperated with others only when they were phenotypically similar; otherwise, they defected. In contrast, defecting individuals played an unconditional strategy and always defected. This model showed that clustering of phenotypes in phenotypic space promoted the evolution of cooperation and is a simple mode of tag-based cooperation which can lead to the evolution of cooperation without spatial structure (Riolo et al. 2001; Traulsen and Claussen 2004; Jansen and van Baalen 2006).

More advanced tag-based models are based on individuals differentiating between cooperates and non-cooperators, with cooperators actively preferring to associate with cooperators using recognition traits (green beard mechanisms) (W. Hamilton 1964b; Dawkins 1976; Keller and Ross 1998; Jansen and van Baalen 2006). Such systems are susceptible to defecting individuals displaying 'recognition traits' and thus exploiting cooperators (A. Gardner and West 2007). Very few examples of green beards have been described, and these are restricted to microbes and one example in ants (A. Gardner and West 2009), where there is a relatively simple link between genotype and phenotype. In contrast, in vertebrates the polygenic nature of most behaviours would allow for the evolution of defectors who displayed the tag but do not cooperate.

There is some support in the literature that cooperative behaviour in real-world populations may be linked to other behavioural traits that could drive the social clustering of cooperative phenotypes. For example, Croft et al (2009) showed for a wild population of guppies in a laboratory experiment that boldness during predator inspection was negatively correlated with shoaling tendency. When the fish were returned to the wild, Croft and colleges found that the social network was significantly assorted by these behavioural traits (Croft et al. 2009). While a number of mechanisms may contribute to this assortment, one could imagine that interindividual variation in shoaling tendency (gregariousness) has a role to play in driving the observed patterns of assortment. Individuals that have high levels of gregariousness will be more likely to occur together in larger shoals.

Spatial and temporal assortment by simple behavioural rules

Theoretical work suggests that simple behavioural rules or heuristics (Hutchinson and Gigerenzer 2005) that are based on conditional movement can lead to the evolution of cooperation by generating assortment among cooperators (Aktipis 2004, 2009). An example of one such model is 'walk away when encountering non-cooperation' (Aktipis 2004, 2009), whereby cooperative individuals leave areas or social groups that contain defectors. Theoretical work has demonstrated that this can maintain cooperation in the absence of complex cognitive abilities by

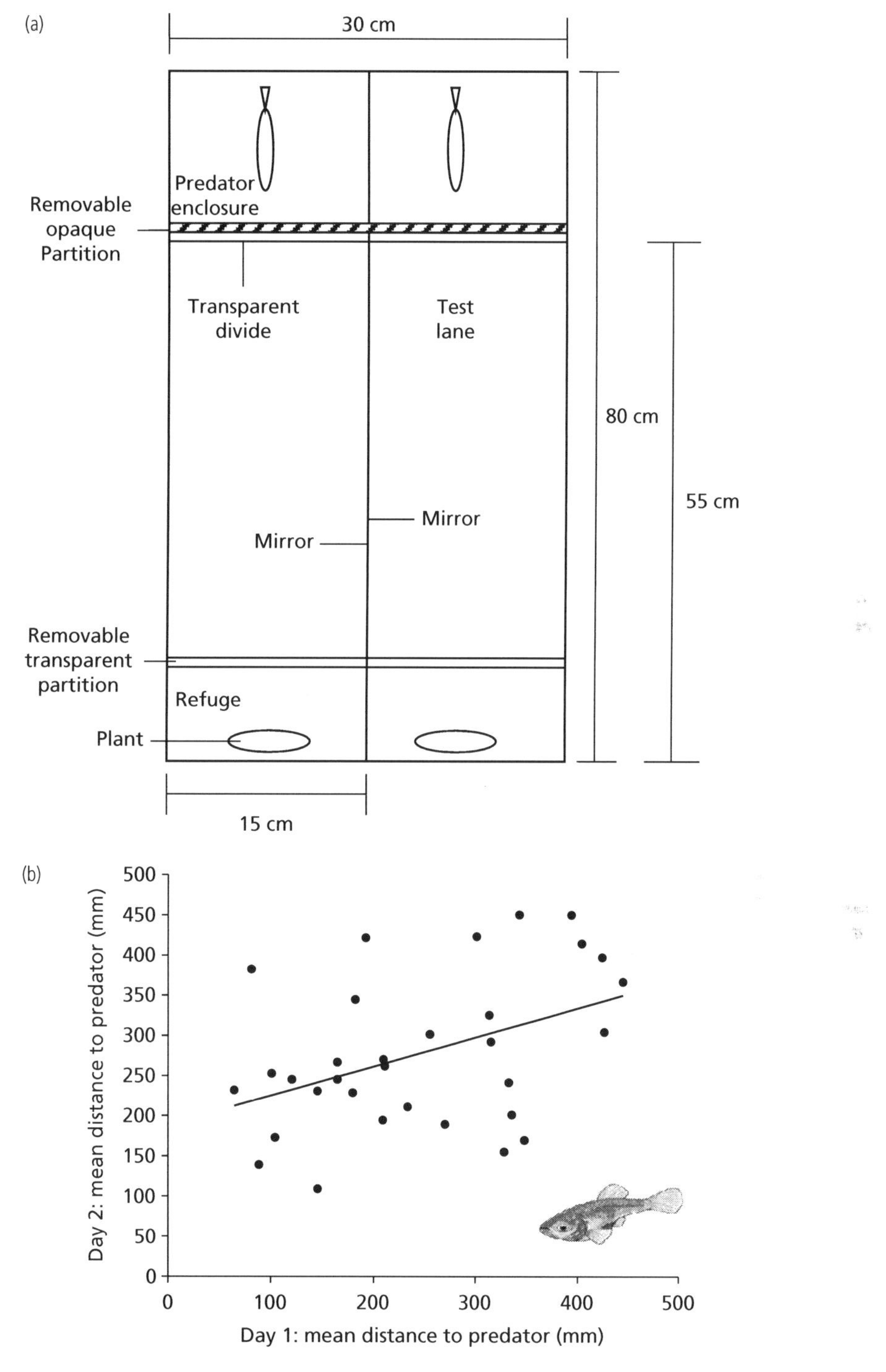

Figure 3.2 (a) Schematic diagram of the test arena used to screen fish for cooperation (two test lanes per arena). (b) Scatter plot presenting mean distance from a predator stimulus (*Aequidens pulcher*) during cooperation assays for female guppies across two days of testing separated by 24 h (Edenbrow, unpublished data). Note that lower scores represent individuals that on average moved closer to the predator stimulus during an assay. Females were significantly repeatable in their cooperative behaviour measured as mean distance to the predator (intra-class correlation coefficient R = 0.401, standard error = 0.145, p = 0.012).

promoting positive assortment among cooperators (Aktipis 2004, 2009).

In the walk-away model, individuals perceive and respond to the level of cooperative behaviour of other individuals they interact with. When a cooperative individual encounters a defecting individual (Aktipis 2004) or a group of defecting individuals (Aktipis 2009), they update their social partner(s) after one interaction with the defector. In this way, conditional movement generates assortment among cooperators. In humans, empirical evidence suggests that partner choice is important for cooperation during economic games (Orbell et al. 1984; Boone and Macy 1999; P. Barclay and Willer 2007). Boon and Macy (1999), for example found cooperation was favoured when the option to leave was available, suggesting simple heuristics akin to conditional movement strategies are active in humans. To the best of our knowledge no studies on non-human animals have investigated the degree to which individuals use such conditional movement rules in the context of cooperation.

Similar to the walk-away model, non-spatial behavioural rules have also been applied to social network theory (Perc and Szolnoki 2010). In such models, individuals strategically maintain and break apart interactions based upon the behaviour of their network neighbours. As outlined above, experimental work in humans suggests that such dynamics are fundamental to the emergence of cooperation (Rand et al. 2011).

Conditional cooperation

In the absence of strong spatial and/or temporal assortment by cooperative phenotypes, theoretical work suggests that conditional cooperation can drive the assortment of social networks by cooperation (see Figure 3.3). Conditional cooperation has generally been modelled using the prisoner's dilemma framework (see Raihani and Bshary 2011 for a recent review). In the basic prisoner's dilemma, each player has a choice to cooperate or defect and when modelled as a one-shot interaction (i.e. not repeated), defection is favoured and the only evolutionary stable strategy (Maynard Smith 1982). However, when interactions are repeated, and the game is played in an iterated context, individuals may adjust their behaviour dependent on the partner's behaviour in the previous round.

There have been multiple models generated within the iterated prisoner's dilemma, and these have in turn generated multiple strategies that can lead to cooperation. Perhaps the most well-known of these strategies is tit-for-tat (and its alternatives generous tit-for-tat (Walker and Herndon 2008), contrite tit-for-tat (Nowak and Sigmund 1992) and Pavlov (A. Foote 2008)), which can lead to the maintenance of cooperation via direct reciprocity ((Axelrod and Hamilton 1981) (Figure 3.3a). Tit-for-tat generates assortment because individuals display cooperation in response to cooperation and defection in response to defection (Axelrod and Hamilton 1981).

Many examples of cooperation have been interpreted as direct reciprocity: for example, predator inspection in fish (Milinski 1987; Dugatkin 1991), the sharing of blood meals in vampire bats (*Desmodus rotundus*) (G. Wilkinson 1984), and reciprocal support in olive baboons (*Papio anubis*) (Packer 1977). In relation to population social structure, early work on social networks in wild populations of guppies illustrated that pairs of fish repeatedly interact in stable partnerships (a prerequisite for direct reciprocity) and that these partnerships are underpinned by individual recognition and active partner choice (Croft et al. 2006). When examining patterns of cooperation during predator inspection, Croft et al (2006) found that cooperation was higher among these stable social pairs than between them. As an important note, while such observations fit with the model of direct reciprocity (i.e. individuals repeatedly interacting and cooperation), there are a number of mechanisms that could lead to such patterns.

Reciprocity can also come in the form of indirect reciprocity, in which individuals help those who have helped others (Nowak and Sigmund 1998) (Figure 3.3b). In cooperation via indirect reciprocity, help is given to others based on their reputation, and this mechanism does thus not require repeated cooperative interactions between the same two individuals. This model of cooperation is particularly attractive in human populations, where cooperative interactions between individuals are often not repeated (Nowak and Sigmund 2005). Indeed,

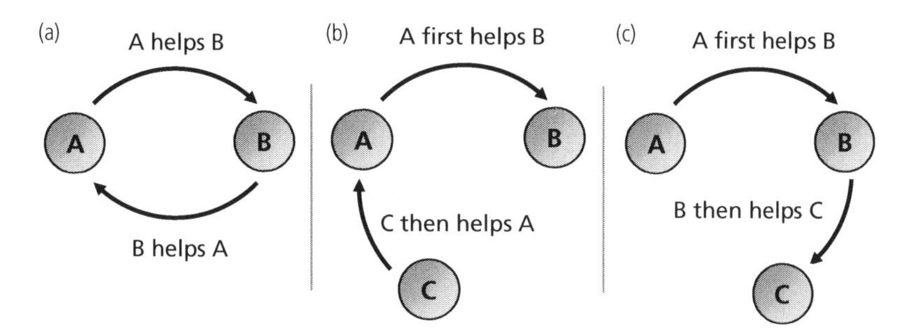

Figure 3.3 Diagrams illustrating the different forms of reciprocity. (a) Direct reciprocity, where A helps B and B helps A. (b) Indirect reciprocity, where A first helps B, C observes A helping B, and then C provides help to A, based on A's reputation as a cooperator. (c) Generalized reciprocity, where A first helps B, and then B, having received help from A, provides help to C. Figure adapted from Nowak and Sigmund 2005; reproduced with permission from *Nature*.

in humans there is empirical support that indirect reciprocity can lead to cooperation with individuals that are helpful, as they are more likely to receive help (Wedekind and Milinski 2000; Milinski et al. 2001; Seinen and Schram 2006).

While this strategy appears to be widespread in humans, there are very few examples in non-human animals (Bshary and Grutter 2006). Indeed, both direct and indirect reciprocity are both 'bookkeeping strategies' that require individuals to remember both the identity of individuals and their behaviour, cognitive abilities that are likely to be beyond the majority of real organisms (J. Stevens and Hauser 2004; J. Stevens et al. 2005). In fact, in humans it has been proposed that selection for reciprocity has played a major role in the evolution of human intelligence (Nowak and Sigmund 2005).

The extent to which reciprocity can explain the evolution of cooperation in non-human animals has been greatly debated (Hammerstein 2002; Pfeiffer et al. 2005; Clutton-Brock 2009). Where direct reciprocity has been proposed as a mechanism underpinning cooperation, other, simpler mechanisms that may generate similar behavioural patterns are often overlooked. One such mechanism is generalized reciprocity (also referred to as 'pay it forward' reciprocity (Fowler and Christakis 2010), upstream tit-for-tat (Brent, Heilbronner, et al. 2013), and upstream indirect reciprocity (Nowak and Roch 2007)), which works on the principle 'help anyone when helped by someone' (Pfeiffer et al. 2005; Nowak and Roch 2007; Rutte and Taborsky 2007) (Figure 3.3c). In contrast to direct and indirect reciprocity, generalized reciprocity does not require advanced cognitive ability and thus may have general applicability for the evolution of cooperation between unrelated individuals in less cognitively advanced species. Generalized reciprocity depends on individuals using information from the previous interaction, regardless of the partner, and thus is a much more likely mechanism in non-human organisms. One potential problem associated with generalized reciprocity, however, is that it is an anonymous strategy and, because defecting individuals cannot be identified and punished, is open to exploitation (Van Doorn and Taborsky 2012).

Recent theoretical models have demonstrated that, even in the absence of punishment or bookkeeping strategies, cooperation via generalized reciprocity may evolve when interactions are heterogeneous (Van Doorn and Taborsky 2012) and may be evolutionary stable when individuals repeatedly interact in small groups (Pfeiffer et al. 2005), when individuals assort within populations (Rankin and Taborsky 2009), or when social clustering and thus network modularity is high (Van Doorn and Taborsky 2012). In addition, generalized reciprocity has also been shown to be evolutionary stable when combined with group-leaving strategies (I. Hamilton and Taborsky 2005), additional forms of complex reciprocity (Nowak and Roch 2007), or partner-updating rules such as, for example, state dependence (Barta et al. 2011). In humans, generalized reciprocity can occur regardless of general positive state or awareness of prosocial norms (Bartlett and DeSteno 2006), is favoured in smaller groups

(Greiner and Levati 2005), and can be stronger than direct reciprocity (Stanca 2009).

Generalized reciprocity has received very little empirical attention in non-human animals. To the best of our knowledge, the only study to have looked at this directly is a study by Rutte and Taborsky (2007), which showed that female wild-type Norway rats (*Rattus norvegicus*) were more cooperative towards an unknown partner if they had previously received help. In that laboratory experiment, individuals had to pull a stick in order to produce food for a partner. It still remains unclear whether generalized reciprocity can act as a general mechanism for the evolution of cooperation across species, and importantly whether it occurs in wild animal populations. However, as outlined above, recent modelling work suggests that the social conditions required for generalized reciprocity to evolve could be widespread (Van Doorn and Taborsky 2012).

Cooperation in social networks: conclusions and future directions

Understanding the mechanisms that maintain cooperation among non-relatives is a major challenge for the twenty-first century. Decades of theoretical effort have been devoted to this question and, as outlined in this chapter, some general predictions have emerged. In particular, the modelling of cooperation using evolutionary graph theory suggests that cooperators are more successful in heterogeneous networks and in networks with low average connectivity (high social viscosity), both of which increase the likelihood and benefits of assortment by cooperation. Moreover, both theoretical and empirical work suggests that network rewiring (i.e. individuals adjusting their social ties) is key for cooperators to succeed.

However, the most fundamental requirement for the evolution and maintenance of cooperation is assortment among cooperators. When cooperators assort, they can prevail, because they interact with one another and thus gain higher fitness payoffs than defectors in the population. The key to unravelling the paradox of cooperation among non-kin is thus identifying the mechanisms that underpin assortment in real-world networks, and new studies on wild populations are essential to advance this field as a whole.

An underlying notion of behavioural ecology is that behaviour is a relatively plastic component of the phenotype (West-Eberhard 1989). Moreover, plasticity forms the foundation for much of our understanding of cooperation, because the option to cooperate or defect is dependent upon the behaviour expressed by a partner. Surprisingly, there is considerable evidence documenting interindividual differences in cooperation (Bergmüller et al. 2010) that is consistent across time (e.g. see Figure 3.2).

While there is strong evidence for cooperative personality across taxa (Bergmüller et al. 2010), we currently have very little understanding of the mechanisms that underpin this variation, and in particular the extent to which variation in cooperation is driven by intrinsic differences among individuals (i.e. personality) or whether this is the result of state-dependent plasticity. Addressing this gap in knowledge provides an exciting area for future research. Moreover, it is well documented that behavioural traits are often correlated in syndromes across individuals in a population (Sih, Bell, and Johnson 2004). For example, relatively bold individuals are often relatively aggressive and exploratory.

A major objective for future research should be to quantify how cooperation correlates with other behavioural traits such as aggression, boldness, and general activity. This is particularly important because, if cooperation covaries with other behavioural traits (e.g., activity levels), then assortment by cooperation may be a by-product of assortment by other behaviours. For example, the costs of disassortment based on activity budgets is proposed to be a major mechanism driving social segregation by sex in ungulates (Conradt 1998).

While studies have started to document the existence of interindividual variation in cooperation, we have very little understanding of how populations are socially structured by cooperative phenotypes (see Croft et al 2009 for an exception). Over the last decade, we have gained a detailed understanding of how phenotypic traits such as size and sex underpin population social structure, and spatial and social segregation of populations by such traits is well documented (Krause and Ruxton 2002;

Ruckstuhl and Neuhaus 2005). New research is needed to extend this body of work to include cooperative phenotypes and quantity, how cooperative phenotypes are distributed in space (for example, along environmental gradients), and how cooperative phenotypes are distributed across social groups (i.e. testing for social assortment).

Key to understating the evolution of cooperation is not just documenting the patterns of assortment by cooperation but determining the mechanisms that drive assortment. Theoretical work again provides testable predictions here. For example, the social partner-updating models such as the walk-away model (Aktipis 2004) suggest that simple social partner-updating can generate social assortment. To the best of our knowledge, however, such models remain untested in non-human animals, and new experiments are needed to test the assumptions and predictions of such models.

In dynamic social systems where social groups regularly encounter one another, social assortment may be underpinned by social recognition. For example, individuals may learn the identity of others and exhibit preferred and avoided interactions based on this information (Ward et al. 2009). Such individual recognition may be context dependent (whereby individuals learn the identity of an individual by either interacting with them or observing them in a particular context, e.g., cooperation) or context independent (whereby individuals learn the identity of an individual independent of associating with them in a particular context). More recent work has shown that animals may also use global habitat cues for social recognition: fish for example, generally prefer to associate with individuals that have the same habitat odour (Ward et al. 2009). Such social preferences are manifested in the social network structure and generate both local and global network characteristics such as clustering and assortivity through maintaining stable social relationships (Croft et al. 2008). While there has been a great deal of interest in the mechanisms of social recognition and the structure of animal social networks, these two research areas have rarely been integrated to determine the role of social recognition in driving social network structure.

In the absence of spatial or social segregation by cooperative phenotypes, conditional cooperation can lead to the evolution of cooperation. Early work on conditional cooperation focussed on direct reciprocity; however, there is a general consensus that direct reciprocity is unlikely to explain the majority of cooperative behaviour observed in non-human animals. Theoretical work over the last decade suggests that other, less cognitively demanding strategies such as generalized reciprocity are more likely to underpin cooperation in non-human animals. Generalized reciprocity has received a great deal of attention in the theoretical literature and has been the focus of a number of empirical studies on humans but has received very little attention in non-human animals. New experiments are needed that can tease apart the effects of direct versus indirect reciprocity in non-human animals and compare the relative importance of these two mechanisms directly.

One area that is particularly exciting for future work is to study the co-evolution of social networks and cooperation (Fehl et al. 2011), and how cooperation can spread through social networks (Fowler and Christakis 2010). In humans, opportunities for reality mining (Komers 1997), which uses online social networks or other technology (e.g. mobile phones) to collect real-world data on human social behaviour, can provide access to enormous datasets that can track changes in social networks in real time. If methods can be developed to integrate information on cooperation with these network databases, then it will provide an exciting test bed for the theoretical predictions regarding the role of social networks in underpinning cooperation.

Acknowledgements

D. P. C. and S. K. D. acknowledge funding from the Leverhulme Trust.

CHAPTER 4

Mating behaviour: sexual networks and sexual selection

Grant C. McDonald and Tommaso Pizzari

Introduction

In non-selfing organisms, reproduction is mediated by interactions between individuals. These interactions represent a social trait, in which the sexual behaviour of an individual has the potential to affect not only its own fitness but also that of other individuals through its influence on mating and fertilization (Pizzari and Gardner 2012). Differential sexual interactions among members of a population often constitute a major source of variance in individual fitness, creating opportunity for selection underpinning rapid and strong evolutionary change (Shuster and Wade 2003). However, studying the fitness implications of sexual behaviour is not trivial.

The outcomes of sexual interactions are often complex, for three main reasons. First, the sexual behaviour of an actor has ramifications which transcend dyadic interactions by influencing the fitness of multiple recipients. Second, both the behaviour and its fitness outcomes can change plastically with the phenotype of the interactants. Third, individuals typically interact non-randomly within a population. This can be due to multiple ecological and behavioural factors, which include spatio-temporal variation in resource distribution, constraints on dispersal, and assortative or disassortative matching.

The goal of this chapter is to discuss how social network theory, by addressing these complexities, can help biologists better understand the evolutionary significance of sexual behaviour. Recent years have witnessed an explosion of studies utilizing social network approaches to study animal behaviour. Our focus is to elucidate how these approaches can be harnessed to complement more traditional approaches and provide novel information on the evolutionary ecology of sexual behaviour. Below, we briefly introduce the principles of sexual selection theory, which is the Darwinian framework for understanding the functional significance of sexual behaviour, and then present social networks within this context (i.e. sexual networks).

Sexual selection

The functional significance of much of the diversity of sexual behaviours can be understood in light of two social mechanisms: (i) competition among members of the same sex over access to reproductive opportunities, and (ii) selection of reproductive partners by members of the opposite sex. Darwin (1859, 1871) intuited that these mechanisms arise from variation in individual reproductive success, generating opportunity for two distinct episodes of selection, namely intra- and intersexual selection. The realization of sexual selection resolved a number of outstanding issues in Darwinian theory: the evolution of extravagant traits, which was unexplained by viability selection, the evolution of sexual dimorphism, and the rapid evolutionary divergence of related species (Darwin 1871; Andersson 1994; Shuster and Wade 2003). Modern sexual selection theory developed this Darwinian framework in two fundamental ways.

First, while the Darwinian view of sexual selection was limited to variation in mating success, Parker (1970) realized that in several insect species

Animal Social Networks. Edited by Jens Krause, Richard James, Daniel W. Franks and Darren P. Croft.

sexual selection can in fact continue after mating because females mate with multiple males (polyandry) in a way that forces their ejaculates to compete for the fertilization of a single set of ova, a process that became known as sperm competition. Others noted that polyandry can also enable females to systematically bias the outcome of sperm competition in favour of the ejaculates of certain males, a process known as female sperm selection or cryptic female choice (Childress and Hartl 1972; Thornhill 1983; Eberhard 1996). Sperm competition and cryptic female choice represent episodes of intra- and intersexual selection, respectively, which arise after copulation from variance in the share of paternity within a single set of eggs. The last decades have demonstrated that polyandry is near ubiquitous and that episodes of postcopulatory sexual selection can be powerful agents of evolutionary change (e.g. Birkhead and Pizzari 2002). Clearly, understanding the operation of sexual selection requires an appreciation of both pre- and postcopulatory processes.

Despite this, however, only recently has a full appreciation for the potential for the interaction between both pre- and postcopulatory processes begun to develop (Collet et al. 2012; Kvarnemo and Simmons 2013; Parker and Birkhead 2013). Collet et al. (2012) investigated the relationship between the level of polyandry and the relative strength of both pre- and postcopulatory selection in red junglefowl (*Gallus gallus*) and showed that increasing levels of polyandry are associated with a decrease in the overall strength of sexual selection but that this decrease was chiefly the result of a reduction in the magnitude of the pre- (rather than post-) copulatory sexual selection. This study highlights how mating dynamics can not only affect the overall strength of selection but also differentially modulate the roles of pre- versus postcopulatory episodes and thus the evolutionary trajectories of populations.

Second, a quantitative framework has been developed to formally define the opportunity for sexual selection and the strength of its episodes on individual phenotypic traits (Lande and Arnold 1983). The total reproductive success of an individual can be decomposed into three multiplicative constituents as follows:

$$T = (M * N * P) + \varepsilon \qquad (1)$$

where M is the number of mating partners, N represents their fecundity, P represents the proportion of eggs fertilized (paternity share), and ε is an error term with zero mean. The variables M and N contribute reproductive success through mechanisms leading up to mating (i.e. precopulatory: intrasexual competition over mates and mate choice), while P captures postcopulatory mechanisms (sperm competition and cryptic female choice). The maximum potential strength of sexual selection is proportional to the variance in T, which is determined by the independent variance in each constituent and their covariances. The opportunity of sexual selection I can therefore be expressed as the amount of standardized variance in each constituent (Shuster and Wade 2003) as follows:

$$I_X = \frac{\sigma^2}{\overline{X}} \qquad (2)$$

where $\overline{X}$ represents the population average of a given constituent of reproductive success (T, M, N, or P) and σ^2 its variance. Therefore, I is proportional to the maximum strength that sexual selection can exert on a given trait, z, which causally covaries with reproductive success.

This approach entails the estimation of the relationship between a measure of individual reproductive fitness and measurements of a phenotypic trait (e.g. tail length, body size, display rate) as follows:

$$\omega_T = \beta_z + \alpha \qquad (3)$$

where ω_T represents reproductive fitness and α represents the intercept of the regression. Selection on such traits can then be formalized as β_z, the slope of the regression of ω_T on z, where ω_T and z are standardized to the mean of the population as a whole (Arnold and Wade 1984). This *selection analysis* provides researchers with an estimate of both the strength and direction of selection operating on the phenotypic traits measured, termed *selection gradients*. These selection gradients can be used to estimate both directional, as well quadratic selection such as stabilizing or disruptive relationships between fitness and the phenotypic trait (Brodie et al. 1995; Hunt et al. 2009). Selection analysis has allowed researchers to determine the targets of sexual selection and understand its relative strength and form on different traits, as well as test how selection varies across populations and fitness components (Kingsolver et al. 2001; Kingsolver and Pfennig 2007; Siepielski et al. 2011).

Sexual selection in structured populations

The implicit assumption of these selection analysis studies, which measure both fitness and trait value relative to the population average, is that competitive and selective contexts are homogenous across the population (Formica et al. 2011; McDonald et al. 2013). However, recent studies taking a more sophisticated approach have undermined this assumption by showing how sexual selection may vary across a variety of ecological axes in both time and space (Cornwallis and Uller 2010). Both spatial and temporal variation in the ecological context of competition can have large effects on the strength and form of sexual selection. Furthermore, there has been increasing focus on how local variation in the social environment determines the force of selection on phenotypic traits and the evolution of sexual strategies. For example, Arnqvist (1992) found that variation in population density across three populations of the water strider (*Gerris odontogaster*) not only changed the strength and form of sexual selection on male phenotypes but also qualitatively changed the targets of sexual selection. Similarly, Gosden and Svensson (2008) investigated both temporal and spatial variation in sexual selection regimes in the damselfly (*Ischnura elegans*) across an ecological gradient and found a complex and variable mosaic of disparate sexual selection pressures on male body size, determined by fine-scale differences in densities of female colour morphs and body size. These studies call into question the assumption of classical population-level approaches that intrasexual competition at the population level (global scale) is representative of patterns of intrasexual competition and the social competitive context at the local scale (i.e. the structure of the competitive environment is homogenous across the population). Using simulations, Benton and Evans (1998) showed how such local structure can result in bias estimates of sexual selection. In populations where females select males following a 'best of *n*' strategy, local grouping of males by attractiveness can blur the relationship between male attractiveness and fitness at the population level. To account for this, several quantitative methods have been developed to explicitly estimate the effect of the local socio-competitive group on individual fitness (Nunney 1985; Heisler and Damuth 1987; Wolf et al. 1999). These approaches, known as multilevel selection analyses, embody a hierarchical extension of classical selection analyses where selection is measured both at the level of individual phenotype and at the level of the group phenotype (Okasha 2004a,b). The main multilevel selection approaches include contextual analysis and neighbour analysis (Nunney 1985; Heisler and Damuth 1987; Okasha 2004a).

The explicit goals of these analyses are to estimate the relationship between individual fitness and the individual phenotypic trait, as well as the effects of the local competitive context on individual fitness. More formally, contextual analysis takes the form

$$\omega_{ij} = \beta_1 z_{ij} + \beta_2 z_j + \mathcal{E} \quad (4)$$

where ω_{ij} and z_{ij} are the fitness and trait values for the *i*th individual in the *j*th group, respectively, Z_j is the trait value for the *j*th group, and ε is the residual variance. The regression coefficients β_1 and β_2 are, therefore, the partial regression of individual fitness on individual character, controlling for group character, and the partial regression of individual fitness on group character, controlling for individual character, respectively. The fitness effects of an individual's local social group on individual fitness are captured by β_2. The neighbour analysis differs from contextual analysis only in the calculation of what constitutes group phenotype (i.e. Z_{jk}; see McDonald et al. 2013). In practice, both methods are likely to give similar results; however, larger differences between the two methods are expected when group size is low (Okasha 2004a; McDonald et al. 2013).

A third common multilevel selection approach, social selection analysis, identifies local groups in the same way that neighbour analysis does (Wolf et al. 1999; Formica et al. 2011). However, social selection analysis differs slightly from contextual analysis in that it also includes a term called the interactant covariance C_{ij}, which is estimated using the Pearson product correlation between z_{ij} and $Z_{j.}$ This term measures the correlation between individual phenotypes and the phenotypes of their local group and is included thus:

$$\omega_{ij} = \beta_1 z_{ij} + C_{ij}\beta_2 z_j + \mathcal{E} \quad (5)$$

When positive, C_{ij} demonstrates that individuals of similar phenotypes tend to share group membership, whereas negative phenotypic assortment results in negative values of C_{ij}. By multiplying β_2, the term C_{ij} translates local group effects into population-level selection on the focal trait. By estimating explicitly the effect of local group on individual fitness, multilevel selection provides a powerful tool to understand how local competitive structure affects the evolutionary trajectories of phenotypic traits. Furthermore, such methods bring to the fore the opportunity for selection to act on alternative social strategies that individuals may use to alter their own selection regimes by manipulating their own competitive context (Formica et al. 2011).

Despite the appeal of the multilevel selection approach, however, its utility relies on the identification of individuals' local competitive groups. This has often posed a non-trivial challenge (West et al. 2002), with local competitive context often approximated by using spatial proximity (Kasumovic et al. 2008; Eldakar et al. 2010). In many systems such measures of 'local' may often require arbitrary decisions and may not reflect the scale of competition.

Difficulties in determining local group have resulted in a bias in systems studied towards more tractable invertebrate (Tsuji 1995; Eldakar et al. 2010) and sessile systems, where an individual's nearest neighbours are used to define local competition (Stevens et al. 1995; Donohue 2003, 2004; Weinig et al. 2007), with only one recent study in vertebrates (Laiolo and Obeso 2012). More complex analyses have been developed that use home-range overlap to infer the extent of competition between individuals (Formica et al. 2010). While these estimation methods may prove powerful, especially in cases with largely indirect competition, they may not be appropriate to accurately identify functional groups in species with complex within-range space use (Formica et al. 2010). In fact, even within groups with home-range overlap, there can be considerable variation in interaction patterns between individuals both within and between groups (Wolf et al. 2007; Marsh et al. 2010).

The logic of sexual networks

Here we argue that the use of network data explicitly built upon interactions between individuals provides a powerful framework in which to characterize intra- and intersexual interactions and can be used to describe which individuals are in local competition. For example, in a wild population of house finches (*Carpodacus mexicanus*) Oh and Badyaev (2010) used network analysis to characterize the local competitive environments of individuals and its effects on male reproductive success. To achieve this, Oh and Badyaev (2010) first used network algorithms to define, within the finch social network, subsets of individuals (communities) which interacted with each other more often than with the rest of the population. The authors then quantified selection on male ornamental traits by standardizing male trait value at the level of the community rather than at the level of the population. The study found that individual males can modify their own selective context, such that males of lower attractiveness were able to increase their relative attractiveness by moving between different socio-competitive groups.

However, when measuring selection, community approaches may provide relatively coarse measures of individual competitive group. This is because community detection techniques do not account for variation in the structure of interactions within communities and do not account for weak ties between communities. An example of a population with a modular community structure is provided in Figure 4.1. In this population, females inspect males and then choose to copulate with males based on their relative attractiveness. If the phenotypic distribution of males within this population were structured such that males within communities were more similar, females would tend only to inspect a biased representation of the population phenotypic distribution, due to the tendency of individuals to interact within rather than between communities.

For this reason, traditional population-level selection analyses provide biased estimates of sexual selection on male attractiveness by ignoring non-random female sampling regimes. Selection

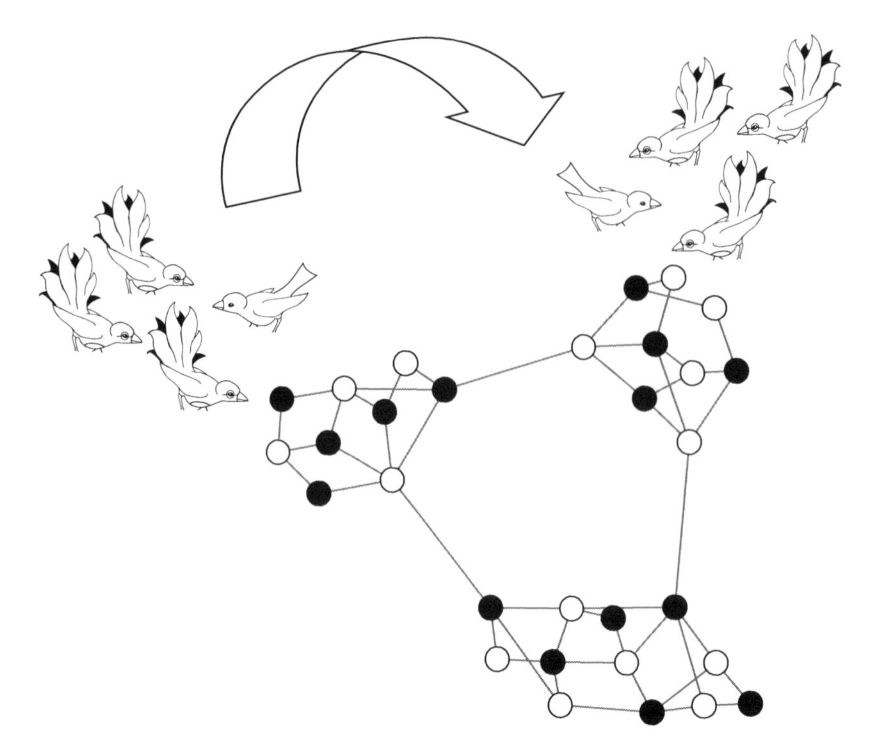

Figure 4.1 A hypothetical modular intersexual network with both males (black nodes) and females (white nodes). An edge (link) between a male and a female node represents courting between these two individuals. The diagram highlights bridge females moving between communities to inspect males.

measures based purely on community membership also ignore the between-community ties that themselves may have important repercussions for the estimation of phenotypic selection. A key benefit of the network approach is that it allows a common framework that can be applied across different episodes of selection, which may operate on vastly different social and spatial scales. In the following sections, we illustrate how sexual networks can be utilized to: (i) measure sexual selection, (ii) characterize patterns of intersexual interactions and their consequences for sexual selection, (iii) develop a fuller understanding of mating systems, and (iv) study the functional significance of individual sexual network measures.

Intrasexual interactions and the measurement of sexual selection

McDonald et al. (2013) developed a quantitative framework based on network data of competitive interactions to allow unique competitive environments for each member of the population. This method utilizes 'competitive networks' to capture the structure of intrasexual competition in sexual populations. Competitive networks can be derived from direct intersexual competitive interactions (e.g. male combat) or via intersexual male–female networks (e.g. male–female courtship or copulations). These networks consist of individuals connected by edges representing unique competitive events. For instance, if two males within a lek display to the same female, both males have competed to mate that female and to fertilize her batch of ova and hence will share a competitive tie in a competitive network. Similarly, if two males inseminate a female sufficiently close in time such that their ejaculates overlap at the time of fertilization, then these males share a competitive tie through sperm competition.

Following this pattern, we can construct a complex web of interactions, building competitive structure as an emergent property of individual interactions. Further resolution can be added to these networks by adding value to the edges of the network such that, if two males compete over two unique females or for two batches of ova within the same female, the competitive edge between these males will carry an edge weight of 2. In Figure 4.2a, we show how intrasexual competitive networks can be derived through the projection of intersexual courtship or copulation networks for dioecious animals. This approach is not limited to dioecious animals and can be extended to encompass sexual competition networks in hermaphrodites (Figure 4.2b) as well as pollen competition networks in flowering plants (Figure 4.2c).

Here, we extend the analysis of McDonald et al. (2013) and explore how the scale of competition may vary for both pre- and postcopulatory selection. We use a simulation approach where males are organized into 'leks', and females visit multiple leks, forcing males to compete both before and after mating within and across leks. Specifically, we generated replicate populations of 500 males and 500 females. Individuals were placed randomly within a circle with radius 0.5.

Each male was allocated a phenotypic trait z, ranging from 0 to 1, which positively influences both his pre- and postcopulatory reproductive success. Male trait z was allocated in direct relation to the position of males on the horizontal axis across the circle diameter, such that males on the left of the circle have lower values for z, and males on the right of the circle have higher values for z. This generated an approximately normal distribution of male trait values for each population.

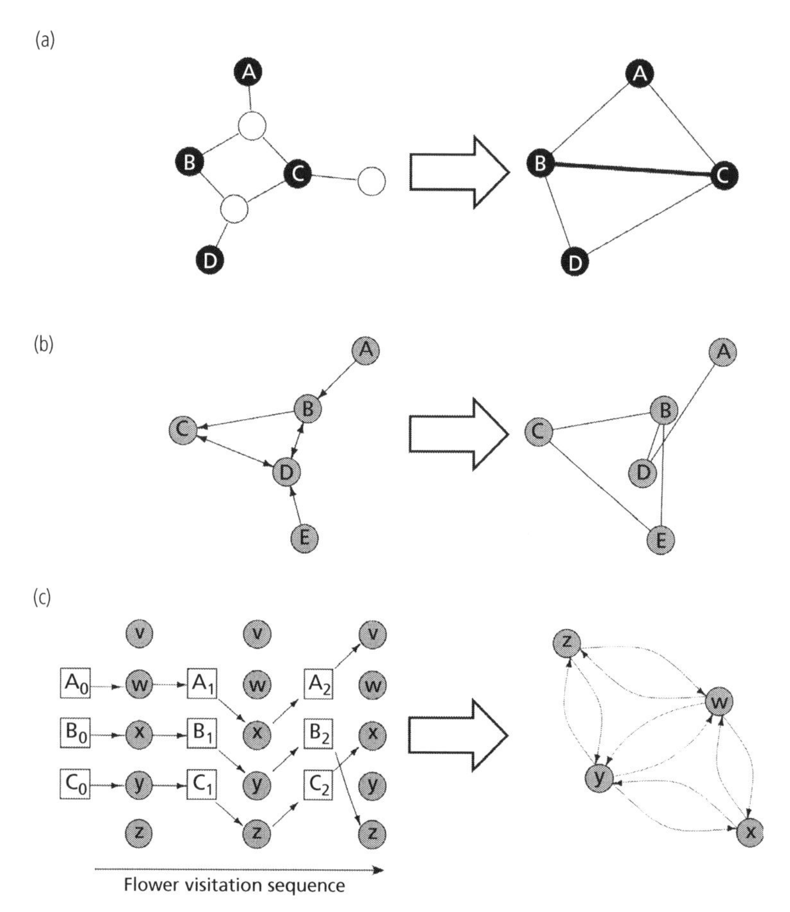

Figure 4.2 (a) Intersexual network with males (black) and females (white) nodes projected into a male–male competitive network. (b) A simultaneous hermaphrodite mating network where edges represent copulations. This network is directed with black arrows showing the direction of ejaculate transfer. This network is then shown projected into a male–male competitive network. (c) Network of simultaneous hermaphroditic plants (circles) and pollinators (squares). Network shows pollinators visiting three successive plants each, and projection represents the pollen competition network between plant individuals.

Males were organized into 100 'leks' of 5 males each. Lek composition was determined with respect to z values such that the first lek contains the five males with the lowest z, and the last lek, the five males with the highest z values. Females visit a total of n leks, inspect all associated males, and mate with the two males with the highest z within each lek, before moving onto the next lek. We generated 1000 replicate populations each, for a range of values of n from 2 to10. This created a scale ranging from low levels of n, where females are limited in the trait space they can inspect, up to $n = 10$, where females could potentially inspect males from across the entire trait space.

Lek composition was fixed for each male across all female movements. This process generated a competitive structure such that males only engaged in precopulatory competition within their lek, but postcopulatory competition could occur within and between leks. For every male, we calculated his mean mating success as the mean probability that an inspection would result in a successful copulation. Postcopulatory reproductive success was calculated as the mean probability of fertilizing ova across all females mated, where the probability of fertilization was calculated as the focal male's z, proportional to the sum of all male z values that mated the particular female.

For every replicate, we generated both pre- and postcopulatory competition networks. We estimated the gradient of pre- and postcopulatory sexual selection on male phenotype (z) using network-based multilevel selection analyses; for each model, we included male trait z and the weighted mean group trait for each male as calculated from intrasexual competitive networks. We show how this approach provides a substantially more accurate representation of selection compared with classical population-based methods, which quantify selection on a global scale.

Precopulatory selection

Across a gradient of number of leks visited by females, classical population-based methods failed to identify directional selection on male trait z, despite strong precopulatory selection imposed by our simulation protocol (Figure 4.3). This is because competition was highly local within the male leks, and lek composition was positively assorted by z values. This obscured the correlation between precopulatory mating success and z at the population level.

Multilevel selection approaches using network information take this population structure into account. By identifying the scale at which competition occurs, both contextual and neighbour analyses identified strong positive precopulatory selection on z across all levels of n. The strong positive selection on male z was also closely mirrored in multilevel selection analyses by the strong negative selection on mean competitive group z values. This indicates strong selection against competing with males that have high relative z values. This result highlights the potential for selection to operate not only on increasing values of z but also on social strategies that may allow males to modify their competitive context, a behaviour which has recently been identified in a number of studies (Oh and Badyaev 2010; Gasparini et al. 2013).

Postcopulatory selection

Similar to precopulatory selection results, multilevel selection analyses showed consistently higher selection estimates on male trait z than global selection analyses. This difference was strongest when females visited few leks and polyandry was low. Global estimates of selection increased with the number of leks visited. This result was due to the combined effect of polyandry and the increase in the breadth of trait values that females sample, increasing the covariance between male trait z and fertilization success. These results are consistent with those obtained by McDonald et al (2013), showing that, for low to intermediate levels of polyandry, multilevel selection analyses measure selection more accurately than classical approaches. As competition becomes more global due to female promiscuity, both multilevel and global estimates of selection on z begin to converge.

Overall, these results highlight how both pre- and postcopulatory selection can operate on vastly different scales, both social and spatial,

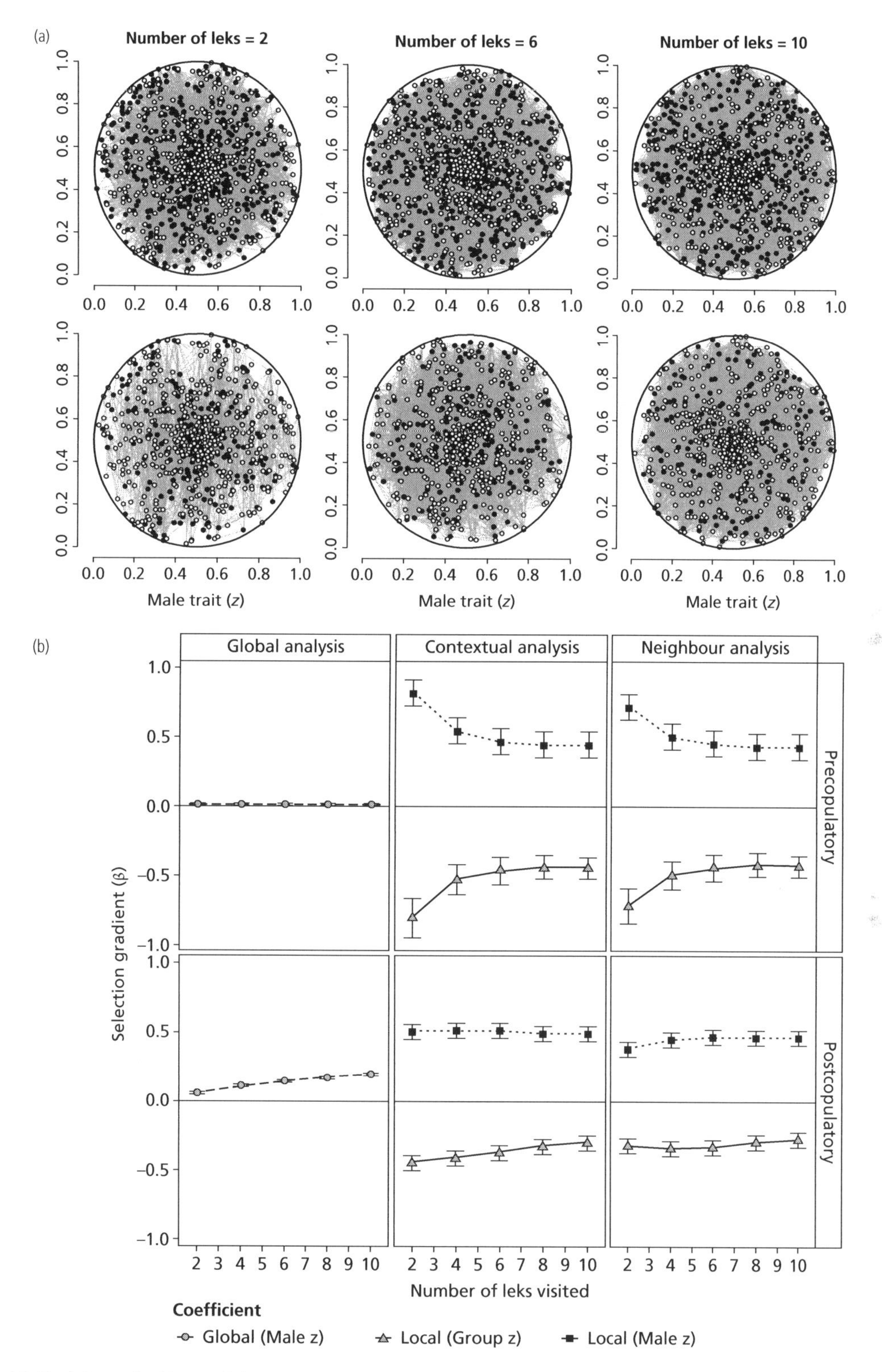

Figure 4.3 Simulation results. (a) Network diagrams show representative simulation results for 6 replicate populations of 500 males (black nodes) and 500 females (white nodes), where females inspect and mate males from 2, 6, or 10 different leks. The top row shows precopulatory networks where edges represent mate inspection events, and the bottom row shows mating networks where edges represent copulations between pairs.
(b) Selection estimates on male trait *z* using population (global) selection analyses and two multilevel selection techniques based on network information (contextual and neighbour analyses). Selection gradient estimates show mean values across 1000 replicate populations ±1 standard deviation.

and that this provides opportunity to misinterpret measures of both pre- and postcopulatory selection. Specifically, the nature of competition through both episodes of selection and the scale at which they occur will determine the extent to which classical measures of selection provide biased estimates of phenotypic selection. Furthermore, these results demonstrate the power of multilevel selection techniques informed by sexual competitive networks to provide a common framework in which to measure sexual selection across disparate episodes of selection operating on different scales.

Although we have focused on sexually competitive traits that result in negative group effects on individual fitness, more complex fitness implications may arise. For instance, competing in highly competitive groups may reduce the fitness of relatively poor competitors, but these individuals may also gain benefits if more females visit groups containing highly attractive males. It may be possible to measure such effects by partitioning the group component of multilevel selection analyses into one component which consists of the focal phenotype *relative* to mean competitor phenotype, and one which is explained by mean competitor phenotype alone. Finally, the methods presented above, if combined with information on genetic relatedness, may provide a useful tool to understand the extent to which related individuals compete directly and how this mediates the potential for kin-selected behaviours.

Intersexual interactions, mating patterns, and the operation of selection

In this section we explore how the mating patterns of a population, the 'mating topology', captured by network analysis can further elucidate the operation of pre- and postcopulatory sexual selection. In a mating network the number of mating partners an individual has is described by its node degree. In species where females do not mate multiply (i.e. no opportunity for postcopulatory selection), node degree may be sufficient to explain individual reproductive fitness (number of offspring). However, when females are polyandrous, a male's reproductive success also depends on the risk and intensity of sperm competition across the females with which he mated (Parker 1998).

Under sperm competition, mating and sperm transfer may not necessarily result in the fertilization of ova (Preston et al. 2001). Copulations with polyandrous females may therefore represent decreasing returns in terms of fertilization success because they are associated with intense sperm competition (Parker and Pizzari 2010). Male reproductive success is thus tempered by the extent to which female partners mate with additional males. How such sperm competitive environments are distributed across males in the network may have strong repercussions for the distribution of reproductive success and the strength of selection.

If the males with the highest number of mating partners also tend to mate with females with the highest number of mating partners (i.e. the most polyandrous females), this may reduce the benefits in terms of fertilization success of high-degree males, as those males will also face on average the highest intensity of sperm competition (Sih et al. 2009). Indeed, if males that mate with many females also tend to face the most intense sperm competition, then the number of mating partners may indeed negatively correlate with reproductive success. Importantly, this pattern would be expected to reduce the strength of precopulatory selection on mating success (i.e. β_M, the 'Bateman gradient'). This is because males with high mating success also on average face high sperm competition intensity, flattening the relationship between mating success and reproductive fitness. In network terminology, this pattern is described as positive assortativity by degree (i.e. nodes of similar degree are positively associated (Newman 2002a, 2003)) (Figure 4.4c,d).

Degree assortment in networks has been explored in a variety of contexts, including human and animal social networks (Newman 2002a; Croft et al. 2005) as well as ecological networks (Abramson et al. 2011). In ecological networks such as that of plant–pollinator or plant–herbivore systems, edges are drawn across different trophic levels or between plants and their pollen vectors. Similar to male–female sexual networks, these networks are bipartite; that is, there are two sets of nodes (e.g. consumers and producers), and edges are only

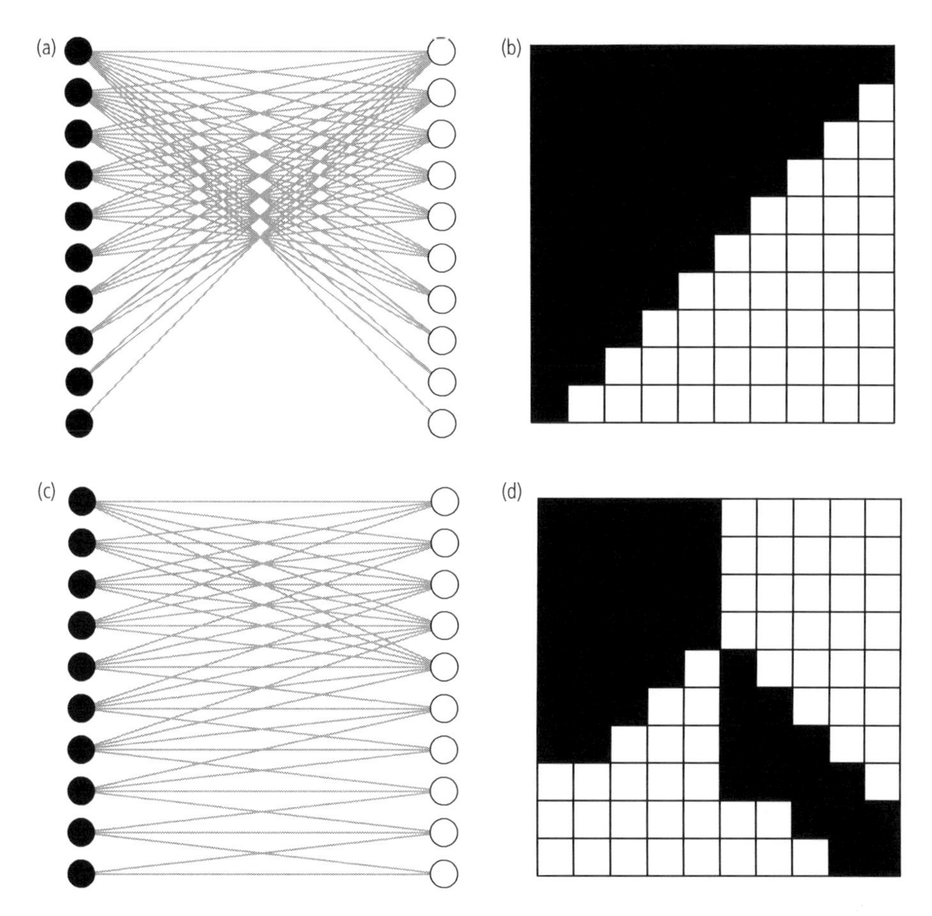

Figure 4.4 Two hypothetical mating networks, each presented in two alternative graphical representations. (a) Network representation of a mating network characterized by positive degree assortment. (b) Matrix representation of a mating network characterized by positive degree assortment. (c) Network representation of a mating network characterized by negative degree assortment (disassortativity). (d) Matrix representation of a mating network characterized by negative degree assortment (disassortativity). Black nodes and rows represent males, and white nodes and columns represent females. Edges and filled squares indicate a pair that has copulated. Males and females are ordered by number of mating partners.

drawn between these sets, and never within. Such networks tend to show negative degree assortativity (disassortativity) where generalist species (those with many connections) tend to be connected to specialist nodes of the other set, and vice versa (Bascompte et al. 2003). These disassortative networks also tend to show a nested structure of interactions where specialist nodes only interact with a restricted subset of more generalist nodes (Bascompte et al. 2003; Bascompte and Jordano 2007; Ulrich et al. 2009; Joppa and Pimm 2010) (Figure 4.4a,b). This can generate a strong core–periphery structure, where individuals within a well-connected core are strongly connected to each other, whereas peripheral nodes are weakly connected to the core and sparsely connected to each other (Bascompte and Jordano 2007; Ulrich et al. 2009) (Figure 4.4a,b).

In the context of mating networks, such disassortative nested mating patterns may have strong implications for variation in the level of sperm competition intensity experienced by different males. This is because males with few partners face on average higher sperm competition than those who have many mating partners and who instead tend to experience higher exclusivity of mating (Sih et al. 2009). This scenario may increase the variation in male reproductive success and the Bateman gradient (Sih et al. 2009; McDonald et al. 2013).

Using existing network techniques, it is possible to quantify degree assortativity in mating networks. Newman (2002a) introduced a quantitative measure of degree assortativity in networks, now termed the Newman assortativity coefficient r. This measure is the standard Pearson correlation coefficient between the degree of a node and the mean degree of its neighbours, and ranges from −1, representing negative assortativity, to 1, demonstrating high positive assortativity, with 0 representing no relationship. Similarly, there has been a proliferation of nestedness measurements (Ulrich et al. 2009). One of the most widely used methods is that of Almeida-Neto et al. (2008), named NODF. This method is based on the rules of percentage overlap and decreasing fill between pairs of rows and columns, and measures the nestedness of the whole network ranging from 100 (perfect nestedness) to 0 (no nestedness).

A strong positive relationship between nestedness and disassortativity is clear (May et al. 2008; Abramson et al. 2011; Gonzalez et al. 2011; Koenig et al. 2012). However, disassortativity is best regarded as a more general property of degree correlations also found in one-mode networks, whereas nestedness is restricted to bipartite (two-mode) networks (Sugihara and Ye 2009). Furthermore, generally positively degree assortative networks can show some degree of nestedness (Abramson et al. 2011). Despite its clear implications for sexual selection however, the way in which such population-level patterns of mating behaviour (the mating topology) generate variation in the level of sperm competition experienced across males has rarely been quantified (Sih et al. 2009).

So far, we have discussed binary ties, where an edge is present or absent (i.e. the pair either mated or not). However, in many systems multiple mating events between the same individuals may be an important determinant of reproductive success (McNamara et al. 2008; Collet et al. 2012). In such cases, the frequency with which a male mates with individual females also needs consideration. This can be achieved using weighted networks, where edges are weighted by the number of repeat mating events between individuals. With respect to disassortativity and nestedness, there has been effort to develop methods that utilize such quantitative interaction data (Galeano et al. 2009; Almeida-Neto and Ulrich 2011).

A perfectly nested weighted network similar to that provided by Almeida-Neto and Ulrich (2011) is shown in Figure 4.5a. In this network, males are represented by rows and females by columns. Individuals are ranked based on the total number of unique partners (node degree) and the total number of copulations across all partners (node strength). In the perfectly nested example, the male with the highest number of mating partners also has the highest re-mating rate with the most polyandrous females, and male re-mating rate decreases with female polyandry for all males, demonstrating that males re-mate with individual females based on their polyandry. Such a pattern would be expected

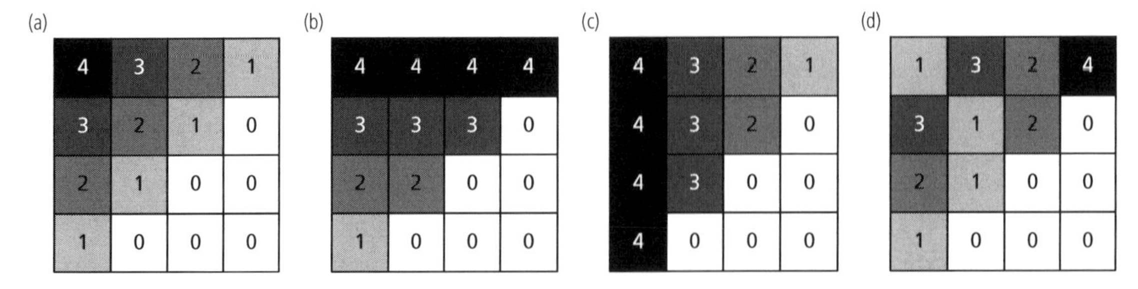

Figure 4.5 Four mating networks; rows represent males, and columns represent females. Filled squares indicate mating between pairs. Numbers and shading intensity represent the number of copulations between pairs. Rows and columns are ordered by number of mating partners and total number of copulations. Networks demonstrate (a) perfect weighted nestedness, (b) weighted row nestedness, (c) weighted column nestedness, and (d) high binary nestedness but relatively low weighted nestedness.

to produce high variation in reproductive success among males, as the males with the highest number of mating partners also have the highest relative sperm investment across females mated.

As in binary nested networks, we can again quantify the extent that the interaction patterns of nodes with few connections tend to represent subsets of those nodes with higher mating rates. Interestingly, the nestedness contribution for males and females can also be calculated separately, to understand whether the pattern of ejaculate transfer is nested across males (Figure 4.5b) or across females (Figure 4.5c). Quantitatively distinguishing between such population-level mating patterns may facilitate tests of optimal sperm allocation strategies across males and the influence of group-level patterns of such strategies for variation in male reproductive success.

More recently, further quantitative methods have examined the nestedness of ecological interaction networks such as host–parasitoid networks (e.g. Staniczenko et al. 2013). This work has revealed that, although networks were binary nested, weighted networks tended to be non-nested, suggesting that interaction strengths between species are organized to reduce competition. Exploring such patterns in mating networks may provide an interesting complement to traditional techniques to understand how mating behaviour contributes to the operation of selection and variation in reproductive success. Furthermore, it may be enlightening to investigate how such competitive structures vary in response to other factors such as group relatedness.

Finally, degree assortativity in the mating topology will also have ramifications beyond sperm competition. For instance, classic research in the patterns of sexual behaviour in humans demonstrated how assortative mixing by degree will affect the transmission of sexually transmitted diseases (May 2013). In negatively assorted networks, epidemics may spread more slowly but reach a higher level of infection than in similar but positively assorted networks (Gupta et al. 1989).

Mating system ecology

Here we discuss how patterns of intersexual interactions captured by network analysis can help biologists understand both how ecology shapes mating systems, and the evolutionary ecological implications of such patterns. Classical characterizations of animal mating systems employed coarse categories based on male and female promiscuity (e.g. monogamy, polygamy) (Thornhill and Alcock 1983; Rubenstein and Wrangham 1986; Shuster and Wade 2003). In their seminal work, Emlen and Oring (1977) linked mating system definitions to the ecological context of competition. This provided a predictive framework for variation in mating systems based on the ecology of the population.

A key prediction is that the spatial and temporal distribution of resources determine the ability of individuals to monopolize access to mates (Emlen and Oring 1977). For instance, where resources are spatially clumped and more easily defendable, a subset of the population may be able to better control access to this resource and therefore access to mates that utilize these resources. Therefore, variation in resource availability may affect the variation in reproductive success and the potential strength of sexual selection among populations. Similarly, the ratio of reproductive males to females (operational sex ratio) as well as the spatial and temporal distribution of mates (Emlen and Oring 1977; Clutton-Brock and Parker 1992) is expected to affect the strength of sexual selection acting in a population by varying the ability of individuals to monopolize reproductive opportunities. However, recent research has highlighted some of the limitations of this classic categorization of mating systems. Bertram and Gorelick (2009) highlighted how vastly disparate mating patterns can result in the same variation in both male and female mating success. An example of this is illustrated in Figure 4.6.

In social network research, tools have been developed to distinguish such subtle but potentially important differences in interaction structure. For example, network clustering coefficients can be used to measure the extent that groups of nodes form smaller groups within the population with strong connections within the group and weak connections between groups (Watts and Strogatz 1998), and have been extended for use with weighted networks (Opsahl and Panzarasa 2009). Furthermore, a variety of community detection techniques allow the delineation of such strongly interconnected

$$\text{(a)}\quad \begin{array}{c} \\ \male \end{array}\overset{\female}{\begin{bmatrix} 10 & 10 & 10 \\ 10 & 10 & 10 \\ 10 & 10 & 10 \end{bmatrix}} \qquad \text{(b)}\quad \begin{array}{c} \\ \male \end{array}\overset{\female}{\begin{bmatrix} 28 & 1 & 1 \\ 1 & 28 & 1 \\ 1 & 1 & 28 \end{bmatrix}}$$

Figure 4.6 Matrices of two polygynandrous mating systems. Rows represent males, columns represent females, and numbers in matrices represent the number of copulations between pairs. In both groups every male and female mates with three partners, so the opportunity for precopulatory sexual selection (I_S) is zero. Furthermore, every male and female has 30 total copulations and an average of 10 copulations per partner. Despite such similarities, however, the system depicted in (a) is highly promiscuous, where every female mates equally across all males, leading to intense postcopulatory competition, whereas in (b) there is clear preferential assortment between mates, generating relatively little postcopulatory competition.

subgroups and have been used in variety of biological networks (Girvan and Newman 2002; Lusseau et al. 2006; Olesen et al. 2007). Such techniques may provide a useful tool to determine how demographic, genetic, and ecological variables shape the non-random associations in sexual interaction networks. For example, Wolf et al. (2007) explored how factors such as demography and territoriality determine the structure of social networks in a colony of the Galápagos sea lion (*Zalophus wollebaeki*). Wolf et al. (2007) found that sea lion social structure was strongly hierarchical, with sex and age determining population-level associations, while community structure was determined by individual space use.

Such methods could be used to understand how mating structures change in fine-scaled ways according to different ecologies. For instance, do different resource or nesting site distributions predictably alter the topology of mating interactions and the distribution of partner phenotypes available to individual males and females, and what are the resulting implications for the spread of sexually transmitted infections? Recent studies have investigated how changes in the availability of important food resources can alter the feeding regimes of different individuals, leading to variation in diet preferences and diet specialization (Araújo et al. 2008). Such preferences may result in consistent differences in associations between individuals (Ward et al. 2004, 2005) and have been shown to play a role in assortative mating (Snowberg and Bolnick 2008).

How such assortativity is organized may have strong implications for the maintenance of alternative phenotypes and reproductive isolation. A recent theoretical treatment showed that whether the pattern of assortative mating acts in a graded fashion across the population or is realized in a more extreme (i.e. modular and stepwise) fashion can have important implications for speciation, as well as for the maintenance of distinct species during secondary contact (Kondrashov and Shpak 1998). Although assortative mating and incipient speciation have been documented (Reynolds and Fitzpatrick 2007), the extent to which this mating is stepwise and modular remains to be investigated empirically. We argue that network analysis provides a tool through which we can directly and empirically address these significant gaps in knowledge.

Individual network measures

In addition to population- and group-level characterizations of sexual structure, network analysis also allows the identification and characterization of the position of individual nodes within networks (Newman 2010). How these individual-level metrics relate to the roles that individuals play within animal societies (Lusseau and Newman 2004), as well as the costs and benefits of individual network position, have begun to be explored (Godfrey et al. 2009; Aplin et al. 2012). Here we discuss the potential for individual-level metrics within sexual networks to shed light on reproductive strategies and on how the position of individuals within sexual networks can affect reproductive success.

The characterization of individual node position, for example in terms of measures of network centrality, has been used to characterize the risk of sexually transmitted infection of different individuals, as well as their role in disease spread (e.g. Corner et al. 2003; De et al. 2004; Christley et al. 2005; Fichtenberg et al. 2009). A number of studies have adopted a similar approach to investigate how individual position within social networks can affect individual mating and reproductive success (e.g. McDonald 2007; Ryder et al. 2008). The extent that such positional roles may represent reproductive strategies has recently been investigated by Oh and Badyaev (2010) in the house finch. Within this population, male social network position was measured in terms of individual *betweeness centrality* (Freeman 1979).

Oh and Badyaev (2010) showed that individual males with the highest betweeness centrality and who thus may be expected to experience the greatest social diversity had higher reproductive success. This is because, as discussed earlier, by moving between different groups, these males were able to modify their socio-competitive environment such that they increased their relative attractiveness to females compared to their competitors. This study provides an excellent example of how individual network measures can capture sexual strategies which would be missed using more conventional methods.

In 'Intersexual interactions, mating patterns, and the operation of selection', we discussed how a nested mating network may relate to group-level variation in the strength of precopulatory sexual selection. In such networks we may expect strong relationships between mating success and the level of sperm competition. Recent methodology developed in ecological research allows the calculation of each individual's contribution to the global level of nestedness in the network (Saavedra et al. 2011). Such methods may provide a useful characterization of sperm competition such that, for any male with any given mating success, his contribution to the nestedness of the network may mediate his reproductive success.

Recent research, however, has shown that such higher-order network properties may be correlates of simpler measures (James et al. 2012). For example, Formica et al. (2012) used path analysis to explain male mating success in light of male position within male social networks in the forked fungus beetle (*Bolitotherus cornutus*) and showed that, although some network metrics independently predicted male reproductive success, measures of centrality were simply correlates of male activity level. Future work should therefore seek to establish how measurements such as individual contribution to nestedness add novel biological information to simpler approaches.

Sexual selection and sexual networks: conclusions

In this chapter we have outlined the diverse and complex array of social interactions arising from sexual reproduction. This complexity provides a challenge to researchers studying sexual selection. With this challenge in mind, we have discussed how network analysis may allow the characterization of the complex web of socio-sexual interactions in both animals and plants. We then demonstrated how such quantitative network approaches may be utilized in combination with traditional sexual selection techniques to provide fertile ground for a more comprehensive view of reproductive behaviours and sexual selection.

Acknowledgements

We are grateful to Darren Croft, Dick James, and Jens Krause for inviting us to write this chapter. We are especially grateful to Dick James and Andy Gardner for helpful discussions about some of the ideas presented in this chapter. G. M. was supported by a Biotechnology and Biological Sciences Research Council CASE scholarship in collaboration with Aviagen Ltd, and T. P. by a grant from the Natural Environment Research Council.

CHAPTER 5

Quantifying diffusion in social networks: a Bayesian approach

Glenna Nightingale, Neeltje J. Boogert, Kevin N. Laland, and Will Hoppitt

Introduction to social transmission in groups of animals

Social learning is often broadly defined as 'learning that is influenced by observation of, or interaction with, another animal (typically a conspecific) or its products' (Heyes 1994). However, the phrase 'influenced by' is unacceptably vague, and we prefer to characterize social learning as learning that is *facilitated by* observation of, or interaction with, another individual, or its products (Hoppitt and Laland 2013). In addition, we prefer the more specific term 'social transmission' to refer to the process by which behavioural traits spread through groups. We define social transmission as occurring when 'the prior acquisition of a behavioural trait T by one individual, A, when expressed either directly in the performance of T or in some other behaviour associated with T, exerts a lasting positive causal influence on the rate at which another individual, B, acquires and/or performs the behavioural trait' (Hoppitt and Laland 2013).

The study of social learning was initially motivated by an interest in the cognitive or psychological mechanisms underpinning social learning (e.g. Galef 1988; Heyes 1994), leading to research conducted in controlled laboratory conditions (e.g. Zentall et al. 1996). In recent years, the focus has shifted to animal traditions and culture, driven by the discovery of group-specific behaviour in a number of taxa, including primates (e.g. Whiten et al 1999; S. Perry et al. 2003), cetaceans (e.g. Rendell and Whitehead 2001) and birds (e.g. Madden 2008). Such group-specific behaviour patterns often appear to be the result of different behavioural innovations spreading through groups by social transmission. Many researchers are now studying the conditions under which novel behavioural traits spread and form traditions in the field or in a captive group context, in which the subjects are free to interact with one another (e.g. meerkats (*Suricata suricatta*) (Thornton and Malapert 2009); vervet monkeys (*Chlorocebus pygerythrus*) (Van de Waal et al. 2010); and humpback whales (*Megaptera novaeangliae*) (Allen et al. 2013)). Such research has motivated the development of novel methods for studying social learning in freely interacting groups (Laland and Galef 2009; Kendal, Galef et al. 2010; Hoppitt and Laland 2013). Laboratory experiments provide valuable insights into learning mechanisms, and other aspects of social interaction, but it remains challenging in a natural or field context to ascertain whether social learning has occurred and quantify its impact without the use of sophisticated statistical methods.

In our use of the terminology, 'social transmission' can be distinguished from 'diffusion', as the latter term refers to the observed spread of a trait through a group, irrespective of the cause of the spread. Therefore, a trait might be said to have diffused through a group without any evidence that this occurred by social transmission. For instance, the diffusion may result from independent asocial learning by each individual, or there may be an unlearned social influence on behaviour, as occurs, for instance, when animals influence each other's movements, as is reported in sticklebacks (Atton et al. 2012).

Animal Social Networks. Edited by Jens Krause, Richard James, Daniel W. Franks and Darren P. Croft.

A promising alternative approach to highly structured laboratory experimentation for inferring and quantifying social transmission in naturalistic group contexts is network-based diffusion analysis (also known as NBDA) (Franz and Nunn 2009). Network-based diffusion analysis infers social transmission when the time (or order (Hoppitt, Boogert, et al. 2010)) of acquisition of the trait by individuals in animal groups follows a social network. Similar models have been used in the social sciences (Valente 2005); their relationship with network-based diffusion analysis is discussed in Hoppitt and Laland (2013). In this chapter, we will first briefly review the uses of network-based diffusion analysis and then explain why a Bayesian formulation is required. We will then present our Bayesian formulation of network-based diffusion analysis and test it using simulated data.

Network-based diffusion analysis

Network-based diffusion analysis infers and quantifies the magnitude of social transmission in a set of diffusion data from the extent to which the pattern of spread follows a social network. There are two versions: time of acquisition diffusion analysis (also known as TADA) (Franz and Nunn 2009), which takes as data the times at which individuals acquired the target behavioural trait, and order of acquisition diffusion analysis (also known as OADA) (Hoppitt, Boogert, et al. 2010), which is sensitive only to the order in which they do so. The former is more powerful but makes stronger assumptions. The greater power of time of acquisition diffusion analysis stems from the fact that order of acquisition diffusion analysis is only sensitive to social transmission if it results in a difference in the relative rate of acquisition by individuals, whereas time of acquisition diffusion analysis is also sensitive to absolute changes in the rate of acquisition and thereby has more data with which to detect social transmission (Hoppitt and Laland 2011).

Despite being a recently developed method, network-based diffusion analysis has already been used a number of times to analyse diffusion data from wild and captive animal populations, usually using an association metric to obtain the social network. For example, Aplin et al. (2012) found strong evidence that the time of and probability of discovering of novel food patches followed an association network in a wild population of great tits (*Parus major*), blue tits (*Cyanistes caeruleus*), and marsh tits (*Poecile palustris*). Likewise, Allen et al. (2013) found strong evidence that the acquisition of lobtailing, a foraging innovation, followed an association network in a wild population of humpback whales. Kendal, Custance, et al. (2010) also applied the method to analyse the diffusion of a novel foraging behaviour in lemurs (*Lemur catta*), although there was no evidence that social transmission followed the network in that case.

In other cases, network-based diffusion analysis has been applied to diffusion data arising when captive groups of animals are presented with a novel foraging task. For example, Boogert et al. (2008) constructed a nearest neighbour association network for three groups of five starlings (*Sturnus vulgaris*) each and then separately presented each group with six tasks. Hoppitt, Boogert, et al. (2010) applied a continuous time of acquisition diffusion analysis to these data and found strong evidence of social transmission, although further analysis (Hoppitt, unpublished data) found that there was little evidence that it followed the social network. Specifically, a model with social transmission following the association network had more support than a model without social transmission, but it had similar support to a model that had homogeneous connections between individuals in each group. Under such circumstances, the evidence for social transmission comes from the observation that individuals were more likely and/or faster to solve the task once other individuals in the group had solved it (Hoppitt and Laland 2013). One drawback of many such captive studies is that animals may have little opportunity to avoid each other, as a result of which network-based diffusions may be less like to be detected, either because all network associations are strong or because network connections become superfluous in confined spaces, where learning between poorly connected individuals is feasible.

Network-based diffusion analysis has also been used to analyse diffusion experiments on groups of captive sticklebacks (Atton et al. 2012; M. Webster et al. 2013). Atton et al. (2012) expanded network-based diffusion analysis methodology by recognizing that individuals can move between multiple states. For example, rather than just moving from 'naïve' (not solved the task) to 'informed' (solved the task) states, individuals move from being 'naïve', to having 'discovered' the task, to 'solving' the task. In principle, social influences might operate on both the discovery and the solving transition, and Atton et al. (2012) found that the social network affected each transition in a different way. Atton et al. (2012) also expanded network-based diffusion analysis to allow for multiple options available to solve the task (cf. Kendal et al. 2010).

While network-based diffusion analysis has, thus far, been used to assess whether the pattern of diffusion follows a measured association network, it could instead be used to compare the support for different hypothesized pathways of diffusion (Franz and Nunn 2009; Hoppitt and Laland 2011). For instance, one could test networks corresponding to different theories of 'directed social learning' (Coussi-Korbel and Fragaszy 1995). For example, the hypothesis that all social transmission is vertical would correspond to an asymmetrical binary network in which all connections lead from parents to offspring (Hoppitt and Laland 2013).

Why do we need Bayesian network-based diffusion analysis?

Network-based diffusion analysis can be expanded such that it quantifies the evidence for social transmission across a number of diffusions (Hoppitt et al. 2010), although care must be taken in the interpretation of such models (Hoppitt and Laland 2011, 2013). This expansion provides a valuable way of combining information arising from diffusions across different groups of animals (e.g. Webster et al. 2013). The expansion of network-based diffusion analysis to multiple diffusions could also be valuable where researchers have repeated diffusions across the same group, or groups, of animals (e.g. Boogert et al. 2008), especially when they only have a limited number of animals, allowing them to obtain good statistical power.

However, a statistical problem arises if they fail to account for the fact that the same individuals are involved in multiple diffusions. To illustrate this, imagine a group of experimental subjects presented with a foraging task and who vary in their asocial learning ability in a way that seems to imply social transmission occurs. For example, the best asocial learner may be well connected to the second-best asocial learner, making it appear that the latter is learning from the former, when in fact they are solving the task independently (i.e. through asocial learning processes). For a single diffusion, this chance possibility is automatically accounted for when assessing the evidence for social transmission.

If multiple foraging tasks are presented to different groups of individuals, and a pattern in the resulting diffusion data consistent with social transmission arises each time, then this adds to the evidence that social transmission is occurring; it is unlikely that a chance pattern of asocial learning abilities, consistent with social transmission, occurs in all cases (although confounding variables are possible) (see Hoppitt, Boogert, et al. 2010 for discussion). If *multiple* diffusions are run on the same individuals, and a similar pattern arises, this too will be taken by the model as being strong evidence of social transmission. However, in this case if a chance pattern of asocial learning ability consistent with social transmission happens to arise over the single group of individuals, it is repeated over multiple diffusions. If a model without random effects is fitted, each diffusion will be unrealistically taken as an independent set of data supporting the hypothesis that social transmission is occurring, thus potentially resulting in a spurious result. In a similar way, a chance pattern in asocial learning abilities counteracting the effects of social transmission would lead us to underestimate the effects of social learning, with an overinflated level of certainty.

By including an individual random effect on the asocial rate of learning, the model accounts for the fact that the same individuals have the same (or similar) asocial learning ability in each diffusion. However, random effects can be difficult to implement using maximum likelihood methods (used to fit network-based diffusion analysis models thus far), especially when the random effects structure is complex, because one has to integrate the likelihood function across all the possible values the random effects could take. It is easier to include random effects in a Bayesian model, using Markov chain Monte Carlo methods (Gelman et al. 2004).

Bayesian methods take a joint prior distribution for the model parameters, quantifying researchers' knowledge about the plausible values those parameters could take before receiving the data, and update this in light of the data to yield a joint posterior distribution. The joint posterior distribution thus quantifies the state of knowledge arising from the data, showing which combinations of parameter values are plausible. The marginal posterior distribution for a parameter (often shortened to 'posterior distribution') is the joint posterior distribution integrated over all the possible values of the other parameters in the model (including all the possible values for each level of each random effect). For relatively simple models, a mathematical expression can be obtained for the exact posterior distribution for each parameter of interest. However, for more complex models, such as those containing random effects, this is not possible, and Markov chain Monte Carlo is used.

Markov chain Monte Carlo is a procedure that simulates drawing values for the model parameters from the joint posterior distribution for all parameters in the model. By drawing a large number of values for one parameter from the joint posterior distribution, one is automatically accounting for uncertainty in the other parameters in the model. Consequently, when using Markov chain Monte Carlo, instead of integrating the likelihood over the random effects numerically, inferences on a given random effect parameter can be made by simulating draws from the marginal posterior distribution of this parameter. For a comprehensive explanation of the use of Markov chain Monte Carlo for random effects models, we refer the reader to Hoff's (2009) book on Bayesian statistics. Here, we develop a Bayesian version of time of acquisition diffusion analysis and then test whether it solves the problems outlined above, using simulated data. To aid explanation of the model formulation, we describe the simulated data first.

Simulated diffusion data

The data were simulated from the time of acquisition diffusion analysis model using the Gillespie algorithm (D. Wilkinson 2012). The social network used to simulate the data consisted of interactions of equal magnitude, set at 1. This allows us to show that the extracted parameter estimates closely match the real ones in a simple case. However, we stress that the approach works effectively for more realistic social networks. Bayesian network-based diffusion analyses for networks with more complex structures can be found in the tutorial posted on the Laland lab website (<http://lalandlab.st-andrews.ac.uk/>).

We simulate data corresponding to ten different tasks performed by the same group of ten individuals. We use a fixed value for the social effect (set at 0.6) and a fixed value for the baseline rate of asocial learning (0.3). The output from each data simulation includes both latency-to-solve times and solving order for the group of ten individuals. Random effects for each individual i were also incorporated in the simulation by assigning a number R_i from a set of numbers with variance 9, and multiplying its rate of asocial learning by a factor of exp (R_i). The incorporation of the random effects at the individual level thus models the heterogeneity arising from variation in asocial learning ability across the ten individuals. For ease of reference, each individual was given a unique name (Ned, Ted, Ron, Wim, Jim, Sue, Fay, Lou, Joe, and May).

Figure 5.1a shows the solve times for the ten diffusions, and Figure 5.1b shows the solve times for the first four tasks, with each task represented by a unique plotting symbol. Finally, Figure 5.1c shows how performance across the tasks varies among individuals. From Figure 5.1c, it is evident that the variation in solve times is highest for Ned, Ted, Jim, Ron, and Wim. For Fay, Lou, Joe, and especially May, the latencies to solve the tasks are much shorter than those of the other individuals, and the variation in times is low (the points on the plot have merged into one).

Previous formulation of time of acquisition diffusion analysis

The time of acquisition diffusion analysis model is based on standard survival models using an exponential distribution. We therefore use survival analysis terminology, referring to the 'hazard function' as giving the instantaneous rate at which an individual acquires the target trait, which in this

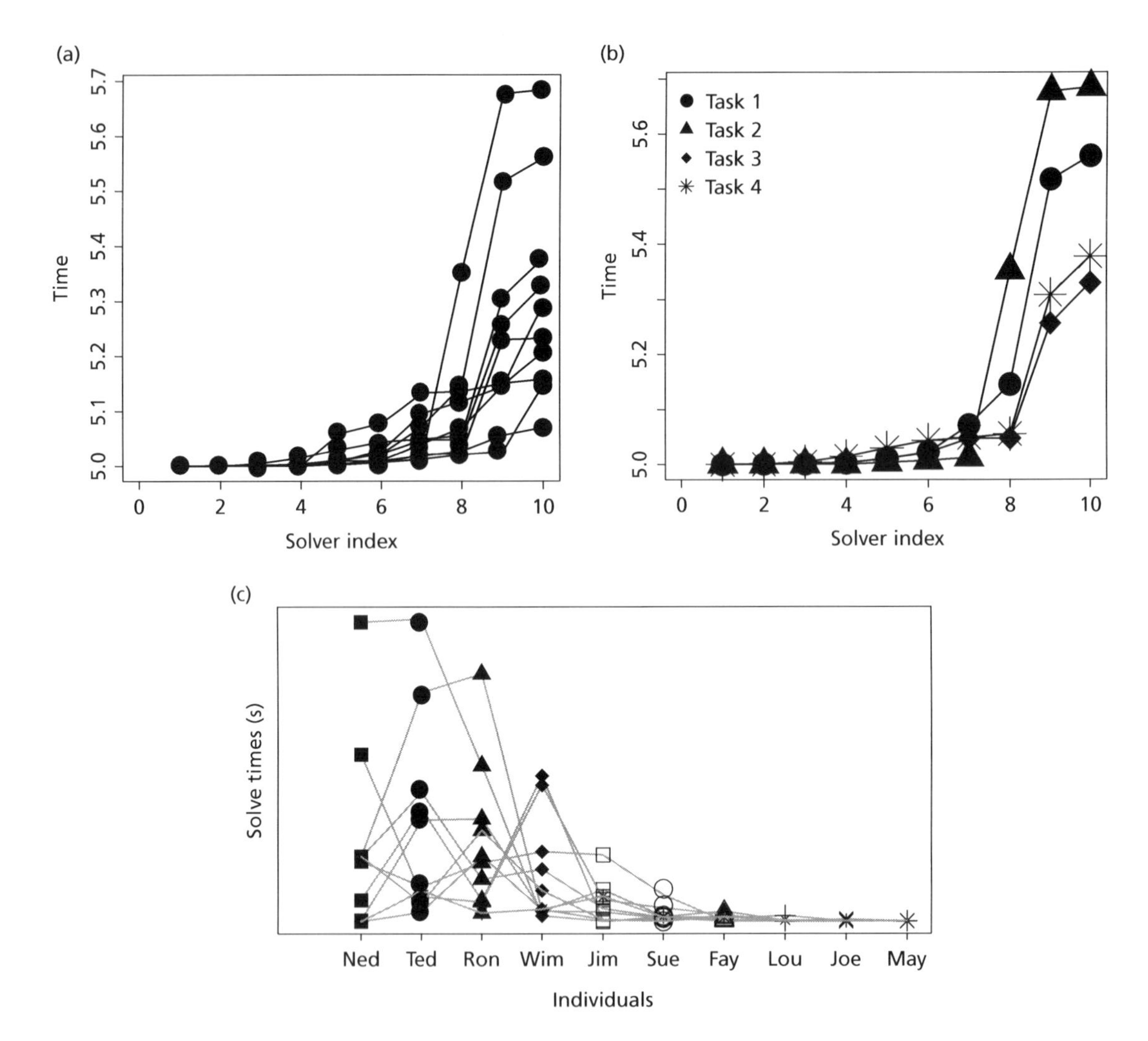

Figure 5.1 (a) Plot showing simulated diffusion times for each of the ten tasks. The points for each diffusion are joined in a separate line and represent a unique task. (b) Plot showing simulated diffusion times for Tasks 1 to 4. Each diffusion represents a unique task and is represented by a unique plotting symbol. (c) Simulated solve times per individual. The points per individual represent the time at which this individual solved the given tasks. Points for each diffusion (task) are joined in a separate line. In cases where the variation in solve times for a given individual is small, some of the points on the graph have merged together.

case is the task solution. There are two parameters of interest in the basic time of acquisition diffusion analysis model: the rate of social transmission between individuals per unit of network connection, *s*, and the baseline rate of trait performance in the absence of social transmission, λ_0. Throughout this chapter, we refer to the *s* parameter as the social transmission parameter, and to λ_0 as the baseline parameter.

The hazard function for the model is expressed as:

$$\lambda_i(t) = \lambda_0(t)(1 - z_i(t))R_i(t) \tag{1}$$

such that

$$R_i(t) = \left(s\sum_{j=1}^{N} a_{ij} z_j(t) + 1\right) \tag{2}$$

where $\lambda_i(t)$ is the rate at which individual i acquires the task solution at time t, $\lambda_0(t)$ is a baseline acquisition function determining the distribution of latencies to acquisition in the absence of social transmission (that is, through asocial learning), and $z_i(t)$ gives the status (1 = informed, 0 = naïve) of individual i at time t. The $(1-z_i(t))$ and $z_i(t)$ terms ensure that the task solution is only transmitted from informed to uninformed individuals (Hoppitt, Boogert, et al. 2010). Previous versions of time of acquisition diffusion analysis allow for an increasing or decreasing baseline rate $\lambda_0(t)$ (Hoppitt, Kandler, et al. 2010). Here, we restrict ourselves to expanding the version for a constant baseline rate (i.e. $\lambda_0(t) = \lambda_0$) (Hoppitt, Boogert, et al. 2010), although the version for a non-constant baseline rate can be expanded in the same way.

The model assumes that the rate of social transmission between individuals is proportional to the connection between them which is given by a_{ij} (see Equation 2). The model is used to generate a likelihood function, allowing it to be fitted by maximum likelihood or analysed using Bayesian methods. Social transmission is inferred if a model including s is better than a model with $s = 0$, using, for example Akaike's information criterion, if maximum likelihood fitting is used, or Bayes factor if Bayesian methods are used (see 'Bayesian formulation of time of acquisition diffusion analysis'). For simplicity, here we assume that a social effect (i.e. $s > 0$) is always indicative of social transmission, although in reality this need not be the case (Atton et al. 2012).

Network-based diffusion analysis can be adapted to include other variables influencing the rate of social transmission or asocial learning that vary across individuals and/or time, by expanding the model for V continuous (or indicator) variables as follows:

$$R_i(t) = \left(s \exp(\Gamma_i(t)) \sum_{j=1}^{N} a_{ij} z_j(t) + \exp(B_i(t)) \right) \quad (3)$$

where

$$\Gamma_i(t) = \sum_{k=1}^{V} \gamma_k x_{k,i}(t) \quad (4)$$

and

$$B_i(t) = \sum_{k=1}^{V} \beta_k x_{k,i}(t) \quad (5)$$

Here $\Gamma_i(t)$ and $B_i(t)$ are linear predictors similar to those used in a generalized linear model.[1] The term $x_{k,i}(t)$ is the value of the kth variable for individual i at time t, β_k is the coefficient giving the effect of variable k on asocial learning, giving the natural logarithm of the multiplicative effect per unit of $x_{k,i}(t)$. Similarly, γ_k is the coefficient giving the effect of variable k on the rate of social transmission (Hoppitt and Laland 2013).

Bayesian formulation of time of acquisition diffusion analysis

In principle, the formulation of the model can remain the same for a Bayesian approach as for a model fitted by maximum likelihood. However, here we wish to include random effects, and reparameterize the model in a way that makes it easier to use in a Bayesian context. Thus, we apply a Bayesian time of acquisition diffusion analysis to the simulated dataset described in 'Previous formulation of time of acquisition diffusion analysis' to assess its performance under different circumstances. To illustrate the importance of both random effects and social transmission, four models were considered based on their inclusion/exclusion. Two of the models (Models 1 and 2) do not include random effects, while Models 3 and 4 do. Likewise two of the models (Models 1 and 3) do not include an s parameter, while Models 2 and 4 do. Please see Table 5.1 for details.

The linear predictors are easily adapted to include random effects. For Models 3 and 4 for example, random effects $\boldsymbol{\varepsilon}$ at the individual level were considered such that $\boldsymbol{\varepsilon} = \{\varepsilon_1, \ldots, \varepsilon_{10}\}$ and the total number of individuals is ten. The term $R_i(t)$ in Equation 2 is therefore expanded to

$$R_i(t) = \left(s \sum_{j=1}^{N} a_{ij} z_j(t) + \exp(\varepsilon_k) \right) \quad (6)$$

where $k \in \{1, \ldots, 5\}$ and depends on which task is involved. The rate of trait performance $\lambda_i(t)$ for individual i, at time t therefore becomes

[1] This general formulation allows the effects of individual-level variables on asocial learning and social transmission to differ. Hoppitt, Boogert, et al. (2010) suggested a constrained additive model constraining for all k, and the multiplicative model by constraining for all k.

Table 5.1 Models considered.

Model	Parameters
1	λ_0
2	λ_0, s'
3	λ_0, ε_i (i = 1:10), σ_ε^2 *i* denotes individual effects
4	λ_0, **s′** ε_i (i = 1:10), σ_ε^2 *i* denotes individual effects

$$\lambda_i(t) = \lambda_0(t)(1 - z_i(t))\left(s\sum_{j=1}^{N} a_{ij}z_j(t) + \exp(\varepsilon_k)\right) \quad (7)$$

To allow us to more easily set a prior distribution reflecting our state of knowledge (see below), Equation 7 is then re-parameterized to obtain

$$\lambda_i(t) = (1 - z_i(t))\left(\lambda_0 s\sum_{j=1}^{N} a_{ij}z_j(t) + \lambda_0 \exp(\varepsilon_k)\right)$$

giving

$$\lambda_i(t) = (1 - z_i(t))\left(s'\sum_{j=1}^{N} a_{ij}z_j(t) + \lambda_0 \exp(\varepsilon_k)\right)$$

where $s' = \lambda_0 s$. The effect of social interactions on the rate of learning s' and the baseline rate of learning $\boldsymbol{\lambda}_0$ are the two parameters of interest. We refer to the re-parameterized s' as the unscaled social transmission parameter, since it is not scaled such that it is quantified relative to the rate of asocial learning, as s is. The full parameter vector $\boldsymbol{\theta}$ is defined as $\boldsymbol{\theta} = \{s', \lambda_0, \varepsilon, \sigma_\varepsilon^2\}$, where ε refers to random effects at the task level. The variance term σ_ε^2 denotes the variance for the distribution of the task-level random effects.

Likelihood function for time of acquisition diffusion analysis

Given the observed data, ω, the likelihood that the nth individual learns the behaviour at time t_n (where t_n is the observed time of acquisition for individual i) is expressed as

$$L(\lambda_i(t_{n-1}) \mid \omega) = \lambda_i(t_{n-1}) \exp(-\lambda_i(t_{n-1}))[t_n - t_{n-1}]$$
$$\prod_{j \neq i}^{J} \exp(-\lambda_j(t_{n-1}))[t_n - t_{n-1}] \quad (8)$$

where $n > 1$, and J denotes the number of individuals in the group during the time period $t_n - t_{n-1}$. This expression represents the product of the probability density of individual i solving the task for the first time and the probability density of the naive individuals ($j \neq i$) that did not solve the task for the first time in the time period under consideration. Note that this expression is used when all the individuals in the study have solved the task within the observation period. Individuals which have solved the task are classified as uncensored individuals. In contrast, an individual that does not solve the task during the observation period is classified as a censored individual. Censored individuals are taken into account in the analysis, using the modification below.

The combined likelihood for all the events $n \in \{1:N\}$ in the observation period is expressed as

$$\prod_{n=1}^{N} L(\lambda_i(t_{n-1}) \mid \omega)\varphi \quad (9)$$

where there are N performance events (i.e. the given task was solved N times), and where φ denotes the probability density of the naive individuals which did not perform the trait during the observation period $[t_1 - t_N]$. In particular φ is expressed as

$$\prod_{j \neq i} \exp(-\lambda_j(t_N)[t_Q - t_N])$$

The terms t_N and t_Q denote the last time of performance and the time of the end of the observation period, respectively. Equation 9 thus represents the likelihood arising from the data for all individuals, both censored and uncensored.

Prior specification

A fundamental requirement for inferring parameter values using a Bayesian approach is that suitable priors are specified for each parameter. The prior represents the researcher's prior knowledge of the distribution of the parameter's under consideration. The prior specified here for the social transmission parameter s' was a uniform prior such that $\log(s') \sim U[-1,1]$. A similar prior is specified for the baseline parameter $\boldsymbol{\lambda}_0$ such that $\log(\lambda_{0)} \sim U[-10,10]$.

These priors are very wide, which would indicate a lack of prior knowledge about the plausible values these parameters could take. Furthermore, uniform priors specified on a log scale express a prior belief that the parameters are more likely to be near zero. These are chosen fairly arbitrarily for the purposes of this simulation. In this section, we discuss how a researcher might set these priors so they represent the prior state of knowledge and discuss the circumstances under which this is important. A hierarchical prior (Gelman 2006; Gustafson et al. 2006) is specified for the random effects $\varepsilon_i \sim N(0, \sigma_\varepsilon^2)$. A gamma distribution is used as a prior for the variance term σ_ε^2.

Generating posteriors using updating methods

The Bayesian approach (R. King et al. 2010; Lee 1989) usually involves the use of an Markov chain Monte Carlo algorithm, which is deployed to generate a sequence of values which converge to the joint posterior distribution of the parameters (see 'Why do we need Bayesian network-based diffusion analysis?') given the data observed. Note that after the simulations are conducted, the properties of the resulting posterior sample (after removing the output from the initial 10% of simulations, called the 'burn in') will reflect the properties of the posterior distribution of the parameters under consideration. There are various methods of parameter updating, two of which are Metropolis Hastings and Gibbs sampling. The parameters in this analysis were updated using a random walk Metropolis Hastings (Gamerman 1997; Gamerman and Lopes 2006; McCarthy 2007) update method.

For the individual model simulations, each simulation consisted of 10,000 iterations and a conservative burn in (initial 10% of the iterations) was removed before obtaining the posterior sample. In addition, we demonstrate the application of Bayesian methods to diffusion analysis to achieve model discrimination. For this analysis, we employ a reversible jump Markov chain Monte Carlo algorithm to discriminate between the four models considered in this analysis.

Model discrimination

For a given dataset, there are typically a number of plausible candidate models, such as shown in Table 5.1, which may vary in the number of parameters they possess. With a number of plausible models for a given dataset, it is desirable that model discrimination is performed so as to determine which model (or, in some cases, which group of models) has (have) more support for the data observed. To achieve this, the reversible jump Markov chain Monte Carlo algorithm allows for posterior model probabilities (the probability that model is the true model, given that one of them is) to be obtained for each model, which is particularly useful when there is a large number of competing models. A common summary statistic related to the posterior model probabilities is the Bayes factor (Lee 1989), which is simply the ratio of the model probabilities for two specified models (so long as the prior model probabilities are equal). This is usually a preferred measure of relative evidence, since it does not implicitly assume that one of the models in the set considered is true. The Bayes factor is used specifically to compare two competing models (hypotheses) for a given dataset.

Mathematically, for model discrimination, we extend the previous Bayesian approach to treat the model itself as a parameter and then form the joint posterior distribution over parameter and model spaces. However, the posterior distribution is no longer of fixed dimensions, since different models have a different number of parameters. Thus, to explore the posterior distribution and to obtain posterior summary statistics, we use a reversible jump Markov chain Monte Carlo approach (Lesaffre and Lawson 2012).

Note that this approach comprises a two-step algorithm which involves the Metropolis Hastings algorithm and a reversible jump step. The first step involves updating the parameters given the model state, and the second step involves updating the model itself. This results in a sequence of model states at the end of the simulation which represent the exploration of model space during the iterations within the simulation. Figure 5.2 illustrates this concept graphically. Model discrimination in the Bayesian context is discussed in the tutorial provided on Bayesian network-based diffusion analysis on the Laland lab website. In addition, a data example and sample R code are provided to illustrate the concept.

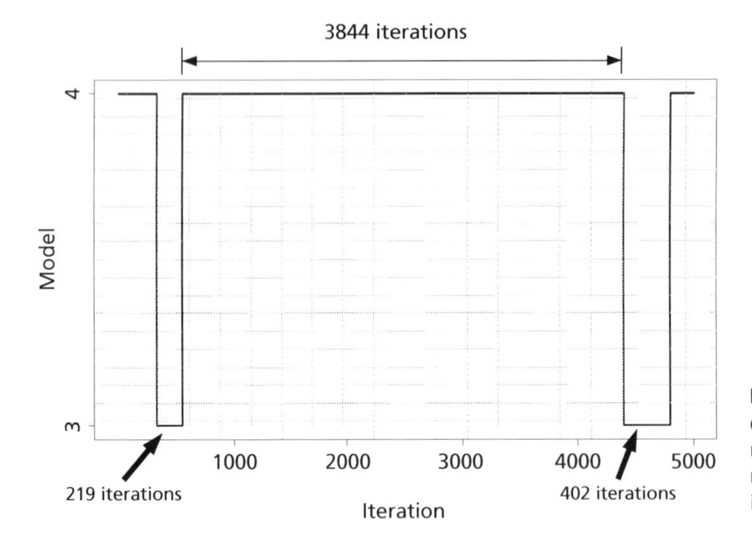

Figure 5.2 Plot illustrating model updating during a theoretical simulation involving two models: Model 3 and Model 4. Note that the model state does not necessarily change at each iteration.

Results

Posterior parameter estimates

The posterior parameter estimates for the parameters in each model are provided in Table 5.2. As noted in 'Simulated diffusion data', we used a fixed value of 0.6 for the social effect and a fixed value of 0.3 for the baseline rate of asocial learning. Below we consider the relative merits of the four models in accounting for the data. However, for the moment we merely draw attention to the fact that the posterior parameter estimates for the model parameters differ between the relevant models. Most strikingly, the parameter likely to be considered of greatest interest to users of network-based diffusion analysis, the social transmission parameter s' differs strongly between Models 2 and 4, which do not, or do, control for random effects, respectively. The posterior estimates for the baseline rate parameter λ_0 showed relatively less variation between models.

From the results, the median posterior estimate of the unscaled social transmission parameter was found to be higher in Model 4 than in Model 2 (the other model which contained this parameter). The log of the estimate in this model (−0.50) is closest to the log of the value (−0.51) used to simulate the data ($s' = e^{-0.50} = 0.61$), whereas the estimate generated by Model 2, which fails to control for random effects, is poor ($s' = e^{-6.62} = 0.001$). Hence, the analysis shows that controlling for random effects is critical to generating accurate estimates of the magnitude of social transmission and that a failure to do so could lead researchers to seriously underestimate or overestimate the social influence.

Figure 5.3 shows the trace plot for the unscaled social transmission parameter s' for the two models. A trace plot is a time series plot which provides a rough indication of how well the Monte Carlo chain has mixed and has explored the posterior distribution. The x-axis represents the iteration number within the simulation, and the simulated values of the parameter are represented on the y-axis. From the trace plots, it is clear that there is more variation in the posterior values in Model 2 than in Model 4, again indicating that Model 2 is poorer.

In both Models 2 and 4, the correlation between the social transmission and baseline rate parameters was negative but small. However, the random effect parameters were found to be correlated with each other and also with the baseline rate (and, to a lesser degree, with the social effect parameter). These correlations explain why the posterior estimates for the baseline rate of solving are so poor, a point to which we return in the final section. For illustration, Figure 5.4 shows the correlation between the baseline rate parameter and the random effect parameters for individuals 9 and 10. The density plots for each random effect parameter are shown in Figure 5.5.

From Figure 5.4, it is clear that there is a negative correlation between the baseline rate parameter

Table 5.2 Posterior parameter estimates are provided in natural logarithms (except the random effects and variance parameter) and are accompanied by symmetric credible intervals for the models considered. For the social transmission and baseline rate parameters, the posterior median and credible intervals are provided. For the random effects and variance parameter, the posterior mean and credible intervals are provided. The median is the preferred summary statistic when the distribution of the parameter of interest is skewed.

Parameter	Model 1	Model 2	Model 3	Model 4
log(s')		−6.62 (−9.81, −2.53)		−0.50 (−0.95, −0.07)
$\log(\lambda_0)$	2.48 (2.27, 2.68)	2.47 (2.25, 2.67)	3.57 (2.38, 5.61)	2.88 (0.78, 4.58)
ε_1 (Ned)			−2.07 (−4.14, −0.72)	−3.65 (−8.59, −0.53)
ε_2 (Ted)			−2.14 (−4.18, −0.83)	−3.94 (−8.12, −0.82)
ε_3 (Ron)			−1.94 (−3.98, −0.62)	−3.29 (−7.62, −0.29)
ε_4 (Wim)			−1.44 (−3.46, −0.16)	−1.82 (−5.48, 0.750)
ε_5 (Jim)			−0.63 (−2.67, 0.79)	−0.06 (−2.14, 2.13)
ε_6 (Sue)			0.48 (−1.61, 1.82)	1.29 (−0.66, 3.42)
ε_7 (Fay)			1.12 (−0.78, 2.46)	1.99 (0.03, 4.11)
ε_8 (Lou)			2.09 (0.07, 3.40)	2.97 (0.99, 5.09)
ε_9 (Joe)			2.89 (0.83, 4.23)	3.76 (1.85, 5.87)
ε_{10} (May)			4.35 (2.32, 5.73)	5.24 (3.23, 7.37)
σ_ε^2			7.29 (3.86, 12.62)	9.79 (5.05, 16.19)

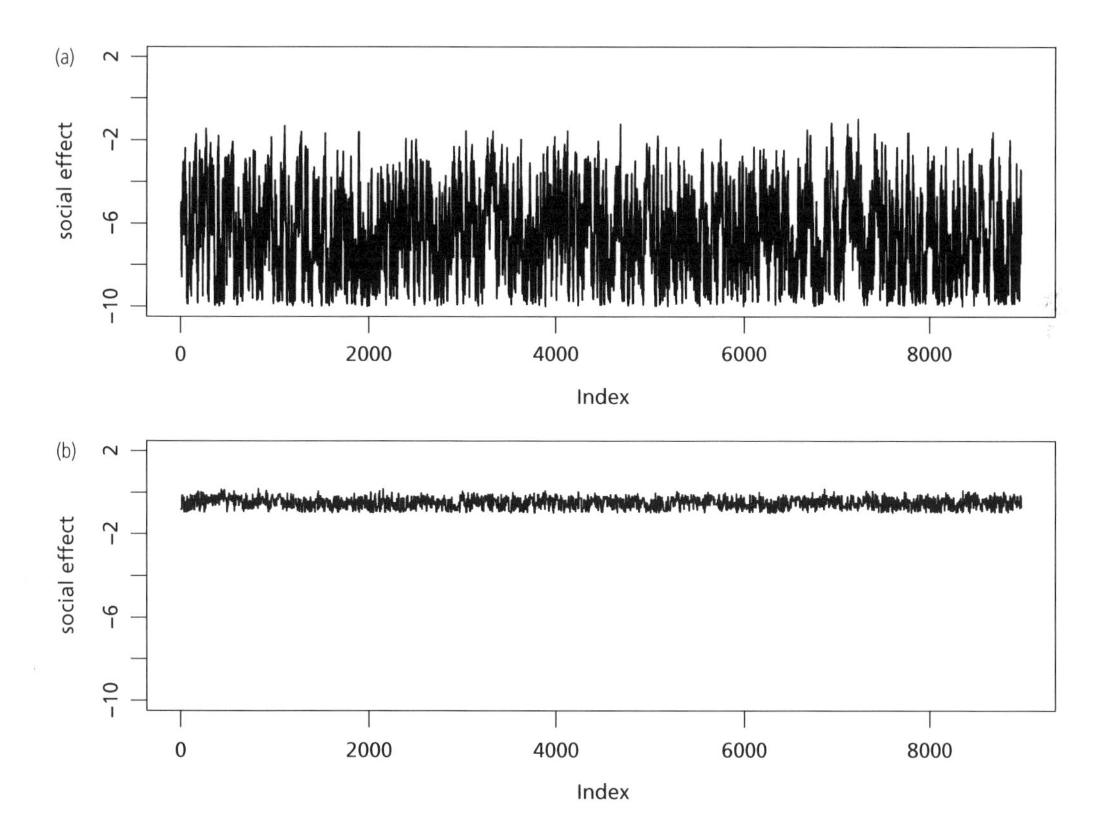

Figure 5.3 Trace plots for the social effect parameters for (a) Model 2, and (b) Model 4. The *x*-axis shows the iteration number, and the *y*-axis shows the log of the simulated value for the parameter. The first 1000 iterations were treated as 'burn in' and were removed before plotting. Identical *y*-axes are used in the plots to highlight the difference in variance of the simulated values obtained under the two models' hypotheses.

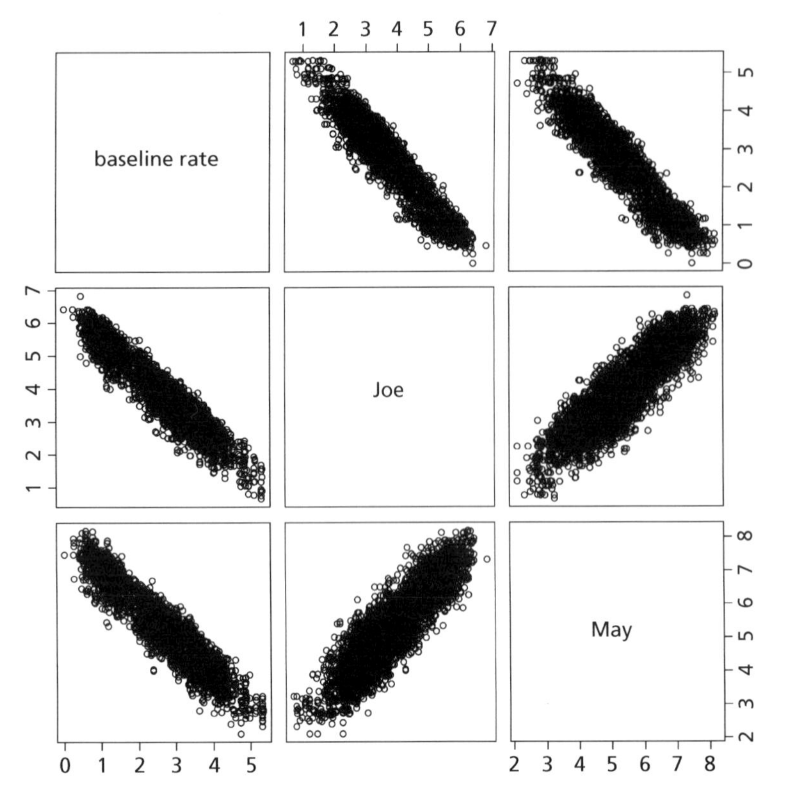

Figure 5.4 Scatter plot showing the correlation in the joint posterior distribution between the baseline rate and the random effect parameters for individual 9 (Joe) and individual 10 (May).

and the random effect parameter representing Joe. A similar effect is observed for May. The interpretation of these posterior correlations would be that, as the posterior value of the random effect parameter for Joe increases, that of the baseline rate parameter decreases. A positive correlation was observed between the random effect parameters for Joe and May. The interpretation in this case would be that, as the posterior value of the random effect parameter representing either of these individuals increases, then that of the other individual would also increase.

The density plots for each random effect parameter are shown in Figure 5a; the density plot for a given parameter can be thought of as a normal approximation to the distribution of this parameter. Each random effect parameter represented in Figure

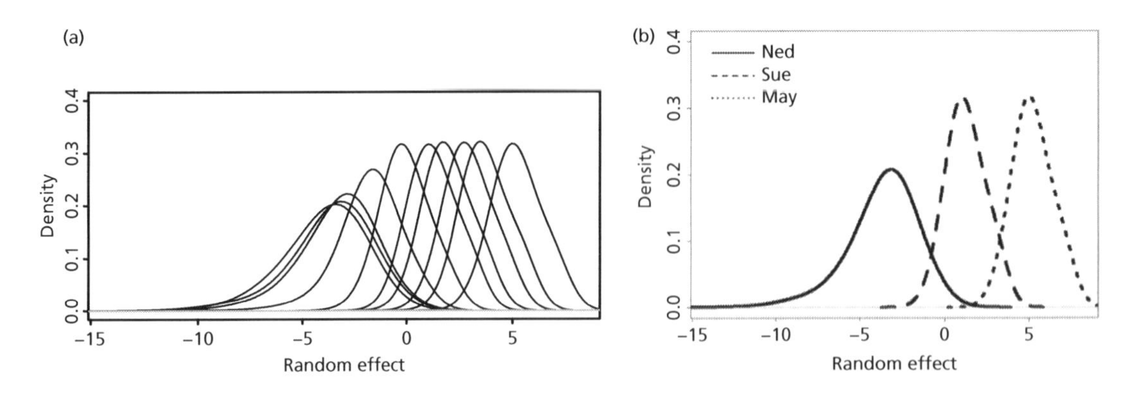

Figure 5.5 Density plots per random effect parameter: (a) ten random effects; (b) three random effects.

5 is associated with a bell-shaped plot, and the width of this 'bell' gives an indication of the spread of the posterior values for this parameter. The x coordinate, corresponding to the apex of the bell, represents the posterior parameter value, which has the highest density and represents the mean (or average) for the distribution. The density plots therefore provide a visual summary of the spread and centre of the distribution of the posterior values for the parameters considered. As expected, for the ten individuals considered, these vary in their effect on the rate of solving, reflecting individual differences in the ability to solve the tasks. This is illustrated more clearly in Figure 5b, where it is evident that May had a positive effect, Ned a negative impact, and Sue had little impact on the rate of solving.

Markov chain Monte Carlo replication

The Markov chain Monte Carlo simulations for Models 2 and 4 were repeated 100 times so as to allow enough time for the respective credible intervals to obtain the values used to simulate the data. From the results of these replications, we note that the credible intervals for the social transmission parameter for Model 4 were observed to be narrower than those for Model 2. In addition, while the credible intervals for the social transmission parameter for Model 4 always contained the value used to generate the data, the credible intervals for Model 2 were not found to contain the parameter value that was used to generate the data. The sole difference between Models 2 and 4 is the inclusion of random effects in Model 4, which thus underlies Model 4's superior performance.

Model discrimination

Model discrimination was performed for three different model comparisons. The posterior model probabilities obtained are shown in Table 5.3. When all four models are considered simultaneously (Comparison 3), there is far more support for Model 4 (the correct model, used to simulate the data) than for any of the other models (i.e. Model 4 was the best model in 94% of the simulations). In addition, when only Models 1 and 2 are considered (Comparison 1; i.e. no random effects) or only Models 3 and 4 are considered (Comparison 2; i.e. random effects), in each case the model that contains a social transmission parameter is by far the best supported model.

Table 5.3 Posterior model probabilities obtained from the application of the reversible jump Markov Chain Monte Carlo algorithm.

Models/ comparison	Comparison 1(Models 1 and 2)	Comparison 2(Models 3 and 4)	Comparison 3(Models 1, 2, 3, and 4)
1	0.11	–	0.0005
2	0.89	–	0.0032
3	–	0.05	0.0550
4	–	0.95	0.9420

These conclusions are reinforced by consideration of the Bayes factor associated with model comparisons, which can be derived by dividing the posterior model probability for the better supported model by the posterior model probability for the alternative. From the analysis using Comparison 3, we note that the Bayes factor in favour of random effects (i.e. (Model 3 + Model 4)/(Model 1 + Model 2)] is greater than 269.5, indicating decisive posterior support for these parameters and their importance in modelling the data observed. The Bayes factor in favour of Model 4, which contains the social transmission parameter, against Model 3, which does not, is 17.2, which suggests that there is very strong evidence for Model 4 (against Model 3). The Bayes factor in favour of the social transmission parameter (i.e. (Model 2 + Model 4)/(Model 1 + Model 3)] is 17.1, which suggests very strong support for the inclusion of this parameter in the model. In this simulated dataset, strong support for s implies a general influence of other individuals on the rate of solving, since here the network is comprised of homogenous patterns of association. However, more typically, support for a model containing s will be indicative of social transmission along pathways of association.

A Bayesian approach to quantifying diffusion on social networks: conclusions and future directions

We have developed a Bayesian version of network-based diffusion analysis as a means to control for random effects that can be generated by individual differences in ability among datasets that repeatedly test the same group or groups. The application of this approach to a simulated dataset clearly illustrates its merits, which we discuss in this section. The incorporation of random effects to account for heterogeneity in the baseline rate of asocial learning in Model 4 yielded a more realistic estimate (0.61) for the social effect on learning (recall that the value of the unscaled social transmission parameter used to generate the data was 0.6). Conversely, when random effect parameters were left out of the model, the social effect was so seriously underestimated that it would have been falsely regarded as negligible. Importantly, we note that the posterior mean (and standard deviation) for the variance of the random effects is 9.79 (2.96), reflecting substantial variation between the rates at which the individuals learn asocially. This is illustrated clearly in Figure, where individuals Joe and May have relatively high baseline rates while Ned and Ted have much lower rates.

The model discrimination exercise indicated that there was decisive posterior support for the random effect parameters, since Model 4 received the highest posterior support, and the Bayes factor in favour of random effects is greater than 100. Of course, to a large extent this is an artefact of the dataset deployed, and different datasets would give greater or lesser support for the models with random effects. However, the result illustrates that at least in some cases it will be necessary to control for random effects and that the Bayesian network-based diffusion analysis is capable of doing this effectively. The exercise also illustrates how a failure to control for random effects can lead to inaccurate estimates for other parameters of interest—most obviously, the magnitude of the social effect.

We note that the median posterior parameter estimate obtained for the social transmission and baseline rate parameters in the model of choice, Model 4, are not precisely equal to the point values used to simulate the dataset. For the social transmission parameter, the 95% credible interval does contain the value used for simulation, and the median is extremely close. However, this was not the case for the baseline rate parameter. The 95% credible intervals for the baseline rate parameter did not contain the value used to simulate the data. The differences between the point values can be attributed to correlations between model parameters in the joint posterior distribution.

This is apparent with the baseline rate parameter, since it appears in the model as a product with the random effect parameter (i.e. $\lambda_0 \exp(\epsilon_i)$). This means that a range of different combinations of λ_0 and $\exp(\epsilon_i)$, for any given i, can explain the data approximately equally well. For instance, a relatively low value of λ_0 and a relatively high value of $\exp(\epsilon_i)$ would explain the data roughly as well as a relatively high value of λ_0 and a relatively low value of $\exp(\epsilon_i)$. This is what a correlation between two parameters in the posterior distribution is telling us.

An alternative formulation for which the correlations between the baseline rate and the random effect parameters are avoided is to estimate individual baseline rates of asocial learning, λ_{0i}, as random effects, which might be appropriate to researchers who wish to ascertain the asocial performance of particular animals. Whether such alternative formulations are warranted depends on the goals of the researcher. In principle, researchers could make this judgement given the nature of the data. However, we suspect that for most applications of network-based diffusion analysis, the primary objective is accurate estimation of s', with λ_0 treated as a nuisance parameter, and hence the formulation presented here suffices. We have found that alternative formulations of random effects for asocial parameters do not generally affect estimation of s'.

Researchers unfamiliar with use of Bayesian methods might be put off using a Bayesian network-based diffusion analysis by the need to specify a prior distribution, quantifying our state of knowledge about the parameters before receiving the data. Often, they will consider themselves to have no solid basis on which to make judgements about the rate of asocial learning, and rate of social transmission prior to collecting data. Here we will discuss the circumstance under which such choices matter, and, where they do, how a prior distribution might be derived.

In some analyses, parameter estimation might be the key focus—probably focussing on estimating s' with 95% credible intervals—with no need for model discrimination. In such cases, one can specify a vague prior for parameters, with a large variance reflecting little prior knowledge, without worrying about exactly how large the variance has to be, or exactly what form the prior distribution should take. So long as the prior is fairly flat in the area that the 95% credible intervals fall, our results will not be greatly affected (Jaynes 2003). A pragmatic approach might be to choose a uniform distribution for s' from 0 to a large value far higher than s' could plausible take (see below), and likewise for λ_0.

In contrast, if model discrimination is the aim (e.g. when trying to decide which of a number of social networks best explains a diffusion), then the choice of priors is important, and the prior should reflect our state of knowledge. This is because the evidence for a given model depends not only on the likelihood of the data but also on how concentrated the priors are in the area in which the model parameters are plausibly located. Consequently, the addition of a parameter for which we have a little prior knowledge will penalize a model more than the addition of a parameter for which we have a lot of prior knowledge (see Jaynes 2003 for an explanation of why this is). For this reason, it is important that the prior distribution reflects our state of knowledge, for model discrimination.

A social learning researcher might protest that we do not know how strong social transmission might be (or indeed how rapid asocial learning might be) before conducting the diffusion experiment. However, we argue that researchers do have prior knowledge about such things, and this can be appropriately incorporated into the analysis. To illustrate this, imagine the researchers are conducting a diffusion experiment on swans: no diffusion experiments have ever been conducted on swans, so, on the face of it, they do not know anything about how fast swans might solve the task, whether it is by asocial learning or social transmission. However, there are possibilities that researchers would consider to be, a priori, impossible. Imagine they got the results of an network-based diffusion analysis on swans which estimated $s' = 1000$ per unit connection per second; this would mean that individuals with a single unit of connection to informed swans would, on average, solve the task in 0.001 seconds. Unless the network was quantified on a very small scale, no researcher would believe this result—they would probably assume something had gone wrong in the analysis (e.g. the time units were days, not seconds).

Such reasoning suggests that researchers do have prior knowledge about how fast social transmission could occur if and when it does occur. If they input a prior that allows a large range of values that are a priori implausible, they are penalizing too heavily against models including social transmission, and the Bayes Factors obtained will not reflect the state of our knowledge. Our suggestion for deriving a suitable prior for s' is to start by setting an upper limit, S_{max}, on how fast social transmission could plausibly occur, when all an individual's associates are informed (this can be estimated as $S_{max} = 1/T_{min}$, where T_{min} is the minimum plausible average time it would take individuals to solve a task under such circumstances). Since s' is the rate of social transmission per unit connection, take the average total network connection over all individuals, k. We then set the upper limit as $s'_{max} = S_{max}/k$. One could then specify a uniform prior $s' \sim U(0, s'_{max})$, which would state that we consider all values of s' within this range to be equally plausible.

This is likely to be a conservative approach, since values at the top end of the range are likely to be, in reality, less plausible than lower values, meaning models with s' included will be penalized more heavily than truly reflects our state of knowledge. Consequently, we suggest that researchers also check the sensitivity of their findings to different priors—perhaps altering the value of S_{max}, and also considering vague priors of different forms. For this purpose, we would suggest that one would need to go through a similar exercise to set a lower limit on s' to obtain s'_{min} where $s'_{min} = \frac{S_{min}}{k}$, $S_{min} = \frac{1}{T_{max}}$, and T_{max} represents the maximum possible time it is believed to take to solve the task at hand. This would result in a prior specification of $\log(s') \sim U(\log(s'_{min}), \log(s'_{max}))$. Priors can be chosen in a similar way for λ_0, by quantifying λ_{0Max} as the highest plausible rate of asocial learning and setting $\lambda_0 \sim U(0, \lambda_{0Max})$. However, this specification is less important than the specification of the prior for s', if the emphasis is on whether social transmission is occurring, since the same prior can be used for λ_0 in models with and without social transmission.

R code is available on the Laland lab website to demonstrate the Bayesian analysis of diffusion data in a simple form, including just the social effect and baseline rate parameters, and also when random effects are incorporated in the modelling process. The code provided is for demonstrative purposes. Development of a more comprehensive package, which can handle more complicated datasets, is currently underway.

Acknowledgements

Research was supported in part by a Biotechnology and Biological Sciences Research Council grant to K. N. L. and W. H. (BB/D015812/1), a European Research Council Advanced Grant to K. N. L. (EVOCULTURE, Ref. 232823), and a Netherlands Organisation for Scientific Research Rubicon Grant to N. J. B.

CHAPTER 6

Personality and social network analysis in animals

Alexander D. M. Wilson and Jens Krause

Introduction to personality and social network analysis in animals

In recent years, there has been a dramatic upsurge of research interest in animal personality and its importance in terms of ecology and evolution (Réale et al. 2007; Sih et al. 2012; M. Wolf and Weissing 2012). In this case, personality refers to consistent differences among individuals in their behavioural responses to ecologically relevant stimuli, both across time and contexts. Such stable interindividual differences in behaviour have been recorded in a wide range of animal taxa and observed across several main axes of behavioural variation including boldness (risk taking), exploration, activity, sociability, and aggression (Réale et al. 2007). These traits can be arranged into syndromes, such that behaviour in one context can be used to predict behaviour in another (Sih, Bell, Johnson, et al. 2004) and can also be used to characterize individuals as belonging to behavioural 'types'. Personality has also been shown to influence, among other traits, dispersal and migratory tendency (B. Chapman et al. 2011), reproductive success (Godin and Dugatkin 1996), response to environmental perturbation and predation threat (Bell and Stamps 2004; Bell and Sih 2007), interspecific interactions and competition (M. Webster et al. 2009), as well as divergence in habitat use and resource polymorphism (A. Wilson and McLaughlin 2007; Farwell and McLaughlin 2009). In group-living animals, personality predicts leadership and social foraging strategies (Leblond and Reebs 2006; Kurvers et al. 2009).

Yet, despite the crucially important role of personality in structuring ecological interactions, comparatively little is known about the role of personality in complex social dynamics and group-level interactions such as those characterized by social networks. Similarly, it is remarkable that given the numerous recent conceptual and empirical advances in both animal personality and social network research, so few studies exist that integrate these two fields of study (Krause et al. 2010; A. Wilson et al. 2013). Typically, personality studies regard social traits on a dyadic or at best hierarchical basis (i.e. individuals are more or less aggressive) and describe such interactions based largely on short term observations. However these behavioural assessments capture only a small part of the overall social complexity and ignore the potential importance of indirect relationships beyond the dyadic interactions observed between focal individuals.

Social network analysis provides a novel array of statistical tools for describing the social fine structure of animal groups and populations in ways that were previously not possible (Croft et al. 2008). Of particular value for personality research, network analysis provides a comprehensive experimental framework for considering not just the intensity (i.e. relative level of response) but also the consistency and frequency of individual-level interactions and preferential associations of individuals in dynamic social environments. As such, social network analysis can provide useful novel insight into the origins and maintenance of individual differences in behaviour by providing a forum for considering personality in the context of whole populations.

Animal Social Networks. Edited by Jens Krause, Richard James, Daniel W. Franks and Darren P. Croft.

The aim of this chapter is to highlight recent advances in the integration of personality and network analysis research and provide an overview of areas of current and future research promise. We will first discuss the importance of key individuals, or individuals with particular behavioural characteristics (i.e. behavioural types) in animal social networks. In the following sections, we will discuss the importance of considering networks and personality over ontogeny and how these phenomena are associated with developmental processes. We will then offer new directions and perspectives for future research on this exciting topic.

Network consistency and 'keystone' individuals

As mentioned previously, most personality studies describing social traits have traditionally been limited to observations of local dyadic associations or interactions. However, the contexts of these relationships extend beyond the dyad of just two individuals. As such, these studies are restricted to making limited assumptions regarding how these observed relationships scale up to more population-based or global social structures. Therefore, the key to integrating personality research and group-level interactions might be found in developing an understanding of individual fitness in the context of social behaviour, or otherwise stated, studying social relationships in the context in which they evolve (i.e. the social group) (Sueur, Jacobs, et al. 2011). This then begs the question, how does one integrate the study of what are generally considered individual traits (i.e. personality) in labile social environments?

The social environment is a powerful selective force, and social experiences from a myriad of ecologically relevant stimuli can impact later behavioural responses and also shape future encounters with conspecifics. For example, numerous studies have demonstrated that previous experience with a predator cue can shape the behavioural response of an individual later in life (Bell and Sih 2007; B. Chapman et al. 2008; Cohen et al. 2008; Edenbrow and Croft 2013). In group-living species, this can be particularly significant due to observed associations based on behavioural type (Pike et al. 2008; Aplin et al. 2013), or for group-level behaviours such as predator inspection (Thomas et al. 2008). Similarly, audience or winner–loser effects based on previous observed or actual encounters are known to have an impact on later behavioural outcomes, particularly with regard to agonistic encounters (A. Frost et al. 2007; Matessi et al. 2010; Dzieweczynski et al. 2012). What remains largely unclear, however, is the temporal consistency of these effects and their implications for the study of personality in networks.

Until recently, observations of animal social networks were typically conducted over brief time periods and lacked the replication needed to examine network positions in terms of personality. An early paper by Lusseau et al. (2004) noted that certain individuals in dolphin social groups possessed higher 'betweenness' scores than conspecifics and suggested that these individuals represented social 'brokers' who played important roles in connecting otherwise distinct social groups. Although such assumptions are possible, this type of categorization is an example of network studies that often characterize individuals as representing behavioural types without checking the underlying assumptions needed for such characterization (i.e. consistency and repeatability of traits) (Dingemanse et al. 2012). Importantly however, such studies demonstrate that particular individuals can indeed have attributes that make them more or less influential in terms of the properties of their given network and thus have significance for group-level properties. Further, they also identify network positions as important and associated with major ecological and evolutionary processes such as disease and information transmission in animal groups (e.g., Laland and Williams 1998; Otterstatter and Thomson 2007; Hamede et al. 2009; Croft, Edenbrow, et al. 2011).

From a personality perspective, understanding how individual consistency in behaviour can be achieved across social contexts is especially difficult. Animal groups are essentially dynamic mixes of individuals with unique behavioural repertoires, so what would consistency in network parameters look like? A first step in this process would be to determine how individuals attain a given position or role within their network. Two potential pathways are possible. First, an individual's behavioural type is relatively fixed, and the group dynamics generated by its interactions with others dictate its position within a network. Second, an individual's behavioural type is flexible and influenced by its interactions with other group members. Both pathways are plausible and likely depend on the individual, the behaviour involved (i.e. vigilance vs. dominance) and the mechanism underlying the association (i.e. genetic vs. experience based).

For example, Sih and Watters (2005) found that both possibilities can co-occur and have profound implications for group dynamics and individual fitness. By mixing individual water striders (*Aquarius remigis*) of known behavioural type into groups based on their average level of activity and aggression, the authors found that, while most individuals generally retained their individual attributes, some individuals did in fact alter their behaviour based on agonistic encounters with others. Further, some individuals that were 'hyper-aggressive' had a particularly strong impact on social and mating dynamics in their respective groups. As such, one might assume that behaviours with a strong relationship to underlying physiological needs (i.e. activity) are more likely going to be fixed over the long term than seasonal or state-dependent traits. On the other hand, giving consideration to the potential impact of these state-behaviour relationships on network position allows potentially interesting, and heretofore untested, opportunities to understand positive/negative feedback mechanisms (e.g. asset protection) associated with trait labiality (M. Wolf et al. 2007; Dingemanse and Wolf 2010, and references therein).

A second step in this process then is to establish the ecological and fitness consequences of different network positions. To ascertain these consequences and their importance for personality and network research, it is necessary to determine some measure of consistency in network positions between individuals and across contexts. Consistency in network position is however, an almost entirely neglected area of network analysis research.

Recently, A. Wilson et al. (2013) presented a conceptual framework and statistical test for characterizing the consistency of network positions in individuals. This allows one to consider social personality traits in the context of their group by examining the longevity and relative importance of consistency in network position relative to other traits. Consistency in network position might have significant implications for behavioural research and potentially, provide insights into the evolution of social phenomena. This is particularly relevant since recent studies have shown that network position can be related to fitness (e.g. D. McDonald 2007; Formica et al. 2010) and might perhaps be heritable (Fowler et al. 2009).

Fitness consequences of network positions

Assuming some level of consistency in the ability of an individual to maintain a network position across ecological contexts, a key challenge then is determining what the potential fitness consequences of such consistency might be and whether selection can shape the evolution of networks. A. Wilson et al. (2013) outlined two plausible scenarios for approaching this challenge. First, individuals in different network positions might achieve different absolute fitness values. For example, in some primate groups, animals are thought to adhere to strict linear dominance hierarchies wherein the highest ranking individuals, based on agonistic interaction data, have the highest frequency of affiliative interactions and therein the most access to potential mates (but see Shizuka and McDonald 2012). In this case, we would be most interested in understanding how these differences arise and are maintained in animal groups and what factors enable certain individuals to have such high fitness and others not? Alternatively, individuals in different network positions may initially be capable of achieving the same fitness, but differences in network position may, for whatever reason, be related to differences in other indirect fitness components (e.g. mortality risk, fecundity profile).

One potentially compelling example of this for personality may be found in a recent paper on wild fungus beetles (*Bolitotherus cornutus*). Formica et al. (2010) measured *strength* (a measure of network centrality via number of connections) and *cliquishness* (as measured by clustering coefficient) and found that network position covaried with copulation success and further that at least some aspects of these fitness pathways were associated with individual-level differences in activity. Generally, males that were active more frequently had higher measures of centrality (greater number of connections), and this result effectively accounted for the observed relationship between *strength* (connectedness) and fitness. As such, the value of considering personality and group-level behaviours via a network perspective clearly provides an important more holistic framework for understanding social evolution.

In general, not many studies have commented on the fitness consequences of network position (Krause et al. 2007), but there are notable exceptions with implications for personality. Silk et al. (2009) demonstrated that 'social capital' gained by female baboons (*Papio cynocephalus*) that maintained more social relationships (i.e. more connections, greater centrality) had more offspring, which in turn had greater survival. In another baboon study, A. King et al. (2011) demonstrated that dominance interactions and affiliative patterns were critical to shaping the observed feeding network. Analysis of network structure revealed that individuals in the group tended to arrange themselves in such a way as to reduce aggression or agonistic encounters and increase foraging benefits.

A number of other studies have also provided evidence for a link between social structure and function and perhaps a pathway for selection to increase (or decrease) the fitness of individual group members. For example, Oh and Badyaev (2010) found that male house finches (*Carpodacus mexicanus*) choose their social background to enhance their attractiveness to female conspecifics. Males with less elaborate ornamentation tended to be more socially labile (showing higher betweenness) relative to more elaborate males (with more colourful plumage). Social labiality was reflected in duller males changing associations between distinct social groups to find groups which best suited their duller ornamentation, and therein provide themselves with the maximum possible fitness advantage by increasing their relative attractiveness to females.

In another study, Ryder et al. (2008) demonstrated that social network connectivity (as measured by degree, eigenvector centrality, information centrality, and reach) predicted the ability of male wire-tailed manakins (*Pipra filicauda*) to become territory holders and therein obtain greater reproductive success via preferential access to potential mates. D. McDonald (2007) found a similar result in that centrality was a good predictor of future reproductive success and adult social status for juvenile male long-tailed manakins (*Chiroxiphia linearis*). However, all of these studies present only correlational evidence and more research is needed to determine whether there is a causal relationship between network position and fitness, and the implications of this in terms of personality research.

Networks and behavioural types

The notion of 'behavioural type' might also be important for understanding group-level properties in other ways. For example, different behavioural types likely have unique network properties that can perhaps play important roles in the formation of social cliques. Pike et al. (2008) quantified individual differences in latency to feed following a fright stimulus as a proxy for boldness in stickleback fishes (*Gasterosteus aculeatus*) and related this to their individual network properties. They found that fish that were bolder had fewer, more evenly spread connections among group members than more timid fish which formed longer-lasting associations based on many interactions with just one or two other fish. This was an important study as it demonstrated for the first time that animal social network structure can be affected by the behavioural composition of the group. Another recent study has found a similar result between network attributes (i.e. centrality) and personality in wild songbirds: Aplin et al. (2013) found that fast-exploring individual great tits (*Parus major*) interacted with greater numbers of conspecifics and moved between flocks more often than slow-exploring birds, which formed more synergistically stable relationships.

The formation of cliques may also be a necessary factor in the evolution of cooperative behaviours that may only develop in such highly interconnected groups of individuals in a population. Later studies have shown that this relationship can also be associated with other traits (i.e. body size) and explain, at least partially, how individuals attain dominance or social status and the establishment of social ties (Schurch et al. 2010). Croft et al. (2009) also found strong evidence for assortment by behavioural type in animal social networks. Using Trinidadian guppies (*Poecilia reticulata*), the authors quantified the behavioural phenotype of individual fish based on predator inspection and shoaling tendency and related that to their network properties. With observations in both the field and the lab, this study demonstrated that individuals in the wild either strengthened or cut social ties in relation to previous social experiences from the laboratory tests (i.e. 'cooperation' during predator inspection). Interestingly, the authors also noted that timid individuals tended to have stronger associations with group members, although also more connections than bold fish.

Another study recently demonstrated that behavioural type and social personality traits can be useful in understanding colonization success and predicting ecological invasions. Using mosquitofish (*Gambusia affinis*), Cote et al. (2011) demonstrated that individual dispersal tendency was influenced by the mean boldness and sociability score of its population. Individuals from bolder, more asocial populations were more likely to disperse, irrespective of their own behavioural type, than individuals from more social populations.

Given that we know that (a) individuals can associate by behavioural type and (b) certain 'keystone' individuals might play important roles with regard to transmission processes (i.e. disease, information), one important question that might be asked is whether it is in fact the individual or the role represented by the individual that is most important. For example, a number of studies have demonstrated that association patterns can be temporally transient depending on ecological context and environmental cues. In free-ranging shoals of fishes, group composition can change within very short time periods (i.e. seconds to hours). Because of this rapid turnover, it has been suggested that self-referent matching of recent prey and habitat use may provide a more parsimonious mechanism of recognition than identification of attributes of the individuals themselves (Ward et al. 2004; Webster et al. 2007; Ward et al. 2009). Similarly, recent evidence suggests that population level attributes might at times be more important for influencing important ecological processes than individual attributes (Cote et al. 2011).

In the context of networks, this might have important implications for understanding the underlying mechanism of observed social associations in animal groups. In fission–fusion groups in particular (reviewed in Aureli et al. 2008), this notion may provide a forum for examining the importance of behavioural type, as animals might not identify individuals per se but recognize and associate with particular 'types' of individuals. For example, individuals may choose to associate with each other based on similar space use, foraging preferences or perhaps with individuals based on their propensity to take risks and evaluate novel stimuli. This result may explain why the networks of fission–fusion groups may be less affected in general by the targeted removal of individuals and further, perhaps, to types of individuals. It may be easier to identify and therein associate with individuals based on their type than to evaluate many attributes of all possible individuals.

That said, many studies have demonstrated that in some species or study systems in particular, individuals do have properties that make them unique and perhaps irreplaceable. For example, using experimental manipulations of network composition in pig-tailed macaques (*Macaca nemestrina*), Flack et al. (2006) found that the removal of high-ranking individuals involved in group policing resulted in a notable decline in group cohesion. Similarly, Bhadra et al. (2009) observed that the colony network structure of a wasp (*Ropalidia marginata*) changed dramatically after the removal of the queen, as agonistic encounters increased among colony members as they vied for centrality and control of colony reproduction. Such studies that demonstrate the importance of particular individuals typically tend to focus on small, relatively closed groups with little immigration or emigration, and often lack replication. It is therefore plausible that by recognizing particular behavioural classes or types of individuals rather than specific individuals, members of large or highly transient groups can temporarily or permanently reduce the costs they experience during group fission–fusion events.

Should individuals recognize behavioural types rather than individuals, this notion may provide a forum for comparing larger scale patterns among animal groups that otherwise differ in fine-scale conditions. For example, comparing group composition or removal of key individuals based on type could perhaps result in similar changes to the social dynamics of populations, irrespective of taxonomic or environmental distinctions. As such, this idea might shed light on the dynamics of how animals socialize, and provide further insights regarding the importance of behavioural types in animal social groups. From an applied perspective, this notion might also be valuable in terms of species conservation and animal welfare practices in captive populations.

Networks and personality from a developmental perspective

One of the great hazards in contemporary behavioural research is to ignore developmental processes and their importance for ecology and evolution. Yet, despite growing awareness of this issue, ontogenetic processes have largely been neglected by personality researchers (but see Bell and Stamps 2004, Sinn et al. 2008, and Edenbrow and Croft 2011 for examples). Groothuis and Trillmich (2011) suggest that the reason for this neglect is primarily due to difficulties associated with reconciling personality (consistent behaviour over time) and ontogeny (changes in many traits over time).

Until recently, personality has most commonly been assessed by observations over short time periods or within a given life-history stage (Stamps and Groothuis 2010). However, there are several inherent problems with this experimental approach. For example, such observations underestimate the consequences of long-term consistency in behaviour and also fail to recognize the potential importance of seemingly maladaptive behaviours later in life. Similarly, behavioural traits may couple with other life-history characteristics over ontogeny, with some such combinations of traits being superior to others in terms of fitness (M. Wolf et al. 2007; Réale et al. 2010).

In many respects, developmental processes offer rare experimental opportunities to evaluate the consistency of personality and social behaviour across a myriad of variables in a natural environment. For example, many species possess complex life cycles or undergo abrupt ontogenetic niche shifts during development. These life-history transitions can involve changes in foraging tactics, sociality, environmental niches and life-history goals, as animals progress from larval or juvenile stages to adulthood.

For example, in amphibians, you might have a herbivorous, social, aquatic larval stage organism geared primarily for growth that metamorphoses into a carnivorous, solitary, terrestrial adult that is geared towards reproduction and dispersal. Metamorphosis, defined as a complex phenomenon incorporating a series of abrupt changes in an individual's morphology, physiology, and behaviour during postembryonic development (H. Wilbur 1980), then provides a rare *in situ* experimental opportunity to study how behavioural personality traits uncouple with other ecological, morphological, and physiological traits over development (A. Wilson and Krause 2012a). Not many studies currently exist that document behavioural consistency over metamorphosis but some do exist for invertebrates (Brodin 2009; Gyuris et al. 2012; Hedrick and Kortet 2012) and, to a lesser extent, vertebrates (A. Wilson and Krause 2012b). That said, interesting comparisons might be made by comparing the network structure of larval versus adult stages of development and teasing apart the underlying mechanistic basis for either change or consistency of social associations and behaviour over ontogeny.

In terms of social networks, the dynamic nature of ontogeny has also largely been neglected. However, the benefits of developmental perspectives are increasingly attracting attention in this field of research. In a landmark study, D. McDonald (2007) demonstrated that the network position of juvenile long-tailed manakins (*Chiroxiphia linearis*) can predict the direction of their later social trajectory. For example, early connectivity in young males was a strong predictor of social rise in the population, with high connectivity being most important for young or low status males and becoming less important upon their reaching a higher level of social and reproductive success. Later studies on this system suggested that the underlying mechanism for this relationship between connectivity and social rise was not explained by kinship and that, as expected, the duration of a male's territorial tenure was a good predictor of his probability of siring offspring (Ryder et al. 2008; D. McDonald 2009).

In another study, Wey and Blumstein (2010) used network analysis to study social cohesion in yellow-bellied marmots (*Marmota flaviventris*). As with the previous example, the authors also found that patterns of direct and indirect interactions changed over ontogeny, with juveniles being involved in primarily affiliative interactions, and older animals engaging in more agonistic ones. However, unlike the previous example, kinship played an important role in explaining affiliative networks in addition to age in this case. While these examples are somewhat limited, they do highlight the central relationship between social network analysis and personality traits (i.e. activity, aggression) and the hazard of basing important ecological assumptions on limited short term observations or those made within particular life-history stages. Since developmental needs and goals might vary over ontogeny, considering network attributes with a developmental perspective is crucial for developing a better more holistic view of social structure and behaviour.

Personality and social network analysis in animals: conclusions and future directions

Until recently, the focuses of many behavioural studies have revolved around *processes* that generate differences between individuals rather than larger scale *patterns*. By including an emphasis on such patterns of variation, and merging social network analysis with personality research methodology, it becomes possible to use individual-level interactions to gain population level insight into biological systems. Such an approach has the potential to provide important new information regarding the evolution of animal personality, while at the same time identifying commonalities between what were previously thought to be distinct social phenomena (i.e. cooperation, aggression, anti-predator behaviour).

In this chapter, we have highlighted how individuals can be important to their network and that this importance can be related directly or indirectly to personality or behavioural types. We have discussed the value of understanding consistency of network positions, and the potential relevance this may have for fitness. We also discussed the possibility that in some social groups, particularly those with high turnover (i.e. fission–fusion groups), individuals might identify and associate with others based on behavioural type or class of individual rather than particular individuals themselves. Lastly, we have highlighted the need to consider personality and social networks from a developmental standpoint, as the significance of different behaviours may change as individuals' progress through ontogeny. While each of these topics has significant potential for network personality research conceptually, experimental solutions represent more of a challenge.

One approach to test many of the ideas presented in this chapter would be to use species or populations which eliminate or at least reduce areas of uncertainty, for example, by studying animals that exist in small- to medium-sized groups with closed membership (not fission–fusion groups), are long-lived, are thought to be capable of individual recognition, and generally inhabit fixed locations. These are all factors that would facilitate the study of individual-level consistency in behaviour and its importance for social network analysis. Such study systems may be represented by captive populations or those species which exhibit these characteristics in the wild (i.e. some primates, birds, and fishes).

For example, many species of damselfish (i.e. *Dascyllus* spp.) found on tropical coral reefs exhibit most of the aforementioned characteristics and are known to exhibit strict dominance hierarchies while inhabiting small coral heads for their entire lives (Booth 1995). Such systems would be ideal for determining the importance of key individuals and the effects of the removal of such individuals on group dynamics and network attributes. Similarly, due to having relatively small home territories, individuals can be tracked for long time periods, allowing evaluation of how behaviour and network attributes (i.e. network position) change over ontogeny. Understanding an individual's role in their network and the potential costs and benefits of retaining or changing their network position during different life phases (with different life-history objectives) represents an important research goal.

On the other hand, using species that are known to possess fission–fusion dynamics may also offer important insight into the evolution of personality and social networks. For example, such populations allow one to test assumptions regarding the identification of behavioural types or classes of individuals versus individuals themselves and the potential advantages of such attributes for highly dynamic or large populations. Understanding any level of consistency in behaviour and interactions between individuals would add crucial insight into the costs and benefits of social associations in dynamic groups and their potential fitness consequences. A crucial first step in this research direction is to understand the biological mechanism of social recognition (i.e. familiarity, environmental cues) and how this relates to similar mechanisms in other comparable taxa. Similarly, while we know that familiarity and other social phenomena can convey a number of benefits, including increased foraging success and reduced aggression as well as enhanced predator escape responses and learning performance (Morrell et al. 2008; Ward et al. 2009), these benefits do not necessarily imply individual recognition.

Logistical issues associated with tracking and identifying all individuals in a network during multiple observation sessions also represent a significant challenge, particularly in wild populations (outlined in A. Wilson et al. 2013). However, recent technological advances (i.e. Encounternet; Rutz et al. 2012) provide a manner by which such problems may be overcome, allowing continuous tracking of large number of individuals over short observation sessions (Krause et al. 2011). Such technological innovations are also valuable in other ways as they provide a manner by which to collect network data in a diverse range of taxa (i.e. birds, reptiles, and amphibians) and in a diverse range of environments (i.e. tropical and temperate forests) that were previously not feasible.

Another important challenge may be found in trying to interpret the relationship between personality and the dynamics of networks from an evolutionary perspective. In a recent review, Krause et al. (2010) highlighted several studies which provided compelling insight into the evolution of network processes (Santos, Pacheco, et al. 2006a,b) and discussed how they might be relevant in terms of personality. These are mostly highly abstracted models and therefore are largely untestable in real populations. As such, few studies have empirically attempted to demonstrate such relationships, but there are exceptions (e.g. J. B. W. Wolf et al. 2011).

That said, in a number of areas, steady progress is being made that allows one to understand the causes and consequences of personality and consistent network positions from an evolutionary standpoint. Such characterizations might be made by examining the fitness consequences of different network positions both directly and indirectly (i.e. via differences in behavioural type). In spite of this progress, however, a more holistic understanding of the fitness consequences of different network positions is still lacking. While there is evidence for the heritability of different network positions, these estimates represent mere snapshots, and long-term measures over several generations are still needed.

The integration of network analysis in the study of social personality traits offers many potentially fruitful avenues of future research. Such integration allows new perspectives on 'old' traits and allows new dynamic models of how animals socialize to be built. Since both personality and social network analysis are associated with important ecological and evolutionary processes, understanding the dynamic relationship between these two phenomena will offer compelling new results and be of substantial value for both behavioural and evolutionary biologists.

Acknowledgements

A. D. M. W. was in receipt of an Alexander von Humboldt Research Fellowship. J. K. acknowledges support from the Pakt programme for Research and Innovation of the Gottfried Wilhelm Leibniz Gemeinschaft.

CHAPTER 7

Temporal changes in dominance networks and other behaviour sequences

David B. McDonald and Michael E. Dillon

Introduction to the analysis of temporal changes in networks

Order is one of those curious English words with multiple meanings and connotations. Here, we will distinguish between its use to describe orderliness, the absence of disorder, particularly with respect to rank order, and its narrower sense of a temporal sequence of interactions that establish a dominance network. In previous studies, Shizuka and McDonald (2012) and D. McDonald and Shizuka (2013) examined order in dominance hierarchies, viewed as social networks, primarily from the perspective of orderliness. A surprising emergent property of these comparative analyses, across 101 published datasets for a wide variety of taxa, is the paucity of disorderly cyclic dominance relations.

A triadic cycle (rightmost triad in Figure 7.1) is a dominance relation in which A dominates B, which then dominates C, which in turn dominates A, meaning that one cannot assign a clear top, middle, and bottom rank to the three individuals in the triad. In contrast, a transitive triad (penultimate triad in Figure 7.1) is one in which A dominates both B and C, B dominates C, and C is subordinate to both A and B. A transitive triad is an orderly set, with a clear top, middle, and bottom animal. Dominance relations are an emergent property of a series of dyadic interactions between animals; those interactions result in a learned submission by a subordinate to a dominant (Rowell 1974; Bernstein 1981). The relation exists only when the subordinate accepts, at least temporarily, its lesser status and reduces or changes its agonistic interactions with the dominant individual. Although dominance interactions can be viewed as fierce struggles whose primary function is to obtain access to scarce resources (e.g. Clutton-Brock and Harvey 1976 as characterized by Drews 1993), a network perspective shows that, in most taxa, dominance hierarchies are surprisingly orderly and places greater emphasis on reduction of conflict (Flack et al. 2006). The orderliness suggests some degree of self-organization (D. McDonald and Shizuka 2013) whereby reduced disorder benefits even those at or near the bottom of the hierarchy, whose acquiescence is the most difficult to explain (Maynard Smith 1983).

In this chapter, we focus on the second sense of the word order—the temporal sequence of events involved in establishing dominance relations (states). We have two primary goals. Our first goal is to ask whether the temporal order in which dominance interactions occur affects the ranks of individuals. One can study many aspects of dominance hierarchies without ever assigning ranks to individuals. Nevertheless, researchers often wish to assign ordinal ranks to the individuals in a group, because those ranks then serve as attributes in other analyses (e.g. Macedo et al. 2012).

At least ten major algorithms exist for assigning ordinal dominance ranks to group members (Gammell et al. 2003; Hemelrijk et al. 2005; Bayly et al. 2006). All but one of the algorithms base the ranks simply on the static, ending total of wins and losses.

Animal Social Networks. Edited by Jens Krause, Richard James, Daniel W. Franks and Darren P. Croft.

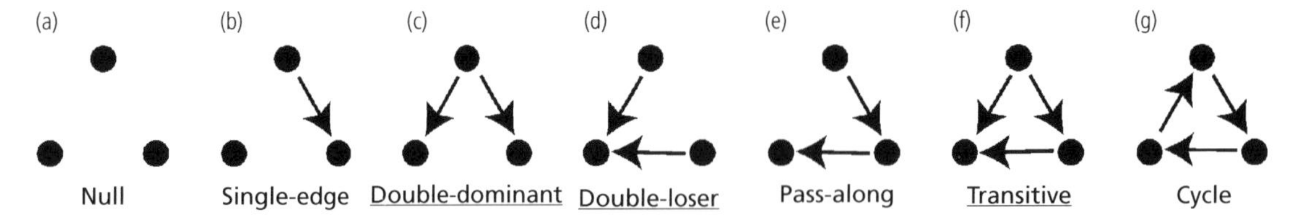

Figure 7.1 The seven types of triads possible when asymmetric edges (one-way arrows pointing from dominant to subordinate) link nodes (animals). A network with *n* nodes has *n*-choose-3 triads (all possible combinations of *n* animals, taken three at a time), distributed across these seven types. The count of triad types is called a triad census. The inherently transitive (orderly) types, (c) double-dominant, (d) double-loser, and (f) transitive, are underlined. If and when the third edge (dominance relation) is established, the two-edge triad type, (e) pass-along, can become either cyclic (disorderly; if the third edge points up) or transitive (orderly; if the edge points down).

The one exception appears to be the Elo ranking method (Elo 1978) recently promoted by Neumann et al. (2011) and explored, from a network perspective, by D. McDonald and Shizuka (2013). For this method, the details of the temporal sequence do affect the assignment of ranks.

As a concrete example, consider two series of three wins (W) and three losses (L). Only under the Elo rating method does the possibility exist for WWWLLL to confer a different rank on the contestant than would the order LWLWLW or LLLWWW. Having established that the temporal order does affect Elo ranking, we will suggest situations in which one might favour the order-sensitive Elo method over an order-insensitive method such as the de Vries (1998; see also Schmid and de Vries 2013) I&SI method, which optimally reorders the rows and columns of the end point dominance matrix.

Our second goal is to assess the shifting nature of multi-actor interactions within a dominance hierarchy, mostly from a social network perspective. That is, does the order in which relations form have predictable consequences for the structure of social organization? From a social network perspective, how does a sequence of interactions affect pattern and process at the level of the node (individual animal), subnetwork (community), or network as a whole? By assessing the network at several temporal stages, we can ask how individual nodes, and the network as a whole, change over time. With the rising use of social network analyses in behavioural ecology (Croft et al. 2006; D. McDonald 2007; Krause, Lusseau, et al. 2009) behaviourists have a rich new toolbox for exploring dominance interactions with quantitative approaches that open the door to insights into the dynamic trajectories of node centrality at the individual level, and network structure at the group level (Pinter-Wollman et al. 2013). By incorporating temporal dynamics, network analyses can help us assess the multidimensional social context, in which relations depend not only on direct dyadic interactions but also on effects between individuals that are not themselves directly linked (at various 'degrees of separation').

Any life is necessarily a trajectory in which individuals assume different roles through time ('one man in his time plays many parts'—Shakespeare). An individual's dominance relations will often form a humped curve, with subordinate status early in life, a peak of status in the prime of life, and increasing subordinance late in life. The middle ranks of most social groups will therefore consist of a mixture of individuals on upward or downward trajectories. One should expect, therefore, that the middle ranks should often be the most turbulent and dynamic, with relatively frequent shifts in dominance relations. The inevitable within-group shifts due to individual trajectories may also be accompanied by, and affected by, changes from outside the group. In many cases, group composition may be relatively stable; over any particular time interval, only a small proportion of individuals may enter, by immigration or birth, and a few individuals may leave, by emigration or death.

Despite such membership changes, previous relations among existing members may be little affected by the entries and exits of other individuals. In some cases, however, novel alliances or third-party effects may mean that the changes have dramatic impacts on pre-existing dyadic relations. Our goal is therefore to contribute to the growing interest in using social network analysis to understand the temporal dynamics of groups (Blonder and Dornhaus 2011; Hobson et al. 2013; Pinter-Wollman et al. 2013). While our focus is on dominance relations, the principles and procedures should apply equally well to other kinds of sequential behaviours, such as singing bouts or sequences of affiliative behaviours.

Network formulation and triad census approach

Published dominance data are almost always in the form of matrices, with the numbers of contests won by an individual in a row, and the number of contests lost in a column. Such a matrix is directly equivalent to a weighted adjacency matrix in network terminology (Wasserman and Faust 1994). D. McDonald and Shizuka (2013) termed this matrix of the raw contests a contest matrix, as distinguished from the 1/0 outcome matrix, described below, that encapsulates the dominance–subordinance relations.

As stated earlier, dominance is fundamentally a learned dyadic relation, in which a subordinate, over a series of contests, eventually learns to submit to the dominant (Hinde 1976; Bernstein 1981). Using a convention such as majority-win designation of dominants (D. McDonald and Shizuka 2013), and the elimination of draws, one can transform the weighted directed network represented by the raw contest matrix into an outcome matrix—a directed 1/0 network in which every dyad has either an asymmetric edge (pointing from the dominant to the subordinate) or a null edge (no relation yet established). The outcome matrix therefore represents the fundamental relation within dyads (states), whereas the contest matrix represents the series of interactions (events) that forge that dominant–subordinate relation and contribute to the emergent social structure. Thus the contest matrix (events), the outcome matrix (states), and the emergent properties of the social network as a whole correspond directly to Hinde's (1976) three hierarchical levels of social organization—interactions, relationships, and social structure.

By eliminating mutual edges (bidirectional arrows pointing both ways between pairs of individuals) one can simplify an important element in the social network toolbox—the triad census (Holland and Leinhardt 1976), a particular form of network motif analysis. The simplified triad census enumerates the seven distinct (non-isomorphic) types of triads (Figure 7.1), ranging from the null triad (no dominance relations established) to the two types of complete triangles—the orderly transitive triangle and the disorderly cycle. This triadic perspective is one example of the network motif approach that has been influential in a variety of network studies (Milo et al. 2002; Faust 2010). The pioneering work of Chase (1982) paved the way for this type of triadic motif view of dominance interactions. Adopting a framework of triadic analysis provides a straightforward basis for assessing temporal dynamics such as the way that single- and double-edge triads progress towards becoming complete three-edge triangles. Further, the triadic structure provides a basis for assessing the interaction between changing network structure and the changing roles of individuals within the network.

Ranking algorithms

We will consider ten different published algorithms for ranking individuals but will focus primarily on the I&SI method of de Vries (1998), and the Elo rating method, derived for chess (Elo 1978). For the I&SI method of de Vries (1998), the contest or outcome matrix is iteratively reshuffled so as to reduce, first, the number of inconsistencies (*I*)—cases in which a lower-ranking individual dominates a higher-ranking individual—and, second, the strength (*SI*) of those inconsistencies—greater strength arises from larger discrepancies between the ranks of the inconsistent relations. Visually, inconsistencies are manifest as non-zero elements occurring below the diagonal of the matrix, and strength by how far away from the diagonal the inconsistencies occur. In Figure 7.2, the lower-ranking bird G dominates higher-ranking B, creating an inconsistency of strength 5, the difference between their ranks.

Schmid and de Vries (2013) recently presented improved algorithms for assessing ranks by the I&SI method. The I&SI method is compelling largely because it strives for a mathematically based internal consistency, similar to the bandwidth problem

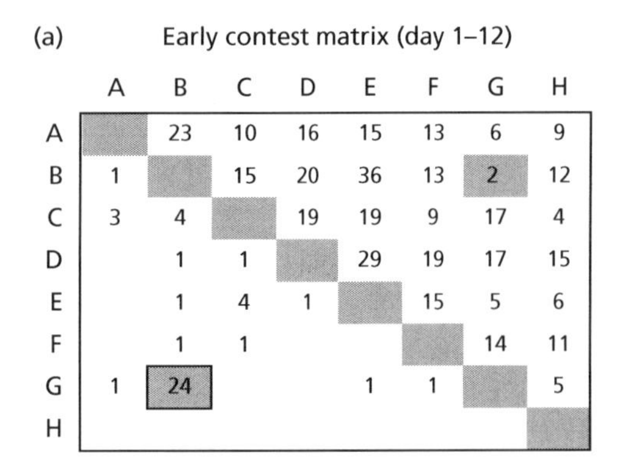

(a) Early contest matrix (day 1–12)

	A	B	C	D	E	F	G	H
A		23	10	16	15	13	6	9
B	1		15	20	36	13	2	12
C	3	4		19	19	9	17	4
D		1	1		29	19	17	15
E		1	4	1		15	5	6
F		1	1				14	11
G	1	24			1	1		5
H								

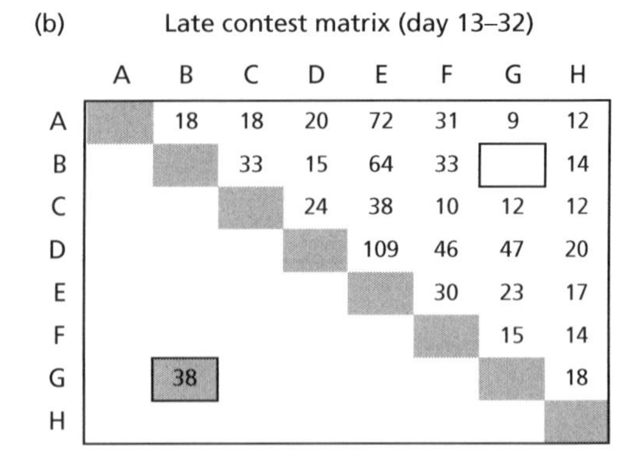

(b) Late contest matrix (day 13–32)

	A	B	C	D	E	F	G	H
A		18	18	20	72	31	9	12
B			33	15	64	33		14
C				24	38	10	12	12
D					109	46	47	20
E						30	23	17
F							15	14
G		38						18
H								

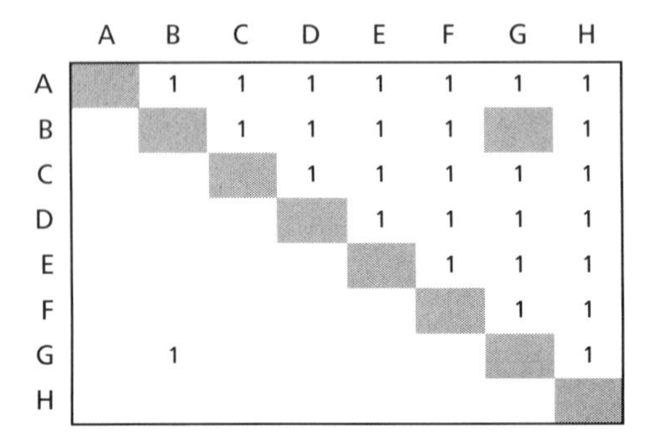

(c) Outcome matrix (1/0 dominant/subordinate relations)

	A	B	C	D	E	F	G	H
A		1	1	1	1	1	1	1
B			1	1	1	1		1
C				1	1	1	1	1
D					1	1	1	1
E						1	1	1
F							1	1
G		1						1
H								

Figure 7.2 (a) Early (Day 1–12) and (b) late (Day 13–32) contest matrices, corresponding to weighted, directed networks from the data of Parsons and Baptista (1980) for eight captive white-crowned sparrows (*Zonotrichia leucophrys*). Despite the numerical differences between the early and late contest matrices, the underlying (c) outcome matrix of 1/0 dominant–subordinate relations is identical, suggesting that dominance relations stabilize early. Note also that, in the late contest matrix, the subordinates cease to 'fight back', including the one inconsistency (sensu de Vries 1998), G > B.

in graph theory and computer science (Chinn et al. 1982). Bandwidth maximizations seek a matrix rearrangement whose non-zero elements are maximally close to the diagonal. The slight twist for the I&SI method is that no penalty accrues for numbers far above the diagonal. That is, domination of the lowest-ranked animal by the highest ranked (the cell in the upper right of the matrix of Figure 7.2) is perfectly acceptable, whereas a non-zero element in the lower left would be the strongest possible and 'worst' inconsistency.

For many kinds of network analyses, a network diagram, or sociogram, is far more visually informative than the corresponding adjacency matrix. In matrix-based demography (Caswell 2001), for example, the life-cycle diagram (a network diagram) is almost invariably more easily interpretable than the corresponding projection matrix. Because they stand out by falling below the diagonal of the matrix (Figure 7.2), inconsistencies and their strengths provide an interesting exception to that rule, with the matrix more visually informative than the corresponding sociogram.

An alternative method for assigning ordinal dominance ranks is Elo rating, as recently promoted by Neumann et al. (2011). Elo rating works from a contest sequence, with scores updated after each contest. The scores of the winner and loser of a contest are incremented or decremented accordingly. Additionally, the scores of the contestants are interpolated across the interval between an individual's current contest and its most recent previous contest. Ranks are then assigned according to the scores at the end of a sequence of contests. For purposes of comparison, and to emphasize the point that ranking algorithms can produce a profusion of distinct rank orders, we will present below the results of rankings generated by the indices proposed by Clutton-Brock et al. (1979), David (1987), Crook and Butterfield (1970), Barlow and Ballin (1976), Kalinoski (1975), Jameson et al. (1999), Van den Brink and Gilles (2000), and Bang et al. (2010).

R scripts for analysing dominance data

We conducted all the analyses with R scripts, many of which are described in D. McDonald and Shizuka (2013) and all of which are available on request. Dominance input data can be of two types—a contest sequence or a contest matrix. Raw data will often consist of a contest sequence: the temporally ordered set of interactions (contests) between dyads. The input file for analysing such a contest sequence is a straightforward three-column file, with rows containing the contest identifier, the identity of the winner, and the identity of the loser. The contest identifier will often be a series of numerals in temporal order but could also be a list of time–date values that could be used to link the analyses to other sorts of time-sensitive analyses. Published studies, however, usually present the data in the form of a compiled contest matrix, where the identities of the animals are the headers for the rows and columns, with wins in the rows and losses in the columns (e.g. Figure 7.2).

The contest matrix is easily compiled from the raw contest sequence and can then be reordered in the I&SI (de Vries 1998) rank order. The contest sequence cannot, however, be reconstituted from the contest matrix. Thus, for analyses such as Elo rating, where the order in which contests occurs matters, one must resort to creating a set of randomly shuffled contest sequences to be able to assess the effects of temporal order. Our main R script begins with a toggle for whether the input data are a contest sequence or a contest matrix. The R scripts also calculate a number of node-based network metrics not presented in this chapter (e.g. flow betweenness), several whole-network metrics (e.g. the global clustering coefficient) and some plots of network diagrams or triad census outputs (e.g. the random, expected vs. observed triad census frequencies, as illustrated in D. McDonald and Shizuka 2013).

Differences among ranking algorithms

Because their assumptions and motivations differ, it is perhaps not a surprise that published dominance ranking algorithms often differ in their rank orders for any particular dataset, as shown for the data of D. Watt (1986) for ten Harris's sparrows in Figure 7.3. For example, for the Watt data, the Elo rating method (Column 3 of Figure 7.3) reverses the third- and fourth-ranked birds relative to the published order, the de Vries I&SI order, and one widely used ranking method, David's score (abbreviated as Dav; Column 5 in Figure 7.3). Note also that concordance is high at the bottom of the rankings and less so for the top and middle ranks.

For some of the 40 datasets used in D. McDonald and Shizuka (2013), each of the ten published algorithms yields a rank order that differs from all the others (thus, to twist Alexander Pope's poem, 'all things differ and none agree'). Furthermore, the rankings may differ, using the same algorithm (e.g. I&SI) between the weighted contest matrix and the 1/0 outcome matrix of relations. Because the underlying relations (states) among the animals in a group are more fundamental than the events that shape those relations, we suggest that rank order is probably best determined from the outcome matrix, in most cases. As it happens, for the sparrow data from D. Watt (1986), the contest and outcome matrices yield identical rank orderings under the I&SI method.

Effect of contest order on Elo ranking

Our analyses demonstrate that the temporal order in which contests occur affects the ending ranks under the Elo rating method. For example, in the case of a set of 200 replicated randomly reordered contest sequences for the dominance data of D. Watt (1986; ten Harris's sparrows) and Allee and Dickinson (1954; ten smooth dogfish) the following points emerge (Figure 7.4). First, in those sequences in which individuals had wins that occurred late relative to their losses, rank increased relative to the mean Elo rank of those individuals across all 200 reordered sequences. In contrast, in

Node ID	I&SI	Elo	Clu	Dav	Cro	B&B	Kal	Jam	van	Ban
A	1	1	2	1	1	1	1	1	1	6
B	2	2	4	2	2	4	2	2	4	1
C	3	4	1	3	3	2	3	3	5	4
D	4	3	3	4	5	3	5	4	2	2
E	5	5	6	5	4	5	4	5	6	3
F	6	8	5	6	7	6	7	7	3	7
G	7	6	7	7	6	7	8	8	8	5
H	8	7	8	8	9	8	6	6	7	9
I	9	9	9	9	8	10	10	9	9	8
J	10	10	10	10	10	9	9	10	10	10

Figure 7.3 Ten different ranking algorithms yield nine distinct rank orders of the outcome matrix for the Harris's sparrow data from D. Watt (1986), as demonstrated by the lack of greyscale alignment. The algorithms are designated as follows: I&SI (from de Vries 1998); Elo (from Elo 1978); Clu (from Clutton-Brock et al.1979); Dav (from David 1987); Cro (from Crook and Butterfield 1970); B&B (from Barlow and Ballin 1976); Kal (from Kalinoski 1975); Jam (from Jameson et al. 1999); Van (from Van den Brink 2000); and Ban (from Bang et al. 2010). The I&SI method (de Vries 1998) yields the same rank order as Dav for this dataset; these concordances, however, differ across different datasets.

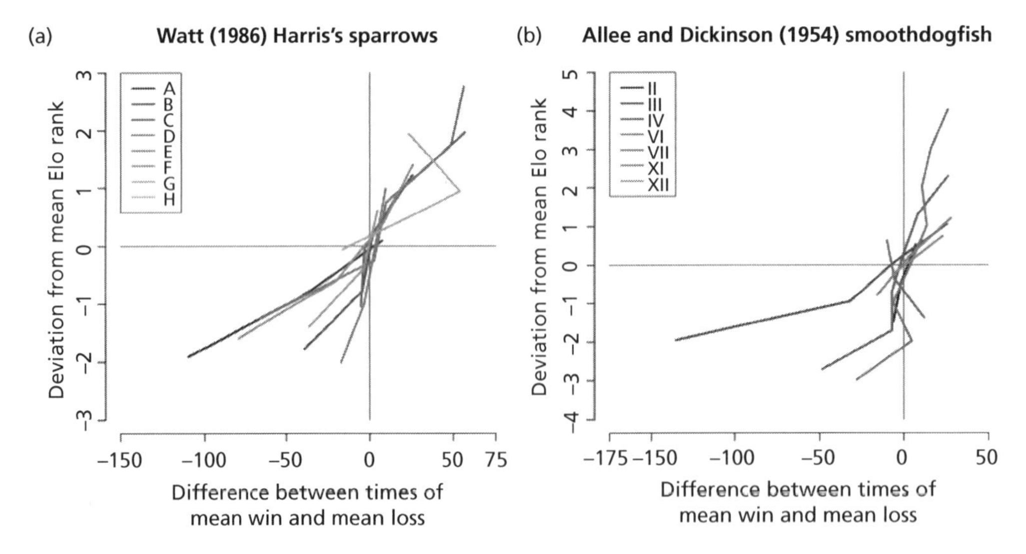

Figure 7.4 The order in which contests occur affects Elo rank ordering. The abscissa represents the difference between the mean win and mean loss for any of the particular simulated temporal sequences of c contests ($c = 540$ for the Harris's sparrows in Figure 7.4a; $c = 267$ for the dogfish in Figure 7.4b). Negative values on the abscissa therefore indicate that wins occurred early relative to losses. Positive values mean that the preponderance of wins was towards the end of the contest sequence. The ordinate is the deviation of an individual's Elo rank in each randomized sequence from its mean rank, over the 200 reordered sequences. Rather than plot all 200 points for each individual, the lines connect the medians of the points for the discrete rank deviation levels of individuals. The top-ranked individuals (darkest lines) can only really deviate downwards, whereas bottom-ranked individuals (lightest grey lines) can only really deviate upwards. (a) Harris's sparrows (D. Watt 1986). The two lowest-ranking individuals never varied in rank and so are not plotted. (b) The smooth dogfish data (Allee and Dickinson 1954) show a pattern very similar to that of the sparrow data, as do plots for other datasets not shown. Individuals attained higher ranks when their wins occurred later than their losses (values in upper right quadrant). In contrast, when their wins occurred earlier than their losses, their ranks were lower than the mean rank (values in lower left quadrant). Elo rating therefore rewards ending on a winning trajectory.

sequences where wins occurred early and losses occurred late, a reduction in Elo rank occurred. The pattern of rank reduction for late losses and rank increase for late wins holds in all of the 40 datasets analysed in D. McDonald and Shizuka (2013); note that, for most individuals, in most of the simulated sequences, their values would fall near the 0–0 centre of the plot, meaning that they attained nearly their mean rank, and that mean win differed little from mean loss. Because few published studies include the original raw sequence of contests, it is an open question as to whether actual contest sequences tend to cluster close to the modal deviation (i.e. close to the 0–0 centre of the plots in Figure 7.4) or whether many datasets have divergent values such as those in the upper right and lower left in Figure 7.4.

Although ranks shifted according to the temporal balance between win and loss times, most individuals remained in the same general echelon. For example, for the sparrow data of D. Watt (1986), although the different iterations of the simulated sequences generated slightly different ranking orders, the broad pattern held of top, middle, and bottom echelons in which most individuals tended to remain, with the greatest uncertainty concerning the second- and third-ranked birds. In a number of the 200 randomly reordered sequences, the Elo rating method ranked individual D (modal rank of 4), above individual A (modal top rank). Figure 7.5 depicts the Elo trajectories for 1 of the 200 randomly reordered win–loss sequences for the 10 individuals in the D. Watt (1986) dataset.

By assessing where the actual ordered sequence falls within a large set of simulated (randomly reordered) contest sequences, researchers could see whether the actual contest order is unusual, relative to a simulated set of reordered sequences. If many datasets tend to show a consistent deviation from the modal order, several intriguing possibilities arise. The first is that datasets might often represent long enough time periods that individuals differ in their trajectories, with some individuals on upwards trajectories as they mature, and others on downwards trajectories as they senesce. Note, for example, that individual D in Figure 7.5 has a rather steep upwards trajectory near the end of the set of 540 contests. That upwards trajectory, combined

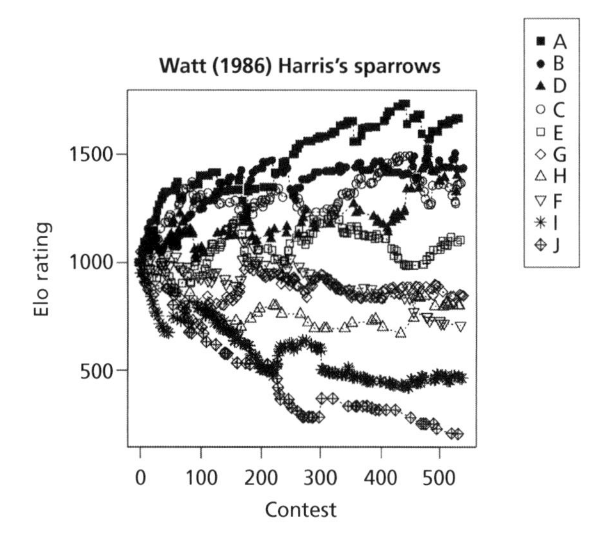

Figure 7.5 Elo rating trajectories for the ten Harris's sparrows in the study by D. Watt (1986, Figure 1, matrix 5). See D. McDonald and Shizuka (2013) for other examples of such plots.

with the downwards trajectory of individual C, results in Elo rankings for those two individuals (Figure 7.3) that differ from the ranks assigned by order-insensitive ranking methods such as the I&SI method of de Vries (1998). Likewise, under Elo rating, individual F's rank suffers, because of a downward-ending trajectory, compared to the other ranking algorithms.

Even more intriguingly, one might hypothesize that dominant individuals should strategically allocate energy devoted to contests to maximize rank order at the point where the payoff for high rank peaks (e.g. just prior to mate selection or territory acquisition). Further, selection might favour avoidance of contests by mid-ranking individuals that might be likely to lose late-occurring contests, which would reduce their rank at the time when rank matters most. Any such strategic avoidance would help explain the sparseness of many non-captive dominance datasets. Active avoidance of others may be strategically valuable for individuals of high rank as well as those of low rank; individuals of high rank may have little to gain by winning but much to lose in the event of chance injuries. A plausible alternative hypothesis to any sort of strategic allocation of contests is that the sensitivity of Elo ranking to contest order has no biological consequences. If order (sensu temporal sequence) does not matter biologically, one might prefer a ranking algorithm that is insensitive to contest order, such as that of de Vries (1998).

In summary, ending on a winning note (upwards momentum) tends to increase Elo rank, while ending on a losing note (downwards momentum) tends to decrease Elo rank. Note that the I&SI method, and each of the other eight other published ranking methods discussed in this chapter, uses only the final summary matrix and is therefore insensitive to the order of wins and losses. Whether it is appropriate that order of winning and losing should affect rank in a dominance context remains an open question. Nevertheless, allowing for the possibility that the order in which wins and losses occurs matters may improve acuity of analysis of both individuals and groups or networks as a whole.

Thus, the order of wins might affect individual trajectories but might sometimes also affect the structure and stability of the network and group as a whole, thereby helping us understand both the causes and the consequences of the structure of dominance hierarchies. For example, in some contexts, we might be interested in assessing the momentum of individuals. In that case, order could easily be crucial; late winners will tend to 'be on a roll'. In other contexts, we might be more interested in the summed lifetime consequences of dominance trajectories—in that case, order might be less important. All individuals should be expected to have trajectories that rise early in life and decline late in life, so that only the sum matters.

Comparing contest and outcome adjacency matrices over time

Few published studies of dominance hierarchies present anything but the final, summary contest matrix. An interesting exception is the dataset of Parsons and Baptista (1980) for white-crowned sparrows (Figure 7.2). Their data include a contest matrix early in the study period (Day 1 to 12; Figure 7.2a) and another for the contests late in the study period (Day 13 to 32; Figure 7.2b).

Several interesting points emerge from analysing the sequential matrices. First, and most striking, the outcome matrix for the early contest matrix is identical to that for the late contest matrix. That is, although the number of contests changed, and contests continued to occur, the underlying relations had formed by Day 12 and did not change thereafter. Second, note that although many new contests occurred in the later period, wins were now confined to the dominant individuals, as demonstrated by the lack of entries below the diagonal in the late contest matrix (Figure 7.2b). That is, the late contest matrix has only one entry below the diagonal, where wins against higher-ranking birds by lower-ranking birds would occur.

The one exception (G > B) was a strong inconsistency (sensu de Vries 1998 I&SI method). In the early contest matrix, low-ranking individual G (a female) won 24:2 against high-ranking individual B (a male). By the time of the late contest matrix, the ratio was even more lopsided (38:0). We cannot resist the temptation to speculate that the interactions were influenced by B's willingness to cede contests to G in return for present or future mating privileges. Similarly interesting and potentially informative inconsistencies and changes in win–loss ratios should characterize many datasets for other taxa and other attributes (e.g. age, size, or personality).

Analysing the contest matrix by quartiles

In principle, one could construct social networks after each new contest and then assess various network metrics for each node, and for the network as a whole. In practice, the network is unlikely to change sufficiently at each step to justify such an approach. Here, we will present results for an analysis of the D. Watt (1986) Harris's sparrow data at the first through fourth quartiles of the set of 540 contests. Note that, because we are working from a published contest matrix, we present data from one of the randomly selected reordered contest sequences; the random sequence used is almost certainly not the true (but unpublished) sequence. Nevertheless, the example will demonstrate the potential for combining multiple network and ranking approaches to temporally ordered dominance data.

Using the contest matrix as the basis for analysis, Figure 7.6 contains information for the ten individuals in the Harris's sparrow study (D. Watt 1986). A group of $n = 10$ individuals has $(n*(n-1))/2 = 45$

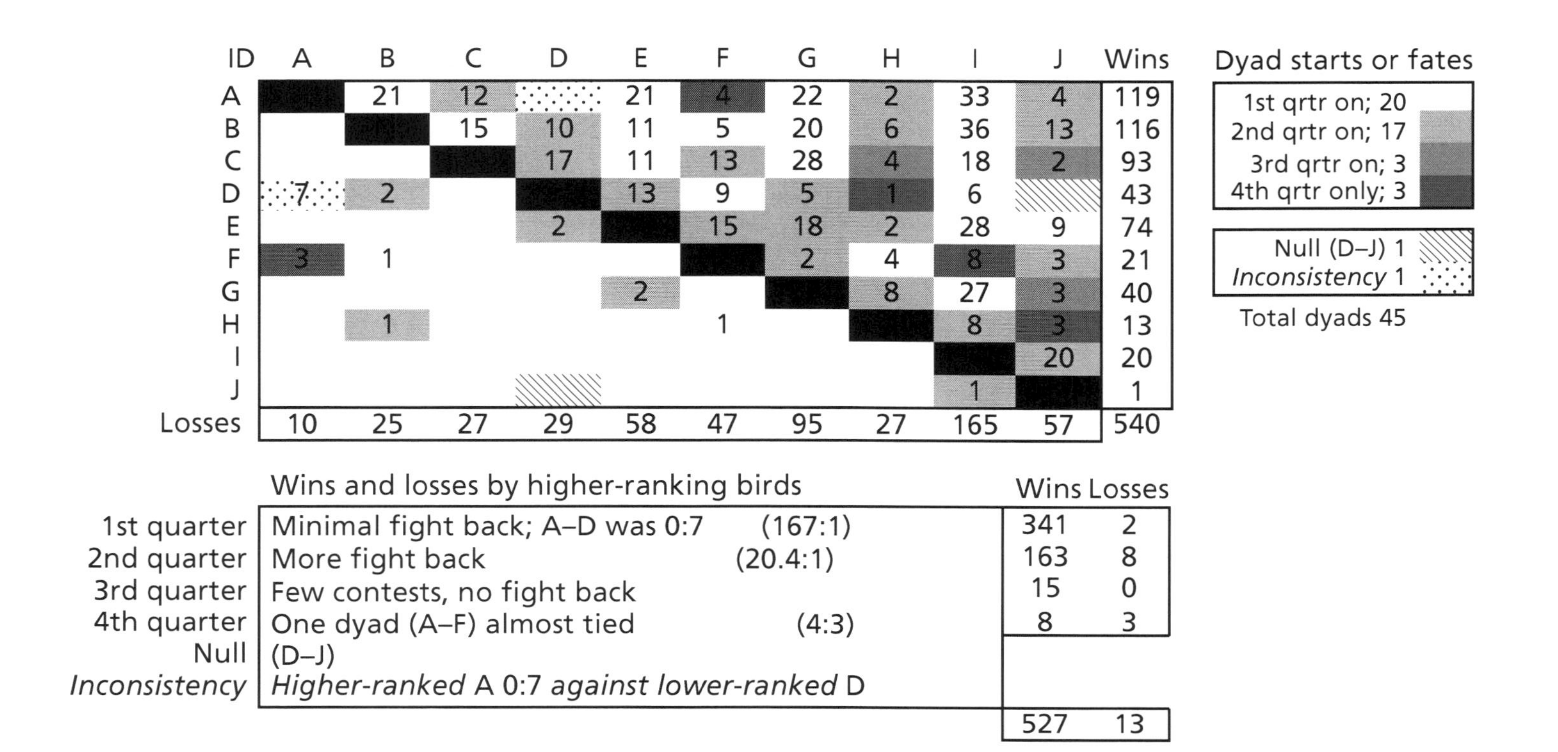

ID	A	B	C	D	E	F	G	H	I	J	Wins
A		21	12		21	4	22	2	33	4	119
B			15	10	11	5	20	6	36	13	116
C				17	11	13	28	4	18	2	93
D	7	2			13	9	5	1	6		43
E				2		15	18	2	28	9	74
F	3	1					2	4	8	3	21
G					2			8	27	3	40
H		1				1			8	3	13
I										20	20
J									1		1
Losses	10	25	27	29	58	47	95	27	165	57	540

Dyad starts or fates	
1st qrtr on; 20	
2nd qrtr on; 17	
3rd qrtr on; 3	
4th qrtr only; 3	
Null (D–J) 1	
Inconsistency 1	
Total dyads 45	

	Wins and losses by higher-ranking birds		Wins	Losses
1st quarter	Minimal fight back; A–D was 0:7	(167:1)	341	2
2nd quarter	More fight back	(20.4:1)	163	8
3rd quarter	Few contests, no fight back		15	0
4th quarter	One dyad (A–F) almost tied	(4:3)	8	3
Null	(D–J)			
Inconsistency	*Higher-ranked* A 0:7 *against lower-ranked* D			
			527	13

Figure 7.6 Annotated contest matrix for ten Harris's sparrows in the study by D. Watt (1986, Figure 1, matrix 5). Most dyads began interacting in the first quarter of the total of 540 contests. Those dyads that started early had the most lopsided win–loss ratios. One dyad (D–J) never interacted, and one dyad (A–D) represented an inconsistency (sensu de Vries 1998) in the corresponding outcome matrix.

dyads. We can evaluate the status of the 45 dyads at each quartile of the 540 total contests.

For these data, 21 of the 45 dyads (including one inconsistency) had dominance relations that formed early and remained consistent throughout. Two of these 21 dyads had one contest won by the subordinate, but the ratio was heavily skewed (341:2) in favour of the dominants. The dyad D–A, an example of an inconsistency (de Vries 1998), exemplifies the difference between dyadic dominance relation and rank order within the group as a whole. Individual D consistently won contests (7:0) over top-ranked individual A, but A, in turn, consistently won against the other individuals in the group.

An additional 17 dyads had relations that formed in the second quarter. Of those 17, 4 had one or two wins by the subordinate, and the win–loss ratio was less skewed (163:8). Three dyads developed their dominance relations during the third quarter, and three dyads did not develop relations until the fourth quarter. One dyad (D–J) was null, meaning that those individuals never engaged in a contest.

Analysing the outcome matrix by quartiles

Using a network diagram to portray the outcome network at each of the four quartiles (Figure 7.7) shows that the roles of individuals change over

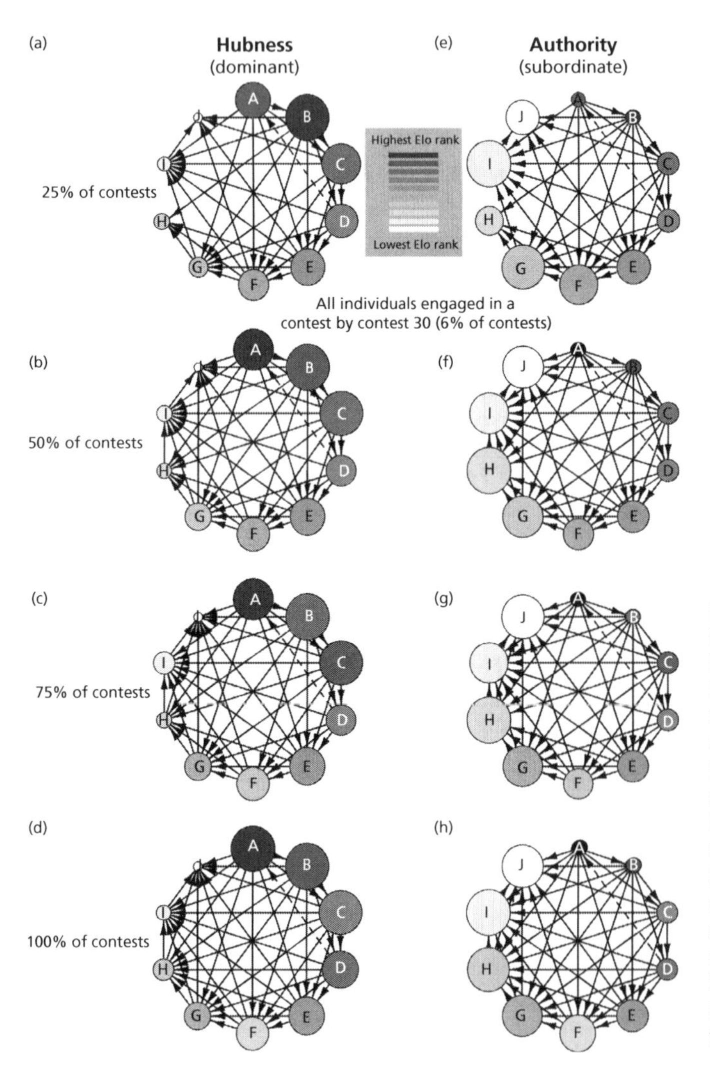

Figure 7.7 The outcome networks at each of the four quartiles of the total sequence of contests, for Harris's sparrow data (D. Watt 1986, Figure 1, Matrix 5). End ranks under the de Vries (1998) I&SI method were in alphabetical order, so nodes are arranged clockwise alphabetically as a fixed visual reference for node placement. Greyscale denotes Elo ranking at each stage. Note that, under the Elo ranking method, B is the top-ranked individual at the 25% stage (a,e), and D is third ranked at the end (d,h). (a–d) Node size is proportional to the hub score metric of network centrality. (e–h) Node size is proportional to the authority score metric. The only inconsistency (sensu de Vries 1998, involving individuals A and D) is shown by dashed arrows. Figure 7.7 is therefore essentially a network perspective on the quartile-based analysis of the contest matrix in Figure 7.6.

time as relations develop. The metrics used here, hub and authority score, are node centrality metrics designed for measuring flow of internet traffic, where hub score (websites often referring to authorities) and authority score (websites often pointed to) are complementary concepts (Kleinberg 1999). The vector of hub scores is calculated, in the iGraph R package, as the principal eigenvector of AA^{T}, where A is the adjacency (outcome) matrix. Hub score measures the tendency of a node (animal) to have directed edges pointing to other nodes; in the context of dominance relations, hub score will tend to yield higher scores for individuals dominant to many other individuals. For the data from D. Watt (1986) for Harris's sparrows (Figure 7.7), the hub and authority scores are highly correlated with dominance and subordinance, respectively, although note that the individual with the consistently highest hub score (B) is not generally the top-ranked animal under either the Elo rating method or the I&SI method. Likewise, the individuals (H and I) with the highest authority scores in Figure 7.7 were not the lowest-ranked animal, whether assessed by their Elo rating or the I&SI method. Thus, the highest hub score will not always or necessarily correspond to the most dominant animal, nor the highest authority score to the lowest-ranked animal.

One way in which such a lack of correspondence might occur is that the top-ranking individual may dominate only one or two other animals, but those animals may be more interactive and dominate many other lower-ranking individuals, giving them high hub scores. A potential problem for these sorts of dynamic assessments of centrality metrics is that some of them (e.g. eigenvector centrality or its generalization, called Bonacich centrality or alpha centrality (Bonacich 1987)) require that the matrix be non-singular. Sparse matrices are much more likely to be singular than are dense matrices. Even at the end of the research period, many dominance networks, especially those with many nodes, will be sparse and will therefore necessarily be very sparse in their early stages. Possible solutions to the problem include dropping individuals from the analysis if they have very few interactions—such individuals will be a major contributor to the problem of matrix singularity.

Experimental and modelling approaches: conclusions and future directions

A potentially important criticism of some social network studies is that they simply describe social interactions without providing either critical tests of hypotheses or novel insights into social dynamics. Further, network analyses can run the risk of shopping for patterns and indiscriminately assessing measures of centrality that may not really be biologically relevant. In considering the suitability of available metrics, it is worth noting that not all metrics are inherently best suited for networks where flow of information or materials is an essential feature of the system (Borgatti 2005). Even where flow is an appropriate property of the system, the nature of the information that flows through the network can affect the appropriateness of the metric. For example, betweenness, because it assesses flow along geodesics (shortest paths; paths which never revisit edges or nodes), is really only suited to indivisible packages, rather than for information that can undergo parallel or serial duplication (Borgatti 2005). Other metrics, such as eigenvector centrality, are appropriate for situations in which flow is along walks (where information can revisit both nodes and edges), or for trails (where nodes but not edges can be revisited).

Not all networks, however, entail flow processes (Borgatti and Foster 2003). Some networks may be primarily structural—what Borgatti and Foster call bond models—and node and network properties may depend on position rather than flow. For example, the network metric "power" (Cook and Emerson 1978) is a bond-model metric, where position, but not flow, determine the value (Borgatti and Halgin 2011). Evaluation of the appropriateness of different network metrics, and the development of new metrics, particularly for bond-model networks, is an exciting avenue into future research on animal social networks.

Despite various pitfalls (R. James et al. 2009), several useful approaches exist for avoiding the twin dangers of simple descriptive studies or over-interpretation of arbitrary or inappropriate metrics. One such approach is experimental manipulation of nodes or edges, including virtual removals (e.g. Flack et al. 2006). Such studies provide the opportunity to make predictions about the expected consequences of manipulations and thereby provide a framework for critical tests of alternative hypotheses. In dominance studies, after first looking at network patterns, one might, for example, remove pivotal animals to test mutually exclusive hypotheses concerning the influence of removals on group dynamics. The network framework can, therefore, provide a rich basis for designing critical experiments that assess both individual and group-level features of the dynamics of social interactions.

Modelling approaches such as exponential random graph models (Snijders 2002; Saul and Filkov 2007) also provide a powerful tool for assessing the most important factors in network dynamics. At some level, exponential random graph models are conceptually similar to logistic regression. One is searching for a factor, or factors, that best recover key features of the structure or temporal dynamics of empirical networks. The resulting best-supported model is therefore explanatory rather than simply descriptive. Comparison of exponential random graph models across disparate taxa or social systems should then help reveal generalities of process, such as those found to underlie social behaviour in the apparently different spectrum from cooperative breeding in birds to eusociality in insects, where degree of reproductive skew is a critical explanatory variable (Sherman et al. 1995). All the various approaches described in this chapter should provide a useful toolkit for addressing the many interesting questions that still remain concerning dominance relations in animal social groups.

Acknowledgements

This work was supported by a National Science Foundation OPUS grant (DEB-0918736) and a University of Wyoming Flittie Sabbatical Award to D. B. M. Comments by Christof Neumann and an anonymous reviewer resulted in several major changes that substantially improved the bond and flow of the manuscript.

CHAPTER 8

Group movement and animal social networks

Nikolai W. F. Bode, A. Jamie Wood, and Daniel W. Franks

Introduction to group movement and animal social networks

Moving in groups is integral to the life histories of many animals. Benefits of group movement can include a reduced risk of predation, an increased capacity to find sources of food, and the improved ability to find and follow migration routes (Krause and Ruxton 2002). The term 'collective motion' is used to describe the synchronized motion of groups of animals such as shoals of fish or flocks of birds that appear to behave as one body, continually changing shape and direction (Sumpter 2006). It is now commonly accepted that the collective motion of animal groups emerges from local interactions between group members (Aoki 1982; Huth and Wissel 1992; Vicsek et al. 1995; Bonabeau et al. 1999; Couzin et al. 2002; Ballerini et al. 2008; Katz et al. 2011).

Computer simulations have demonstrated that realistic collective patterns can emerge from simple local behavioural rules (e.g. Couzin et al. 2002; Hemelrijk and Kunz 2005; Bode et al. 2011). Typically, these models assume that individuals have a sensory range that is limited to a fixed number of nearest individuals or to a perception region of fixed extent. Individuals react to the movement of other individuals that are within their sensory range. Interactions often depend on the distance between individuals and can include collision avoidance at short distances, alignment at intermediate distances, and attractive tendencies to maintain group cohesion at long distances. Each individual in the group follows the same behavioural rules and from the local interactions of many individuals the group dynamics emerge. Empirical work supports the basic assumptions and mechanisms suggested by these simulation models (Ballerini et al. 2008; Herbert-Read et al. 2011; Katz et al. 2011).

The particular appeal of the conceptual framework of local interactions leading to group movement dynamics is that we not only learn about the behaviours governing animals by studying their collective motion, but we can use similar ideas to study human crowds (Helbing et al. 2000) and to design interaction algorithms for teams of robots (Liu et al. 2003). Petit and Bon (2010) suggest that collective movement corresponds to a sequence of events, including a pre-departure period, initiation, and subsequent group movement. We limit our scope to the manifestations of the locally aligned, locally synchronous, and continuous movement of one or more groups of interacting individuals (Bode et al. 2011b). We include the movement of multiple groups and therefore group fission and fusion processes in this definition, as they are important in moving animal groups, and, in our definition, interactions do not have to occur between all individuals involved.

Social preferences between individuals exist in many animal species and could be the preference of large fish to shoal with other large fish, the preference for familiar individuals, or the preference of offspring for a parent rather than a stranger. Evidence suggests that these social preferences can have profound effects on the movement decisions and dynamics of animals. For example, the majority of pedestrians walk in small social groups

Animal Social Networks. Edited by Jens Krause, Richard James, Daniel W. Franks and Darren P. Croft.

(Moussaïd et al. 2010), guppies (*Poecilia reticulata*) prefer to shoal with familiar conspecifics (Griffiths and Magurran 1999), and group fission in macaques (*Macaca tonkeana* and *M. mulatta*) is strongly influenced by individual affiliations (Sueur et al. 2010). This suggests that social preferences could determine how populations move in, split up into, and form separate groups and, conversely, that collective movement could change social preferences by creating social ties that did not exist previously and maintaining existing ties.

Social preferences and interactions between moving individuals evolve at different timescales. While interactions between individuals in moving groups have to change frequently to avoid collisions or to maintain group coherence, for example, social preferences often change slowly (e.g. familiarity, kinship). However, both interactions between moving individuals and social preferences can be expressed in terms of network theory (Croft et al. 2008; Whitehead 2008a; Newman 2010).

In these networks, individual animals are represented by 'nodes', and the connections between them are called 'edges'. Edges can either take binary values (they exist or not) or weighted values (representing the strength of the connection), and they can be undirected (connection between two animals) or directed (connection from one animal to another). The theory of social networks in collective motion rests on the preferences and interactions encoded in two different networks: social networks and networks of interactions, which we call 'interaction networks'. We previously used the term 'communication networks' (Bode et al. 2011b), but to avoid confusion with structured interactions between signallers and receivers, which have also been referred to as 'communication networks' (e.g. McGregor 2005), we have changed our notation here.

We use the term 'interaction networks' to refer to the structure of the flow of information between individuals while they are moving in a group. An edge between two individuals, A and B, could thus indicate that both A and B can observe the spatial position of each other at the current time point. Although there is currently no consensus on the precise structure of interactions in moving animal groups, it is generally assumed that individuals can or do only interact with a fraction of the group in larger groups (Ballerini et al. 2008; Bode et al. 2011; Katz et al. 2011). Assuming a sensory range (e.g. a region of fixed size or a fixed number of nearest neighbours) implies that individuals can move in and out of the sensory zones of other group members. This results in rapid changes in the structure of interaction networks over time, with each configuration of the interaction topology capturing the flow of information at one instant in time. The response of individuals to the information they perceive is not encoded in interaction networks. Figure 8.1a illustrates an interaction network based on the extent of the sensory zones of individuals at one instance in time. For simplicity we show binary and undirected networks.

Using social networks, the social organization of animals can be represented at all levels (individual, dyad, group, and population) and for different types of interaction (Krause et al. 2007). One important difference between social networks and interaction networks is that the former are not limited by communication or spatial positions; two individuals might share a highly weighted social network connection despite not currently being able to perceive each other or being a large distance apart. Therefore, if two individuals are connected in a social network, this does not necessarily imply that they share an edge in an interaction network at one instant of time, as edges in interaction networks only persist for as long as the two individual can perceive each other. However, if all individuals can perceive each other, then a weighted interaction network could express the social preferences of individuals to interact with each other in the weights of its edges. Figure 8.1b shows an example for a social network for the same group as in Figure 8.1a.

Here we investigate the relationship between these two types of networks in the context of collectively moving animal groups. Our focus is primarily on theoretical work, but we relate models and their predictions to empirical examples whenever possible. In the following, we begin by considering the importance of social networks on the movement dynamics within populations. In other words, we review the effects social networks could have on the movement of and between distinct groups. Subsequently, we consider the level of groups by

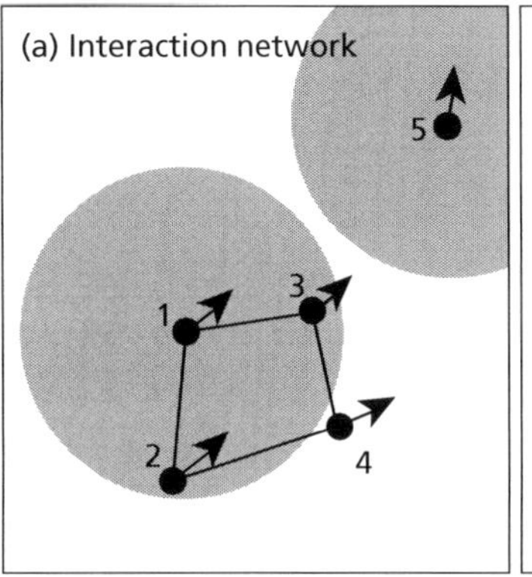

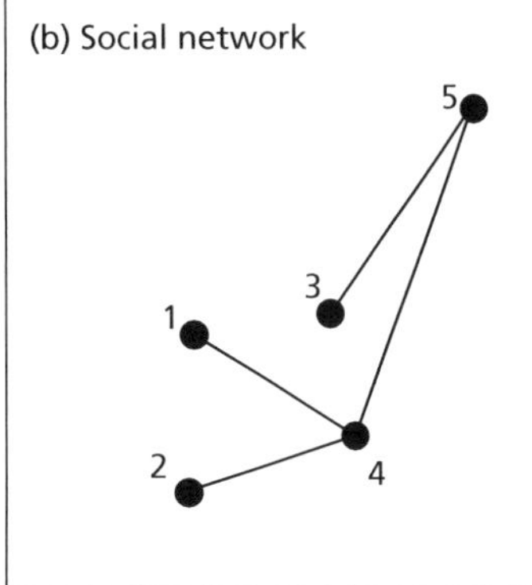

Figure 8.1 The difference between interaction networks and social networks (redrawn from Bode et al. 2011a). The positions of individuals are marked by black circles, with arrows indicating their direction of motion. Connections (edges) are marked by black lines between individuals. (a) Illustration of an instantaneous interaction network. The extent of the sensory zones for individuals 1 and 5 are marked by grey regions. Edges in the interaction network are based on which individuals can perceive each other and can therefore exchange information. (b) Example for a social network, indicating strong social preferences that could underlie the group of individuals in (a). Note how this network contains a connection between individuals 4 and 5, for example, which is not the case in the interaction network in (a). Limited perception can therefore restrict the interaction network to a structure that is different from that of the underlying network of social preferences.

investigating the role of social networks on within-group movement dynamics. Then, we highlight the importance of considering mechanisms at the individual level; and finally, we discuss the challenges in collecting empirical data to test the hypotheses created by theoretical approaches to date.

Population level

In Figure 8.2, we illustrate how underlying social structures could affect the movement dynamics of or between groups at the population level. It is useful to first consider the two extremes: (1) when social interactions are absent and individuals move independently from each other (Figure 8.2a) and (2) when the social preferences are equal between all individuals and the population forms one cohesive, collectively moving group (Figure 8.2c). Intermediate cases are likely to include populations that split into a number of more or less stable groups which occasionally join or split and gain or lose members through between-group movement (Figure 8.2b). This last example is commonly called 'fission–fusion system', and an example for such a system can be found in guppies *(Poecilia reticulata)*, small freshwater fish (Croft et al. 2003). During fission–fusion events, individuals may actively choose to move preferentially towards certain neighbours (e.g. those of the same sex) to whom they have a stronger social affiliation, and the composition of animal groups could therefore be determined by social aspects, not only in the initial formation of groups but also in encounters of moving groups.

We have developed an individual-based model for collective motion in which social preferences between two individuals, A and B, translate into increased probabilities of A reacting to the movement of B, and vice versa. The type of interactions (collision avoidance, alignment, or group cohesion) is not affected by social preferences. Simulating the movement of populations with social subgroups in which all group members have strong social preferences to all other group members but only weak social ties to others in the population results in scenarios in qualitative agreement to the one shown in Figure 8.2b (Bode et al. 2011b). This suggests that social preferences alone can in theory account for fission–fusion events.

Not all populations display frequent fission–fusion events as a result of social preferences. For example, Moussaïd et al. (2010) observed that more than two-thirds of pedestrians in crowds of different densities moved in coherent and stable social groups of two to four individuals. While the

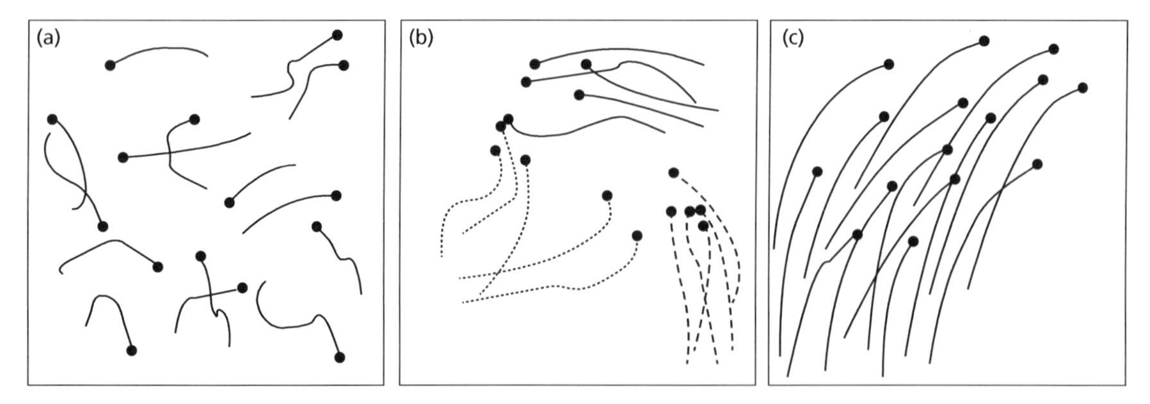

Figure 8.2 Examples for population-level effects of social networks on collective motion. Dots show the current position of individuals, and 'tails' mark the movement paths over a short time period. Panels (a) and (c) show extremes. (a) Example where there is no social interaction between individuals. There are no edges in the underlying social network; individuals move independently and do not interact. (b) Intermediate case: the population is split up into subgroups that move coherently. In this fission–fusion system, individuals may move between groups. (c) Example where individuals have the same social preference for all other individuals. All edges in the underlying social network have the same non-zero weight.

focus of this research was on the internal structure of these social groups, which we discuss in 'Group level', Moussaïd et al. (2010) and others (Braun et al. 2003; Qiu and Hu 2010) have reproduced this group formation using social force models in which social preferences are translated directly into attractive forces leading to stable subgroup formation. At low population densities, such frameworks would result in dispersed and independently moving small social units. An intriguing possibility for interactions between such far-apart groups could be that social preferences may forge long-range communication links that go beyond local information exchange. For example, African elephants (*Laxodonta africana*) mostly move in small social units but are capable of long-distance vocal communication with elephants in separate social units (McComb et al. 2000). Buscarino et al. (2006) extended a simple model for collective motion (Vicsek et al. 1995) to include random long-range interactions in addition to local interactions. This showed that long-range interactions could lead to higher global alignment of the entire population. While this research is primarily related to efficient and sufficient communication between individuals, it demonstrates that long-range communication could impact on or even facilitate the collective motion of spatially separated social groups.

It is not always the case that social structure is reflected more or less directly in the movement dynamics of populations. To study the formation and requirements of ad hoc mobile networks (e.g. networks of wireless, hand-held devices), researchers have combined aspects from social theory and collective motion in models (Musolesi et al. 2004; Borrel et al. 2009). In both studies, the mechanisms of group behaviour are implemented on similar lines to what has been discussed above, and social preferences are included as fixed dyadic weights between pairs of individuals. Borrel et al. (2009) studied the distribution of intercontact durations (based on spatial proximity) for random underlying social networks and found power law distributions (with cut-offs) in their simulations, similar to empirical evidence published elsewhere. In other words, underlying social networks with randomly generated structure resulted in contact networks with power law distributions of edge weights. Therefore, this analysis suggests that the structure of contacts does not necessarily reflect the underlying social network directly.

In the field and in much empirical research, the predominant method for quantifying social structure in populations is based on contacts between individuals, by sampling instances of spatial assortment of individuals in groups in an approach termed 'gambit of the group' ('Whitehead and

Dufault 1999; for other methods see Whitehead 2008a and Krause, Lusseau, et al. 2009). In this approach, associations are recorded between every pair of individuals that are members of the same collectively moving group, where individuals are typically assigned to the same group if they are within a fixed distance from each other. Cumulative networks, constructed by including all recorded associations from a number of 'gambit of the group' censuses, are analysed for non-random features (Lusseau and Newman 2004; Cross et al. 2005; Lusseau et al. 2006; Lusseau 2007a; Croft et al. 2008). The work by Borrel et al. (2009) suggests that we have to choose carefully what aspects of the contacts or communication between animals we study if we want to infer information on the social preferences of these animals.

Models combining social networks and collective motion have already proven to be useful tools in developing hypotheses (fission–fusion can be a result if social preferences alone) and to highlight possible pitfalls in the analysis of association networks. Simulation models allow us to investigate the effect of different large-scale social network structures on population-level summary statistics, such as the number of groups formed, the degree of alignment, and the size of the largest group (Bode et al. 2011b). An area that has not yet been investigated theoretically is the importance of particular individuals in maintaining cohesion of otherwise separate social groups within a population. For example, it has been suggested that the removal of individuals with high betweenness (a measure of importance of individuals in a network to the flow of information between others) could jeopardize the efficient flow of information in a bottlenose dolphin network (Lusseau and Newman 2004). Network analyses commonly presents a static situation, but collective motion and the adaptability of groups might mean that networks reconfigure after the loss of individuals.

Group level

In this section we investigate the effect underlying social networks could have on the movement dynamics in one cohesive, collectively moving animal group. More specifically, we consider the effect of social preferences on the positioning of individuals within groups, the spatial structure of formation adopted by groups, leadership and decision making in groups and finally we consider circumstances under which the maintenance of social ties could lead to a cost–benefit trade-off in animal groups.

It has long been suggested that the spatial position of individuals within groups could have important fitness implications at the individual level (W. Hamilton 1971; Mooring and Hart 1992; Morrell and Romney 2008). For example, fish are subject to different risks of predation depending on their position within groups (Krause et al. 1998), and a theoretical study has predicted central positions for dominant individuals (Hemelrijk 2000). Individuals balance the costs and benefits associated with their position in a group and theoretical and empirical work has shown that individuals displaying differences in behaviour (e.g. different speeds, Couzin et al. 2002; Wood 2010), motivational state (e.g. hunger, Krause and Ruxton 2002) or body size (Hemelrijk and Kunz 2005) occupy different spatial positions. It is not only the spatial position relative to the rest of the group (e.g. at the front) that features in individuals' cost–benefit considerations but also who an individuals' neighbours are (e.g. same size, kin, etc.). Many animal groups have distinctive, often hierarchical, underlying social networks (Croft et al. 2008; Whitehead 2008) and social ties between individuals may affect their interactions. Socially mediated changes in behaviour, due to individuals' social network position, could therefore affect the spatial position of animals within groups.

In addition to investigating the effect of body size on individuals' positioning in groups, Hemelrijk and Kunz (2005) split individuals into two categories: familiar and unfamiliar. Individuals were given higher social preferences for familiar individuals and assigned a higher weight to reactions to familiar individuals. For example, if individual A is attracted to a familiar individual B and an unfamiliar individual C in a simulation, then the attraction of A to B and C is biased towards B as a result of the social preference. Simulations showed that while the group remained cohesive as a whole, individuals spatially clustered with familiar conspecifics within the group (Hemelrijk and Kunz 2005). Similar theoretical results have also been obtained independently in the context of pedestrian crowds (Fridman and Kaminka 2007) and cell sorting (Belmonte et al. 2008). This demonstrates that substructures within collectively moving groups could be explained by social networks.

In a different approach, we simulated social ties as interaction preferences (model introduced above) and investigated the spatial positioning of individuals within cohesive groups as a function of the number of strong social connections versus the number of weak connections individuals maintained (Bode et al. 2011b). We found that higher numbers of strong social connections led to positions closer to the centre of the group (see Figure 8.3a for an example). This result can be explained by the observation that individuals with many strong social connections bias their interactions towards a larger proportion of the group instead of a small subset of the group. If such individuals are situated at the periphery of the group, their average direction of attraction will be towards the centre of the group. Likewise, if such individuals are already positioned close to the centre of the group, the attraction tendencies towards individuals on the periphery on the group will balance, on average. These within-group movement tendencies only hold for individuals that do not bias their interactions towards a small subset of the group (Bode et al. 2011b).

Animal groups can have well-defined and consistent internal structures, such as a positive density gradient from the core to the edges of the group in starlings (*Sturnus vulgaris*) (Ballerini et al. 2008) or from the back to the front of the group in shoals of roach (*Rutilus rutilus*) (Bumann et al. 1997). Perhaps the most striking manifestations of internal structure are the 'V-shaped' formations adopted by flocks of migrating birds (Bajec and Heppner 2009). These internal structures may be a direct result of physical features of the animals involved, or particular behaviours in response to predation pressure; however, the role of social networks in shaping internal group structures needs to be explored.

Qiu and Hu (2010) suggest a model to examine the role of social network structure on the formations of collectively moving human crowds. Their implementation of the effect of social ties on individual-to-individual reactions is based on directed connection strengths that weigh the reaction of one individual to another, similar to what has been introduced above. Qiu and Hu (2010) implement different social structures and model simulations to show that this could give rise to distinct formations in which groups move. For example, a linear network topology (a 'chain' of connections) leads to a linear group formation. A network in which a number of followers are only connected to a 'group

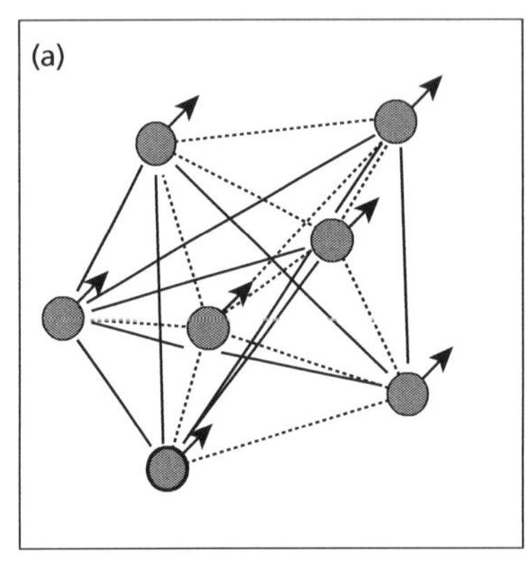

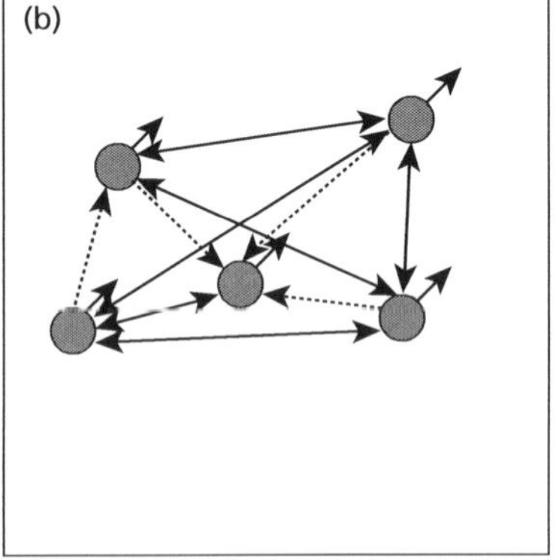

Figure 8.3 Examples for group-level effects of social networks on collective motion. We show individual positions (circles) and movement directions (black arrows), as well as strong and weak preferences in the underlying social networks (dotted and solid lines/arrows, respectively). (a) Example where edges are undirected (social preference between individuals), and individuals with many strong social connections take positions in the centre of the group (Bode et al. 2011b). (b) Example where one individual has additional information about a target (marked in darker grey) and concentrates a number of strong directed edges (social preference from one individual to another) on itself. In this case, the informed individual leads or steers the group from a position that is not necessarily at the front of the group (Bode et al. 2012).

leader' results in a compact formation quantified by a low average spatial distance of group members to the centre of the group, in contrast to the linear scenario. This work suggests group formations can be the result of the underlying social network.

Motivated by detailed empirical observations, an interesting theory specific to the movement and formations of social groups within pedestrian crowds has been developed (Moussaïd et al. 2010). Based on the observation that more than two-thirds of pedestrians moved side-by-side in coherent and stable groups of two to four individuals, Moussaïd et al. (2010) developed a social force model that assumed the communication needs of individuals within small social groups to dictate their relative positions. More specifically, individuals position themselves in such a way that they can see their social partners without having to twist their head too much. This model predicted and empirical results confirmed that as pedestrian densities increase, the linear formations bend forwards and adopt a V-shape, with the open part of the formation pointing in the direction of movement. The empirical observations present strong evidence for the effect social networks have on formations of collectively moving groups, and the research by Moussaïd et al. (2010) demonstrates the potential of comparing and informing models of collective motion and social structure with real-world data.

The mechanisms by which animal groups make decisions have been studied intensely, as group decisions can have direct fitness consequences on individuals and because similar mechanisms could underlie decision making in human groups (e.g. Couzin et al. 2005; Conradt et al. 2009). In general, group decisions can be classified as being 'democratic', that is to say, all group members have a share in making the decision, or they can be 'despotic' or 'oligarchic' and only a subset of the group, often termed 'leaders', makes the decision (Conradt et al. 2009). For example, empirical work has shown that dominant beef cows (*Bos taurus*) have more influence on herd movement than more subordinate cows (Šárová et al. 2010), and it can be argued that leader–follower relationships translate into a social network affecting collective motion. In the context of social networks and collective motion, we need to understand whether certain structures of underlying social preferences facilitate either of the two decision-making mechanisms introduced above.

Some analytical studies have investigated leadership in collective motion using a rudimentary form of social network (Jadbabaie et al. 2003; Liu et al. 2003; Hu and Hong 2007; Consolini et al. 2008). In these models, the leaders dominate group movement decisions by simply pursuing their own goal (typically a preferred movement direction) while the rest of the group moves collectively with equal preference for all individuals (leaders and non-leaders). In this extreme social network structure, all individuals are socially connected, with the exception of the leaders, who have only incoming (directed) social connections.

We investigated the role of social networks on 'despotic' or 'oligarchic' group movement decisions in more detail (Bode et al. 2012). We found that the most successful strategy for leading a group (based on steering the group in a desired direction without it breaking up) requires leaders to balance their own goals with paying attention to the rest of the group. If all non-leaders have a strong preference to react to the leader, the precise balance of behaviours of the leader becomes less important to lead the group successfully. Interestingly, we found that successful leaders in socially dominant positions steered from within the group, rather than the front (see Figure 8.3b for an illustration).

Furthermore, we investigated the case when there was a conflict of interest between two subgroups, one of which had a majority over the other (Bode et al. 2012). Our simulations showed that when all social preferences were equal, majorities dominated the group movement decisions. If, however, members of the minority subgroup occupied influential positions within social networks, the minority was able to dominate group movement decisions (Bode et al. 2012).

While the preceding examples assume leaders to have specific goals or movement preferences which they impose on the group, we also investigated whether individuals can dominate the movement dynamics of groups based solely on social preferences (Bode et al. 2011b). This was motivated by research on pigeon flocks, which revealed that pairs of birds showed consistent leader–follower relationships based on delays in the correlation of their flight directions (Nagy et al. 2010). We found similar movement dynamics in simulations based solely on social preferences and in the absence of other individual differences (Bode et al. 2011b).

The same consistent leader–follower relationships in pigeons were also compared to dominance hierarchies recorded separately from interactions during feeding (Nagy et al. 2013). This approach is particularly interesting, as it presents one way of establishing social relations between individuals (dominance hierarchies) and then testing whether these relationships are preserved in the movement dynamics of the group. Nagy et al. (2013) found that the dominance hierarchy networks did not correlate with networks encoding the leader–follower relationships. This result could arise from different scenarios. The social affiliations and preferences of individuals could depend on the context, or the generative mechanisms for the leader–follower relationships could be independent from social preferences, for example.

One well-known example for democratic group movement decisions is the 'many wrongs' principle (Simons 2004; Codling et al. 2007). In this model, groups seek to navigate towards a target (e.g. during migration) and manage to reduce the navigational error of the group below the individual navigational error by pooling information through individual-to-individual interactions. Using the theoretical framework introduced above, we showed that, in such situations, underlying networks of interaction preferences could result in even more accurate decisions by facilitating long-distance communication in groups (Bode et al. 2012).

A topic that has to date not been covered in detail is the impact of social networks on the movement of social groups in confined spaces. Work by Braun et al. (2003) shows that social preferences could potentially have catastrophic consequences for individuals. Braun and co-workers simulated the evacuation of human crowds from buildings and compared crowds without social ties to crowds with family groups in which members of families were interacting via distance-dependent attractive forces. Crowds without social groups evacuated a single room with one door faster than crowds with social groups. Assuming that in evacuations taking longer increases the risk of injury to individuals, this hints at fitness trade-offs for social ties. It would be particularly interesting to further investigate whether members of social groups evacuate slower than independent individuals and to consider this scenario in the context of an animal collective with a predator in pursuit.

In this section, we have seen that theoretical work has already made much progress to investigate the possible effects of social networks on group structure, formations, and decision making. In contrast to the previous section on population-level effects, empirical data is largely missing at the group level (with notable exceptions: Moussaïd et al. 2010; Nagy et al. 2013). We suggest and discuss further below that the big challenge for this type of research is to make empirical progress.

Another interesting avenue for future research that has not been investigated in detail is the question of cost–benefit trade-offs for maintaining social connections. It is reasonable to assume that maintaining social connections comes at a cost to individuals (e.g. time investment). Throughout this section we have highlighted several cases where certain positions in networks could be beneficial (e.g. positioning or in decision making). Genetic algorithms could provide a tool to investigate what type of social networks we could reasonably expect to find in animals facing different evolutionary pressures.

Individual level

At the individual level, the ability to form and maintain social preferences requires the ability of individuals to perform some kind of selective cognitive process. This can reach from individual recognition (e.g. in primates (Pascalis and Bachevalier 1998)) to sex discrimination (e.g. in fish (Croft et al. 2004)). Studying the cognitive abilities of animals at the individual level can therefore allow inferences on the type of social network we can reasonably expect to be present in different species.

In addition to understanding the cognitive processes animals are capable of, and perhaps more importantly for the topic of this chapter, we need to consider the mechanisms by which social preferences are reflected in the interactions between individuals. We have seen that researchers adopt a number of different approaches. In some models, social preferences are directly translated into attractive forces (e.g. Braun et al. 2003). In other models, individuals react to the average direction and positions of conspecifics around them, and social preferences are reflected in the weighting different individuals are given in computing these averages (e.g. Hemelrijk and Kunz 2005; Qiu and Hu 2010). We developed the concept of individuals interacting with a weighted average of individuals' positions further by converting interaction weights into probabilities for pairs of individuals to interact with each other (Bode et al. 2011b; Bode et al. 2012; Bode et al. 2012). Moussaïd and co-authors (2010) present an entirely different concept in which socially connected individuals adopt relative positions that facilitate eye-to-eye contact and communication. It is not clear to what extent the results reviewed here could be affected by choosing different mechanisms for translating social connections into interactions. Rather than simply assuming that social preferences directly translate into attractive, repulsive, or aligning tendencies—as in most of the literature we review—it may be worthwhile to consider the precise mechanisms of how this could work in animals.

One interesting approach could be to adopt a neurobiological perspective and to explicitly consider neurological pattern recognition. This concept has, to some extent, already been developed for modelling collective motion (Lemasson et al. 2009). Studying the nervous system of animals may provide clues as to how individuals can perceive others. In particular, this approach could allow conjectures or insights into a possible situation dependence of social interactions. For example, some animals are capable of differentiating between genders using their sense of smell (e.g. Bouchard 2001). It could be that the presence of strong global olfactory signals could limit individuals' ability for sex discrimination. Additionally, when moving at high speeds, individuals may not be able to distinguish between different group members, and socially motivated interactions may be overridden by reacting reflexively or instinctively to others. Taking this one step further, it may be possible that the influence of social connections on interactions within moving animal groups could depend strongly on the context. Individuals may change or switch off entirely their social preferences during predator attacks or when foraging. Investigations of these aspects of the role of social networks in moving animal groups are currently missing, and we suggest they could be fruitful avenues for future research.

Group movement and animal social networks: conclusions and future directions

We have reviewed theoretical work on the effect underlying social networks could have on the movement dynamics of animals at the population, group, and individual level. Models have generated many hypotheses, and we have suggested models could be useful to scope potential pitfalls in setting up field studies. We have highlighted areas for future research such as considering mechanisms at the individual level and more explicitly investigating the consequences of cost–benefit trade-offs for social connections in moving animal groups. Most importantly, we have attempted to provide a simple conceptual framework by developing the concepts of the interaction between two networks: social and interaction networks.

We discussed how social and interaction networks evolve at different timescales and that it is likely that in most cases interaction networks change faster than social networks. All of the work we have discussed in the previous sections adheres to the view that social networks affect collective motions. The scenario of collective motion affecting social networks has remained largely unstudied. A possible mechanism for this case could be as follows: theoretical work has suggested that small differences between individuals, such as differences in speed, can affect the positions of individuals within groups (Couzin et al. 2002; Wood 2010). If such differences between individuals persist over some time, this may result in spatial sorting within moving animal groups (see e.g. simulations in Gueron et al. 1996) and subsequently in increased familiarity between similar individuals. Increased familiarity between individuals could result in changes to the underlying social network (e.g. increased preference for familiar individuals). This scenario is supported by empirical work on guppies (*Poecilia reticulata*) showing that individual fish prefer to shoal with conspecifics with whom they are familiar (Griffiths and Magurran 1999).

Some models have been developed in which the social connections between individuals are updated based on their relative spatial positions, but the analysis remains either entirely qualitative (e.g. Musse and Thalmann 1997), or is constrained to the engineering problem of forcing groups to adopt particular predefined formations (e.g. Wessnitzer et al. 2001; Sahin et al. 2002; Trianni and Dorigo 2006). Perhaps the biologically most relevant approach rewards simulated individuals for maintaining preferred distances to others, where the preferred distances could be viewed as a form of social preference (Quera et al. 2010). Successful agents are rewarded by manipulating their preferred distances to other, which in effect changes the social structure of the group. Quera and co-workers (2010) performed simulations to study the formation of stable groups and the emergence of leaders within groups.

In theory, there may be interesting mechanisms for changes in the social structure of groups. For example, animals with few and weak social ties may occupy peripheral positions in groups, where they may face higher predation. As a result, predation pressure on moving animal groups may result in or even select for denser and more homogeneous social networks. It is inevitable that the social structure of animal groups changes through ageing and mortality. In how far the relative movement alone could impact on social structure in a non-trivial way remains to be explored.

Currently the biggest challenge in the field is to link the theory of social networks in collectively moving animal groups to empirical data. In the section 'Population level', we have seen that movement data or relative spatial positions of individuals are typically used to construct networks of social affiliations. Theoretical work has shown that in some cases this could be a reasonable approach (Bode et al. 2011b), but we have also highlighted the need to investigate potential pitfalls.

If we intend to investigate highly synchronized collectively moving animals at the group level, inferring social affiliations from spatial proximity may be inappropriate. An additional problem could be that, with increasing group size, individual behaviour characteristics could be suppressed (Herbert-Read et al. 2013), making it even more difficult to infer social structure in moving groups. In brief, a rich theory with many explicit hypotheses on the interplay of underlying social networks and the movement dynamics of animal groups is in the process of being developed; but because the social structure of groups can typically only be inferred, empirical validation of this theory is currently difficult. To make progress, innovative approaches are needed.

One approach could be to ask more general questions. For example, instead of trying to infer detailed social structures with differently weighted pair-wise preferences, we investigated the question of whether it is possible to distinguish between groups moving towards a target in which individuals interact socially from groups in which animals move independently (Bode et al. 2012). The difficulty of this question arises from the fact that, if many independently moving individuals move towards a common target at the same time, they aggregate, and their movement dynamics can appear similar to the movement of socially interacting groups. One could imagine that this approach could be extended. A good starting point could be the temporal variation of internal structures in moving groups. If tight social groups exist, we might expect them to maintain cohesion within larger groups. This could be tested in human crowds by comparing crowds of commuters which contain few social groups to crowds of museum visitors, which are likely to contain more social groups such as families.

Another approach could be to investigate species in which social structures can be inferred in other ways than by studying spatial association. Non-human primates show a wide range of social behaviours, such as grooming and aggression, which have repeatedly been used to infer the social structure (e.g. Cords 2002; Thierry et al. 2004). Some species aggregate and move in large troops and, while the role of social networks in group movement initiation has been studied (e.g. Sueur et al. 2010), modern technology should allow collecting data while groups are moving. We have mentioned the example of comparing dominance hierarchies during feeding to leader–follower relationships in flying pigeon flocks (Nagy et al. 2013). Recording genetic relationships between individuals could provide another useful proxy for potential social affiliations. In groups where the social structure can be partially inferred, such as in identifying dominant individuals, manipulative experiments could provide another possibility to study the effect of social networks on group movement. For example, the collective motion of groups could be compared before and after dominant or subordinate animals are temporarily removed from the group.

Yet another approach for empirical work could be to turn to genetics. Knocking out certain genes has been shown to reduce the ability in mice to discriminate genders (Stowers et al. 2002). While mice rarely move collectively in the sense that is the focus of this chapter, it is not unreasonable to suggest it could be possible to manipulate behaviour in such a way in other species. Of course, side effects of genetic differences on behaviour would have to be carefully considered, but in principle genetics could allow the controlled manipulation of the social abilities/structures in animal groups.

In conclusion, we suggest that models have already been useful to investigate the role of social networks in collectively moving animal groups and that they will become increasingly important to aid the design of field studies, to highlight potential pitfalls in data collection, and to develop testable hypotheses. We have argued that the most pressing issue for this type of research is to develop ways to validate the theory empirically, and we have presented a number of ways forwards in this respect. Moving in groups is integral to the life histories of many animals. To fully understand the reasons behind group living and perhaps ultimately to understand how social behaviour evolved, we need to understand the role of social networks in moving animal groups.

Acknowledgements

N.W.F.B. gratefully acknowledges support from the AXA Research Fund.

CHAPTER 9

Communication and social networks

Peter K. McGregor and Andrew G. Horn

Introduction to communication and social networks

There are compelling reasons to consider that communication occurs in networks of several animals within communication range of one another and that communication has evolved in such a network context (e.g. McGregor and Dabelsteen 1996; McGregor 2005). Examples of communication networks are ubiquitous in nature, occurring in all taxonomic groups and all sensory modalities (e.g. McGregor et al. 2000). A working definition of a communication network is several individuals within signalling and receiving range of one another (McGregor and Peake 2000). This is strikingly similar to the definition of a social network used by the editors of this volume—'any number of individuals interconnected via social ties between them' (Section 1, this volume). The parallels are obvious and important, given the role of communication in establishing and modifying social ties of all sorts, especially those that underpin critical social behaviours such as mating, foraging, and aggression.[1]

While communication could be considered a social tie in its own right, both the physics of signal transmission (which restrict the range of signal influence) and the generally ephemeral nature of signals mean that communication is different from social ties determined by relatedness or social standing. This means that the relationship of communication networks to social networks is not straightforward. We explore this relationship in this chapter: the aim of this chapter is to identify the main features of communication networks, to discuss briefly their relationship to other networks (information, social), to identify examples where a network approach has increased our understanding of communication behaviour, and to point out some of the ways in which future work can more closely link communication networks to social networks.

Communication and network approaches

To date, most considerations of communication in a network have been informal discussions and thought experiments of how the features of an assemblage of signallers and receivers within communication range of one another may result in behaviours and strategies in addition to those expected in a signaller–receiver dyad (e.g. McGregor and Dabelsteen 1996; Dabelsteen 2005). However, given that formal network approaches have been developed and applied to animal social networks with considerable success (e.g. DeDeo et al. 2010; Flack and Krakaeur 2006; and as evidenced in this book), it is important to explore how amenable communication is to these more formal network approaches. The first stages are to identify how communication differs from other social interactions

[1] By contrast, the use of the term communication network to describe information inherent in the relative spatial positions of groups of moving animals (e.g. Bode et al. 2011a) seems to us to share few if any features with the topic of this chapter and the established use of the term communication network in the literature on animal communication (e.g. McGregor and Dabelsteen 1996; McGregor 2005). Thus, there is the potential for confusion in the developing literature, as modellers familiar with the Bode et al. (2011a) usage of the term communication networks address actual communication networks.

Animal Social Networks. Edited by Jens Krause, Richard James, Daniel W. Franks and Darren P. Croft.

and how communication networks differ from social networks. We begin by recapping basic, but still somewhat contentious, features of communication.

Signals, information, and communication

In the sense used here, communication is any interaction between animals that employs signals, where a signal is 'any act or structure that alters the behaviour of other organisms, which evolved because of that effect, and which is effective because the receiver's response has also evolved' (Maynard Smith and Harper 2003). Many other social behaviours have been shaped by natural selection to affect other animals (receivers). However, signals do so not by physically forcing a change in the receiver's behaviour, as the blows and parries in a fight do, but instead by causing the receiver to change its behaviour on its own, as the threatening wave of a weapon might.

Thus, a signal is a stimulus like any other; but unlike other stimuli (often termed 'cues' (e.g. Maynard Smith and Harper 2003)), a signal is an adaptation whose function is to serve as a stimulus for other animals.[2] For example, roe deer (*Capreolus capreolus*) wrestle each other with their antlers but also signal by shaking their antlers to threaten rival males (Hoem et al. 2007). Wrestling and threatening are both social interactions, and both are accomplished by the use of antlers. Only threatening is communication, however, because in threatening the antlers serve as a visual stimulus that dissuades the approach of rival males, whereas in wrestling the antlers transfer physical force that displaces or gores the rivals (we set aside the grey area, likely true for most physical contests, that wrestling is a test of strength (e.g. Hoem et al. 2007)).

Since, by definition, signals achieve their functions by serving as stimuli, then like any other stimulus, they must, on average (and with frequent and important exceptions spelled out below), serve as a useful source of information for receivers to base their behaviour upon; otherwise, receivers would gain no selective advantage in responding to signals, and thus would evolve to ignore them. Put another way, signals must carry information about the sender or the environment. Indeed, signals have been defined as specialized information carriers (e.g. McGregor 2004).

Information exchange and communication networks

Viewing communication merely as the exchange of information does not encompass all relevant features of communication networks. While such a perspective is not as misguided as some recent critiques have proclaimed (as explained in general by Stegmann 2013 and for communication networks by Horn and McGregor 2013), an information exchange perspective does not make apparent two fundamental features of communication that are particularly important in structuring communication networks.

First, the association between a signal and the information it conveys is probabilistic. A signal's information is a key agent that selects for a response by receivers, but it is not necessary for the functioning of the signal on any given occasion. Whether a given signal actually achieves its adaptive effects on any given occasion depends on many factors besides the presence of the state that it signals, including contextual sources of information (W. Smith 1977) and, most importantly, the selective value of the information conveyed by the signal (Bradbury and Vehrencamp 2000; Koops 2004). For example, a ground squirrel might respond quite strongly to another squirrel's call, even if that particular call is only weakly predictive of a hawk's approach, provided that the selective cost of not responding when there really is a hawk present is sufficiently high—instant death, for example. Conversely, in cases of mimicry, a bird might be repulsed by the colour pattern on an edible butterfly, provided that the cost of eating the toxic species is sufficiently high. While the information a signal conveys is an ultimate (evolutionary) cause of its effectiveness, it need not be a proximate cause of its effectiveness for any given signalling interaction.

[2] Signals used in auto- or self-communication, such as echolocation, are often considerably more directional than advertizing signals. However, there is still scope for them to be received and acted upon by others, so their use has to be considered in a network context. For example, bats eavesdrop on the feeding buzzes of other bats to detect areas of high prey availability (e.g. Barclay 1982; Fenton 2003).

Second, the evolutionary interests of signallers and receivers do not necessarily coincide. Thus signallers may not always release reliable information but might withhold or even misrepresent information. In sexual advertisement, for example, signallers often exaggerate their signals to overcome the sales resistance of receivers, while in parental care, offspring may exaggerate their needs to extract more care from parents. Indeed, in cases of interspecific signalling, such as between prey and predators, there may be no common interests between senders and receivers, and a preponderance of misleading signalling, such as mimicry. How informative signals are is the outcome of a complicated balancing act between the differing interests of senders and receivers. The evolutionary pressures that bring such arms races to a stable signalling equilibrium (or do not) are the subject of a huge literature (concisely reviewed in Botero et al. 2010; see also Searcy and Nowicki 2010). For our purposes, this balancing is particularly important when one considers the diverse perspectives and interests among the receivers that any signaller in a network faces, as discussed in 'Receiver diversity and communication in networks'.

The upshot of these two aspects of communication networks—that information is predictive rather than causative, and that senders might withhold or misrepresent it—is two-fold. First, the exchange of information through a network may be, and indeed should be, viewed as at least conceptually distinct from the interactions involved in exchanging it. Indeed, this property of communication networks, that they are networks of information exchange, distinguishes them from social networks based on other types of interaction. Second, the 'codes' that link signals and information are not set in stone; rather they are provisional outcomes of signaller–receiver interactions (Botero et al. 2010). As we will see below, this provisional aspect of communication has particularly interesting implications for communication networks.

Receiver diversity and communication in networks

By definition, any signaller in a communication network is potentially within range of several receivers. More importantly, these receivers can be of several types (e.g. conspecific competitors and mates, heterospecific predators and prey; see Figure 9.1). By analogy, conversing with a friend in a crowded market might involve not just the friend, but also a companion of theirs who you know is listening, as well as many acquaintances and strangers within earshot that might also listen (Dynall 2011).

Various terms have been used to distinguish between the different types of receiver (see e.g. Peake 2005; Bradbury and Vehrencamp 2011; Dynall 2011), the common theme being the importance of identifying which receivers are 'intended' in the evolutionary sense. For example, the intended receiver of a signal (synonyms include 'addressees' and 'proper recipients') for mate attraction would be an individual of the same species and opposite sex, and the signal and signalling behaviour would show design features that maximize effectiveness in attracting suitable mates. Of course other receivers, such as predators or parasites, could also detect and respond to the signal even though they are not intended receivers. Such receivers impose other selection pressures on signal design and use, as discussed below.

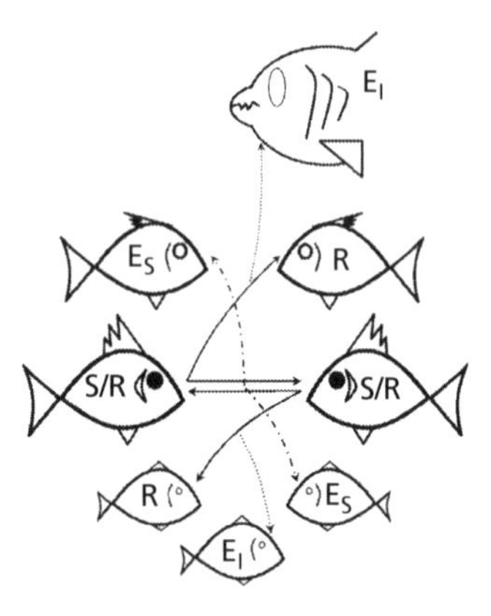

Figure 9.1 A schematic representation of a communication network to illustrate the diversity of roles and patterns of information flow. Most of the individuals shown are males and females of the same species. Males have larger bodies and fins than females. Displaying males have raised dorsal fins, raised opercula, and darkened eyes. An individual predator is shown top right. Four different roles (explained in text) are shown: signallers (S), intended receivers (R), interceptive eavesdroppers (E_I), and social eavesdroppers (E_S). Patterns of information flow are shown by lines, with arrows indicating the direction of flow; signals from signaller to intended receiver are shown by a solid line, signalling interactions by paired lines, signal interception (interceptive eavesdropping) by a dotted line, and social eavesdropping by a dot-and-dash line.

One of the reasons that a unifying terminology has proved elusive is that usually helpful dichotomies, such as whether the receiver is the same or a different species from the signaller and whether the response elicited by the signal has a cost or a benefit for the receiver, have several important exceptions (e.g. Peake's 2005 discussion of eavesdroppers). However, regardless of the exact terms used, it is clear that the diversity of receivers in communication networks is not found in social networks based on association alone. The effect of receivers of various types is also modified by context. Examples include signals that might be occurring at the same time, and a receiver's previous interactions with the signaller. These factors will affect how signals and information propagate through the network.

Understanding how the inherently local process of communication, such as local neighbourhoods of honest or dishonest signallers, affects social behaviour more globally, such as the overall stability of honesty (Botero et al. 2010; Searcy and Nowicki 2010), is a strong reason for using a network approach. A further important aspect of receiver diversity is highlighted by communication networks involving several different species (e.g. Magrath and Bennett 2012), where each species will differ in signal perception and the likely impacts of costs and benefits between the species involved. Even if we do not know these features in detail, we can expect them to differ and therefore to add complexity to communication network structure.

Empirical successes of the communication network approach

In common with other areas in behaviour, formal network approaches seem to be used rarely when considering communication, perhaps reflecting the complexities introduced in the previous section. However, qualitative considerations of communication in a network have produced a number of insights into features of communication, including directing signals in a network (e.g. McGregor and Dabelsteen 1996) and signal timing to chorus or to avoid overlap (e.g. Grafe 2005). Two further notable successes, as well as one feature that deserves more investigation, will be discussed in more detail in the following sections on eavesdropping. These sections are not intended as reviews; rather, they highlight evidence for factors that would be useful to include in formal network approaches to communication.

Eavesdropping

One advantage of considering communication in a network context is that it clearly identifies the opportunity for receivers to eavesdrop (see Figure 9.1), that is, to use information in signals for which they are not the intended receivers (in an evolutionary sense; see 'Receiver diversity and communication in networks'). Eavesdropping is a receiver behaviour that is only possible in a communication network. Two types of eavesdropping can be distinguished (Peake 2005): interceptive and social.

Interceptive eavesdroppers intercept a signal that is targeted (in an evolutionary sense) at individuals other than the eavesdropper. One type of interceptive eavesdropping is predators locating potential prey by the prey's mate attraction signals (e.g. fringe-lipped bats (*Trachops cirrhosus*) preying on calling male frogs (e.g. Tuttle and Ryan 1982; further examples in McGregor 2009). By contrast, social eavesdroppers gather information from the signalling interaction between others. An example is the great tit (*Parus major*), which directs territorial defence (males) and extra-pair behaviour (females) on the basis of aggressive singing interactions between males (e.g. Otter et al. 1999; Peake et al. 2002).

As Peake (2005) points out, it is important to distinguish between social and interceptive eavesdropping because of differences in the information transferred, with consequences for the likely selection pressures imposed by eavesdropping. While it is true that social and interceptive eavesdropping differ in other respects, such differences are not universal, nor are they definitive in identifying the two types of eavesdropping (Peake 2005). For example, interceptive eavesdroppers are usually different species from the signaller and commonly impose a cost on the signaller, whereas social eavesdroppers are usually the same species as the signaller, and the effect on the signaller can be zero, negative, or positive. The consequences for networks of such differences between social and interceptive eavesdropping are explained below.

Interceptive eavesdropping

It has been long recognized (e.g. Otte 1974) that widely broadcast signals, such as those used in mate attraction and resource defence, are vulnerable to eavesdropping. The same is true of extensive, persistent scent marks (e.g. Nieh 1999). There are several examples of predators eavesdropping on prey and of parasites eavesdropping on hosts (e.g. Peake 2005; Schmidt et al. 2005; McGregor 2009), with clear costs to the prey/host. Potential prey also eavesdrop on predators' signals, generally with clear benefits to the potential prey (but see N. Hughes et al. 2012) and potential costs to the potential predator (e.g. Deeke et al. 2005). For example, ground nesting ovenbirds (*Seiurus aurocapilla*) and veeries (*Catharus fuscescens*) react to playback of calls of a nest predator, the eastern chipmunk (*Tamias striatus*), by nesting further from loudspeakers broadcasting chipmunk calls (Emmering and Schmidt 2011).

There is growing evidence that plants eavesdrop within and between species on herbivore-induced release of plant volatiles (e.g. Arimura et al. 2010), and they may also eavesdrop on herbivorous insect signals (e.g. tall goldenrod (*Solidago altissima*) reduces insect damage by increasing defence in response to the sex pheromone of a key herbivore, the gall-inducing fly (*Eurosta solidaginis*) (Helms et al. 2013)). In the context of mate attraction, potential competitors for mates eavesdrop on mate attraction signals; for example, male fiddler crabs (*Uca tangeri*) produce a high-intensity waving signal on detecting an approaching female, and Pope (2005) has shown experimentally that other males eavesdrop, that is, males that cannot see a female nevertheless produce high-intensity waves in response to high-intensity waving by nearby males. In the context of scent marking to assist foraging, it has been suggested that eavesdropping, by favouring the more private exchange of information in the hive rather than as publically available extended scent marked foraging trails, was one of the selection pressures favouring the development of a referential communication system in bees (Nieh 1999; Nieh et al. 2004).

It could be argued that examples such as those described above are comparable to non-signal sources of information (often termed cues) on the presence of predators, prey, parasites, hosts, or mates (i.e. signs of presence). They are also comparable in that selection can act to make information less detectable to eavesdroppers in the same way that, for example, selection can act to make predator approaches stealthier. However, there are several examples that illustrate differences between eavesdropping and detecting signs of presence, and in which interceptive eavesdropping appears to require more complicated processing of information than detection of presence.

In some instances, the signal does not lead to an attack on the signaller; rather, the prey are individuals drawn to the signal. Such 'satellite predation' has been reported in Mediterranean house geckos (*Hemidactylus turcicus*), which intercept female decorated crickets (*Gryllodes supplicans*) approaching calling males rather than preying directly on calling males (Sakaluk and Belwood 1984). In other instances, prey vary their response to different classes of predators, based on information in predators' signals or alarm calls of other potential prey. For example, harbour seals (*Phoca vitulina*) discriminate between the calls of transient, mammal-eating killer whale (*Orcinus orca*) pods and those of resident, fish-eating pods (Deeke et al. 2002). Red-breasted nuthatches (*Sitta canadensis*) use information on the size and risk posed by potential predators that is encoded in the variations of the single mobbing alarm call of black-capped chickadees (*Poecile atricapillus*) (Templeton and Greene 2007).

In social networks eavesdroppers may be required to integrate information; for example, male baboons (*Papio hamadryas ursinus*) use the temporal and spatial juxtaposition of other individuals' vocalizations to achieve sneaky matings (Crockford et al. 2007). A further level of complication has been reported in field voles (*Microtus agrestis*), which compete for resources with sibling voles (*Microtus rossiae meridionalis*) and eavesdrop on their scent marks. Least weasels (*Mustela nivalis*) hunt voles by scent but prefer sibling voles to field voles. Field voles reduced the number of visits to sibling vole scent marks if least weasel scent marks were present in the area, indicating field voles are sensitive to the increased risks of eavesdropping on interspecific competitors when such competitors attract predators (N. Hughes et al. 2010).

Social eavesdropping

Social eavesdropping is defined by the source of information—the signalling interactions between others; therefore, experimental demonstrations are based on providing interactions which contain information likely to elicit a difference in behaviour, while ensuring other features of the interaction remain constant (Peake 2005). There are several experimental demonstrations of social eavesdropping, including visual signals in captive fish and acoustic signals in free-living birds, in the contexts of resource defence and mate choice (e.g. McGregor and Peake 2000, 2004; Peake 2005; McGregor 2009). Of particular relevance to an increased understanding of animal social networks is the type of information that can be gathered by social eavesdropping and how this can underpin relationships between individuals in a social network.

It is becoming apparent that social eavesdropping and therefore social networks are unlikely to be restricted by taxon (see Section 3 of this volume). For example, social eavesdropping has been demonstrated in a crustacean, in a mate-choice context comparable to some of the earliest demonstrations of social eavesdropping in birds and fish. Female crayfish (*Procambarus clarkii*) were observed to visually and chemically eavesdrop on fighting males and then to visit the dominant male first and interact more frequently with the dominant than with the subordinate male (Aquiloni and Gherardi 2008). The authors of this study suggested that this was an unusual discrimination ability for an invertebrate; however, it is worth pointing out that there appear to be no comparable experiments on other invertebrate species, particularly eusocial insects.

It is clear from playback experiments with birds that information gathered by social eavesdropping can be combined with that gained by direct experience (e.g. Peake et al. 2002). Moreover, information from several eavesdropping opportunities can be combined and applied to novel interactions (e.g. Toth et al. 2012). These findings, together with evidence in fish that male cichlids (*Astatotilapia burtoni*) can infer hierarchies of relative dominance from social eavesdropping on pair-wise fights between rival males (Grosdenick et al. 2007), emphasize the relevance of social eavesdropping to social networks.

Audience effects

The ubiquitous nature of communication networks, combined with the opportunities these offer to eavesdroppers, leads naturally to the consideration of the selection pressures eavesdroppers and other receivers in a network exert on signallers. These are often referred to as audience effects (e.g. Matos and Schlupp 2005). Such effects can occur because signallers detect that an audience is currently present and/or because such an audience has commonly been present in the evolutionary past (referred to by Matos and Schlupp (2005) as apparent and evolutionary audiences, respectively).

An audience of interceptive eavesdropping predators or parasites imposes a strong cost on prey/host signallers. Therefore, it is not surprising to find adaptations to an evolutionary audience to make signals more difficult to detect and locate (e.g. Klump et al. 1986). Similarly, reports of behavioural adaptations to an apparent audience are common and usually involve the cessation of signalling.

The selection pressures imposed on signallers by an audience of social eavesdroppers are less easy to characterize. It has been argued that the phenomenon of quiet song in birds (i.e. song that differs from advertising song in amplitude and frequency in ways that should make it more difficult to detect and/or locate) is an adaptation to restrict the spread of the signal through a network (reducing the audience effect of social eavesdroppers) in contexts of high aggression and imminent mating (Dabelsteen et al. 1998). However, there are also examples of signalling that appear to be adapted to help ensure widespread reception (e.g. Dabelsteen 2005), and a fruitful area for consideration is the benefit to interacting individuals of widespread availability of information on the outcome of the interaction—for example, this might be expected to differ between interaction winners and losers (but see the 'good losers' hypothesis in Peake and McGregor 2004). The example of conspicuous signals produced by individuals interacting at close range is often linked to the idea of handicaps (e.g. Zahavi 1979) and honest signals (e.g. Searcy and Nowicki 2010), often without explicit consideration of the possible benefits to signallers of information transfer to a wider audience. Design components that appear to be adaptively ostentatious in this way include the victory displays (Bower 2005) and redirected aggression (Kazem and Aureli 2005) shown in many species, especially birds.

Alarm call spread

Many species produce signals in the presence of predators and other individuals of their own and different species (e.g. the passerine seeet call; Marler 1955). It has been suggested that the propagation pattern of such warning or alarm calls through a network contains information on the likely path of the predator (e.g. McGregor and Dabelsteen 1996; see Figure 9.2). A study by Bower (2000; Bower and Clark 2005) allowed some of the ideas suggested by McGregor and Dabelsteen (1996) to be tested. The study used an acoustic location system based on an array of four microphones to map the location of alarm calls of a population of red-winged blackbirds (*Agelaius phoeniceus*) as a Cooper's hawk (*Accipiter cooperi*) flew over. Fortunately, the silent hawk was closely pursued (i.e. within 1 m) by an eastern kingbird (*Tyrannus tyrannus*) that was producing scolding calls, effectively allowing the hawk to be tracked acoustically.

As the hawk entered the area, the red-winged blackbirds switched from mainly song to alarm calls (Figure 9.3a). The location of alarm calls relative to the hawk (Figure 9.3b; see also Figure 5 in Bower and Clark 2005) shows that most calling came from red-winged blackbirds ahead of the predator's path and that they fell silent as the hawk passed

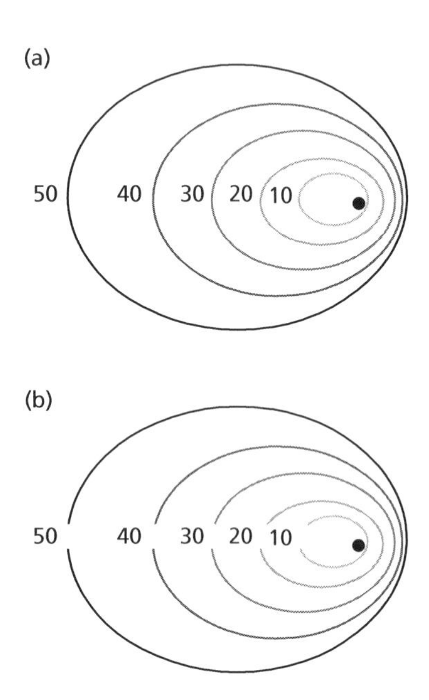

Figure 9.2 Hypothetical change in alarm call patterns in time and space in a communication network, to illustrate how such patterns can provide information on predator location. In the figures, the predator is travelling through the middle of the figure from right to left. Location of the first alarm call elicited by the predator is shown as solid dot. The location of individuals subsequently alarm calling in 10 s periods is shown by curves with the associated time from the first alarm call (numbers) and shading of curves (darker curves are more recent). (a) Pattern generated if individuals call in response to individually detected presence of the predator and to other alarm-calling individuals, with the modification that the probability of calling in response to others decreases with distance from caller, and individuals apparently in the predator's path are more likely to call in response to others. (b) Pattern generated if individuals do not call when close to the predator.

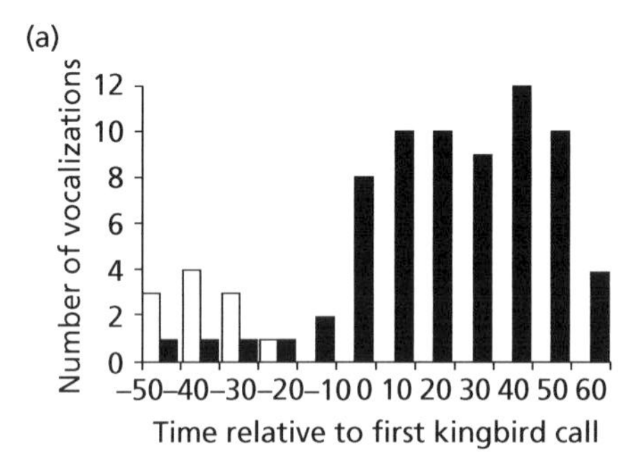

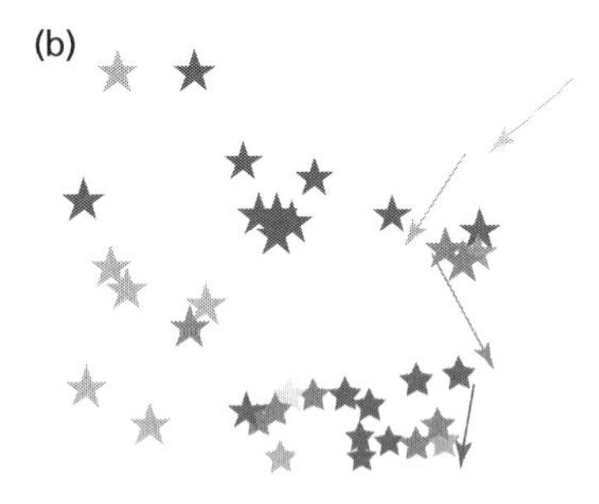

Figure 9.3 (a) Time course of vocalization by red-winged blackbirds as a Cooper's hawk passes through the area. Time is relative to the first Eastern kingbird scolding call. Numbers of songs (open) and alarm calls (solid) bars are shown in 10 s periods. Based on Bower (2000); further details can be found in the text. (b) Location of alarm-calling red-winged blackbirds (stars) in relation to the movement of a Cooper's hawk (arrows). Time (in 10 s periods) is indicated by shading, with darker shades being more recent. Based on Bower and Clarke 2005; further details can be found in the text.

overhead but called again when it was past. This example shows that alarm calls do not propagate evenly through the network from a source (Figure 9.2a). Unlike the ripples from a pebble dropped into a pond, the propagation pattern resembles Figure 9.2b; that is, the form of the propagation pattern contains information that could be used by potential prey in the network to reduce their vulnerability to the predator. A similar result has been reported for Richardson's ground squirrels (*Spermophilus richardsonii*) (Thompson and Hare 2010). In addition, there are similarities in the spatio-temporal nature of alarm call spread information and that of the information used by baboons to achieve sneak matings (Crockford et al. 2007).

Linking communication networks to social networks

The approaches used in studies of communication networks and of social networks are likely to become more integrated in the coming years. In this section, we highlight what we see as key areas where this integration is likely to occur or is already underway.

Signals as methodological tools for studying social networks

Signals and signalling interactions are likely to become increasingly important tools for field studies of social networks. Because signals have evolved for the express function of interacting, through the process of sender–receiver co-evolution described in 'Information exchange and communication networks', they are particularly revealing of social relationships—much more so than mere spatio-temporal association, which is currently the commonest tool for defining links (Croft et al. 2008; Whitehead 2008a). Moreover, as traits designed for social interaction, signals highlight characteristics of social networks that have been shaped by natural selection and thus help to reveal the adaptive structure of societies more generally.

Signals can also be manipulated in ways that are intractable with other behaviours, so they are particularly valuable for experimental approaches. How often a vervet monkey (*Chlorocebus pygerythrus*) grooms a particular individual in its troop is hard to manipulate experimentally, but how often it calls to warn other members of its troop can easily be manipulated by playing back that individual's warning calls (Seyfarth and Cheney 1984). The role of different individuals in the patterns of information flow through the network, as it relates to their reliability, distance to the receiver, and readiness to respond to others, can be assessed through observations followed by such playback experiments (Blumstein et al. 2004; Magrath et al. 2009). In principle, such playback experiments could map the spread of information quite precisely, for example by playbacks of sounds with key encoding features removed—a refinement of the coarser and more disruptive 'knock-out' experiment (Wey et al. 2008) in which social networks are mapped by removing particular individuals.

At the same time, using signals to study social networks has its challenges. One challenge is attaching signals to particular individuals, and more particularly, recognizing the roles of different individuals in the network when many individuals are interacting at once. Positioning information, such as that derived from acoustic location systems (e.g. Blumstein et al. 2011) (Figures 9.2, 9.3), can help to a large degree, as can individually distinctive signals (e.g. Terry et al. 2005). Parsing out the roles of different individuals, however, requires teasing apart the rules of engagement for a given system, which is painstaking but has seen notable successes, especially for acoustic signals (e.g., Blumstein et al. 2004; Foote et al. 2008).

Another important challenge is that many important communication interactions only occur at critical moments, so signalling events that reveal the structure of a communication network may be short-lived, rare, and thus hard to measure. Most obviously, a predator may only rarely fly over a woodlot, and its flight may be brief (about 60 s in the example shown in Figures 9.2, 9.3), but to the birds that may lose their entire lifetime reproductive success to it, the information about its passing is critical. Thus, the effectiveness of alarm calls in reaching receivers, and the attentiveness of receivers to those calls, may be crucial.

Indeed, rapidly changing interactions driven by current needs (e.g. hunger, mating opportunity) are likely as important as long-standing social relationships (e.g. in a stable dominance hierarchy) but may be hard to characterize and manipulate experimentally. Moreover, short-lived interactions may be critical in establishing long-term relationships and effects. Many territorial birds, for example, establish boundaries and dominance relationships early in the breeding season, while those relationships are maintained by subtler displays later on. Similarly, playback experiments lasting a few minutes have affected female extra-pair behaviour (Otter et al. 1999) and subsequent brood paternity (Mennill et al. 2002). While work on communication networks is likely to focus on tonic displays used in ongoing interactions—such as vigilance for predators through sentinel calls, rather than alarm calling per se, or territorial maintenance through the dawn song chorus, rather than territorial disputes—short-term signalling events cannot be ignored, despite their difficulties.

Table 9.1 Examples of studies in which network links were defined by signalling interactions.

Species	Signal	Reference
Daffodil cichlid (*Neolamprologus pulcher*)	Aggressive displays,* submissive displays	Dey et al. 2013
Wire-tailed manakin (*Pipra filicauda*)	Joint display	Ryder et al. 2008
Asian elephant (*Elephas maximus*)	Play, affiliative tactile displays	Coleing 2009
Rhesus macaque (*Macaca mulatta*)	Grooming	Brent, MacLarnon, et al. 2013
Pig-tailed macaque (*Macaca nemestrina*)	Silent bared-teeth display	Flack and Krakauer 2006
Yellow-bellied marmot (*Marmota flaviventris*)	Affiliative displays*	Blumstein et al. 2009
Bottlenose dolphin (*Tursiops* sp.)	Headbutting, mirroring	Lusseau et al. 2007b
Meerkat (*Suricata suricatta*)	Aggressive displays*	Madden et al. 2009

* Pooled with non-signal behaviours or association.

Mapping communication networks as social networks

Social networks are defined most simply by spatio-temporal association, so if association is defined by interaction distance, signals, being the most far-reaching ways of interacting (imagine bird or whale song), are central to defining all social networks (Croft et al. 2008; Whitehead 2008a). Relatively few studies have defined network links using signalling interactions (see Table 9.1 for some examples), but that is likely to change. The potential of doing so is nicely illustrated by two attempts to predict and contrast the structure of communication networks in a range of social systems, such as territorial versus lek breeding aggregations (Matessi et al. 2005; Bradbury and Vehrencamp 2011).

Signals with readily identifiable messages, such as alarm calls and nestling begging, are particularly well suited for studies of information flow, whereas signals that have more ambiguous messages can be particularly important for revealing the structures of social networks. Indeed, the design of the latter class of signals can be baffling if they are viewed as devices for information exchange but makes sense when they are viewed as signals that function primarily or solely to maintain social bonds. For example, the frequent repetition and precise coordination of seemingly arbitrary actions in songbird duets and nest site reunions might well serves as unbluffable signs that the participants have a history of interactions (i.e. a relationship) (Hall and Magrath 2007). Similar design requirements might account for the form of otherwise perplexing joint displays, such as group marching in flamingos (*Phoenicopterus* spp.) and carnivals in chimpanzees (*Pan troglodytes*) (Maynard Smith and Harper 2003). The list of such signals is likely to grow, as the network approaches advocated in this book revive interest in social relationships and the behaviours that maintain them.

Communication networks and information flow

Studying how information flows through a social network is a particularly effective way to reveal the varying roles of individuals and the value they place on the relationships between them. One way to study information flow is to examine how signals vary across different individuals. Not every member of any given population shares exactly the same set of signals, nor has the same contextual information with which to interpret (i.e. extract information from) signals. These local signalling codes cannot be observed directly, but a readily accessible surrogate is the pattern of signal sharing across a social network.

This point is well illustrated by song dialects, qualitatively distinct variants of learned songs found in many songbird species. Song dialects have been studied for decades (Kroodsma 2004), but such patterns of signal sharing are worth re-examination from the context of social networks. For example, they reveal locally shared codes (dubbed 'idiolects' by Mundinger 1982), whose spatial distribution is the result of relatively stable patterns of signalling interactions (Planqué et al. 2014).

A second window on patterns of information flow through a network is to examine the roles of different individuals, or nodes, more directly (Botero et al. 2010; Aplin et al. 2012). Particular individuals may be privy to information or may be more prone to broadcast reliable signals, making them keystone sources or distributors of information through the network. Other individuals may be less reliable or may actively withhold information, making them sinks to information flow. Still others may be more or less inclined to simply pass on information (e.g. by simply repeating what they hear without evaluating the information themselves). Roles can be predicted based on the ecological and behavioural traits of individuals (Goodale et al. 2010; Aplin et al. 2012) and can be tested using observational and playback studies (Blumstein et al. 2004). The effects of these roles and their distribution on information flow has been a rewarding area of research in studies of human communication networks, suggesting that similar rewards will stem from emulating the approach (with due caution; R. James et al. 2009) in studies of non-human animals (e.g. Botero et al. 2010; Dávid-Barrett and Dunbar 2012).

A net result of information flow studies is likely to be that the pattern of information exchange between senders and receivers can shape the structure of the social network as a whole. For example, theoretical work might predict that reliance on key individuals as information sources or a high value of locally shared information yields more hierarchically structured social networks (e.g. Dávid-Barret et al. 2012). Such predictions are readily tested in the field by studying signalling interactions, as we have suggested. For example, observations and playback experiments with bird song can show a high premium of precise matching of song structure at a local scale (e.g. Akçay et al. 2013), and mapping of song sharing 'idiolects' can reveal social structure at the level of large networks of singers on neighbouring territories.

The complexities of information flow through communication and social networks can be integrated under one perspective by viewing information as an ecological resource analogous to energy or nutrients. Animals must eat, but they also must continually consume information that will guide them to optimal behavioural decisions, and other animals provide variably useful sources of this resource (Dall et al. 2005; Valone 2007). This likening of interspecific information webs to food webs is not merely metaphor but rather a shift in how we think of the importance of animal signals in structuring communities and ecosystems (Seppänen et al. 2007; Schmidt et al. 2009; Goodale et al. 2010). That information should be on a level playing field with other resources is perhaps most obvious in predator–prey relationships, in which a prey's likelihood of being eaten is contingent on its ability to use information about predation risk. Indeed, one recent model shows how actual predation rates can be less important than information flow in determining community structure and predator–prey population dynamics (Luttbeg and Trussell 2013).

Although this book is mainly about animals, plants are free of any mentalist connotations of information concepts that might strain analogies between information and food webs. Thus, we will use plants to illustrate the potentially rich ecological implications of information webs. Many plant species release volatile compounds when they are damaged by herbivores (e.g. Kaplan 2012). Eavesdropping on such chemicals within and between plant species appears to be beneficial (see 'Signals as methodological tools for studying networks') by allowing eavesdroppers to prepare defences against herbivory.

Another suggested benefit is the attraction of parasitoid wasps that attack herbivorous insects (e.g. Turlings et al. 1990). However, hyperparasitoid wasps that attack the parasitoids are also attracted by herbivore-induced volatiles, with the effect that the parasitoid attack on the herbivorous insect is less effective (Poelman et al. 2012). Therefore, it is not clear that plants will always benefit from the release of volatiles. This example involves information with fitness consequences passing across four trophic levels; and while it may be rather extreme, it does give an indication of the sort of complexity that will need to be incorporated when modelling some systems—especially as damage-induced volatiles are now considered as part of integrated pest management of food crops (Poelman et al. 2012).

Communication and social networks: conclusions and future directions

In this chapter, we have described the main characteristics of communication networks and have tried to illustrate how their study can be integrated into the study of social networks more generally. We hope that, in ten years' time, this sort of isolated discussion of communication networks will no longer be needed, because network thinking will be fully integrated into the mainstream of the study of animal communication. Signs that communication networks are fundamental to communication include recent studies showing that attentiveness to social information is a basal function of the vertebrate brain (Fernald and Maruska 2012) and a key input to signal function (e.g. Grieg and Pruett-Jones 2010). For the graduate students of today's graduate students, such findings will be as much a part of animal communication as song control centres in songbirds and habitat acoustic effects on signals were to us as graduate students. It is also likely that future graduate students will not have to be introduced to networks of any sort via dyads (cf. Chapter 2) but will be ready to conceptualize them as a network through their familiarity with social media networking (e.g. Twitter, Facebook).

There is also need for an integrated network perspective on animal communication in at least two applications of communication. The first stems from the unprecedented increase in the intensity and scope of anthropogenic noise. Such noise, which also includes signalling modalities in addition to sound, is known to be affecting apparently pristine marine and terrestrial environments as well as extensively man-modified urban areas (e.g. McGregor et al. 2013). The second stems from the role of communication and social interactions in animal welfare of both farmed and captive animals (e.g. Grandin 2001, 2003). Future basic research on the complexities of communication networks that we have outlined here might well provide the tools needed to prevent or control welfare impacts and damage to the ecology of information, much as basic research in the late 1900s helped curtail the damaging effects of pesticides on the food chain.

CHAPTER 10

Disease transmission in animal social networks

Julian A. Drewe and Sarah E. Perkins

Introduction to disease transmission networks

Infectious diseases can have major impacts on populations, from global pandemics to localized outbreaks. It is no surprise therefore that the study of disease transmission is one of the earliest examples of the application of network theory (Bailey 1957). Modelling of infectious diseases using a network approach, however, only gained momentum in the 1980s, with the onset of the HIV pandemic in humans (Klovdahl 1985; Danon et al. 2011). Social network analysis is inherently suitable to understanding the causes and consequences of contact patterns on the dynamics of infectious diseases because contact frequency is at the heart of parasite transmission (May 2006). The connections between individuals (or groups of individuals) that allow an infectious disease to propagate naturally define a network, while the network that is generated provides insights into the epidemiological dynamics (see examples in Godfrey 2013).

Historically, the basic dynamics of infectious diseases have been captured using compartmental models that assume homogenous mixing of individuals within the population, termed 'mean field' (Anderson and May 1991). This mathematical assumption allows for tractable analysis that can predict epidemics on a broad scale but often cannot adequately capture disease dynamics unless the model is made more complex by the addition of epidemiological relevant variation. An example would be an increase in mixing patterns, and subsequent pathogen transmission, brought about by students returning to school during term times (e.g. measles, Bjørnstad et al. 2002). Mean-field models are most suitable when host populations are nearly homogenous, or where individual level variation is likely to be low. However, most animal populations (including humans) are likely to have a form of biological organization (spatial and/or social structure) which results in non-random mixing. In addition, populations also vary in their susceptibility to infection due to epidemiologically relevant host parameters such as sex, body mass, and behaviour (Moore and Wilson 2002; Perkins et al. 2003; Hughes, Border, et al. 2012). As a result, contact patterns and variation in transmission rate are usually highly heterogeneous, meaning that some individuals play a much more important role in transmission than others, which has considerable implications for the emergence, spread, persistence, and control of disease (Newman 2002; Lloyd-Smith et al. 2005; Perkins et al. 2013). Models that assume population-level mixing is random are unlikely to adequately capture heterogeneous host–parasite interactions, and, in these situations, infectious disease may be better explored using network models (Bansal et al. 2007)—an observation that applies to both human and animal diseases.

Constructing transmission networks poses a challenge, as both contacts with infected individuals, or agents, must be quantified while simultaneously monitoring infectious status. Network studies have recently benefited from an increased resolution of contact pattern information using data derived from technologies such as Bluetooth connection through mobile phone networks (e.g. Bengtsson

Animal Social Networks. Edited by Jens Krause, Richard James, Daniel W. Franks and Darren P. Croft.

et al. 2011). Human infectious diseases have seen the advent of 'digital epidemiology'—the field of tracking epidemics by leveraging the widespread use of the Internet and mobile phones (Salathé et al. 2012). Such technologies have given rise to 'big data', which brings new challenges in data storage, accessibility, analysis, and interpretation.

However, just because a dataset is large does not mean it is accurate! These datasets surpass those collected from animals in terms of sample size, but mobile phone data is likely to be a poor proxy for the number and duration of physical contacts—a crucial component of infectious disease transmission—plus, these data lack information on an individuals' infection status and are only relevant when transmission is via contact with other individuals. Studying animal transmission networks may offer some benefits over those generated from human data, as improved diagnostics offer real-time reporters of infectious disease status (Gavier-Widen et al. 2012) while technological advances in ecological data collection used for quantifying contact patterns or animal locations—for example, proximity loggers, GPS collars, and miniature video cameras (Krause, Wilson, et al. 2011)—allow increasingly high resolution contact data to be collected.

In this chapter, we provide an overview of parasite transmission processes in animal social networks and illustrate how using the network approach has helped our understanding of the central biological issue of disease transmission. We address points pertinent to transmission networks, including sampling of the host–parasite network, and ecological methods of specific importance for host–parasite interactions, and explore how host–parasite interactions, animal behaviour, and disease transmission networks are inter-related. We conclude the chapter by discussing how the application of network analysis to our understanding of infectious disease transmission is likely to develop in the future.

The use of animal social networks to study infectious disease transmission

Parasite transmission is composed of two key elements: the probability of acquiring infection (including the infectious dose and route of transmission) and the contact rate (both frequency and duration) between infectious and susceptible hosts. Contact is crucial. No matter how infectious a parasite may be, if an uninfected animal never contacts an infectious host or parasitic stage then it will not become infected. Note that for disease transmission to occur, a 'contact' need not involve a direct interaction and that indirect contacts, such as two animals visiting the same waterhole at different times, may lead to transmission through contamination of the environment, for example with a hardy pathogen such as *Mycobacterium bovis*.

Several studies across a range of species have found that indirect interactions occur much more frequently than do direct contacts. For example, camera traps placed on a reserve in south central Spain revealed that direct contacts between wildlife (wild boar [*Sus scrofa*] and red deer [*Cervus elaphus*]) and livestock (cattle and pigs) were very rare (just 10 contacts recorded in 12 months), whereas indirect interactions—visits by wildlife and livestock to the same waterholes, feed troughs, or pastureland within a specified time window—were two orders of magnitude more frequent (9000 contacts) over the same time period (Kukielka et al. 2013). Similarly, when Drewe et al. (2013) studied the behaviour of European badgers (*Meles meles*) and cattle in the UK by using proximity loggers to record interactions between the two species, they found that direct contacts (interactions within 1.4 m between badgers and cattle at pasture) were very rare (4 out of >500,000 recorded animal-to-animal contacts), despite ample opportunity for interactions to occur. Indirect interactions (visits to badger latrines by badgers and cattle) were 2 orders of magnitude more frequent than direct contacts: 400 visits by badgers and 1700 visits by cattle were recorded within the same 12-month period (Drewe et al. 2013). As such, when constructing a transmission network, the shared resource or space must be taken into account when quantifying transmission-relevant contacts.

The behaviour of some macroparasites means that their transmission networks may also be best captured using indirect contacts. Ticks and helminths, for example, have a 'sit and wait' strategy whereby the parasite is deposited in the environment by a previously infected host; the infective stage develops (sometimes for months) and, due to a lack of motility, transmission will only occur when a susceptible host passively picks up the parasite in passing. As such, space-sharing networks have been used to produce and better understand transmission networks for parasites with this form of transmission strategy (e.g. helminths (Perkins et al. 2009) and ticks (Godfrey et al. 2010; Fenner et al. 2011)).

Networks and disease management

Understanding the structure of transmission networks can provide insight into the options for the practical management of wildlife diseases by elucidating the role of individuals in perpetuating infection (Godfrey 2013). Highly connected individuals, for example, have the potential to become 'superspreaders' of infectious disease because, by the nature of their behaviour, they may have a high chance of not only becoming infected but also of transmitting infection (Lloyd-Smith et al. 2005). Studies of animal social networks and disease transmission have found frequently that the majority of individuals have few connections, with a small minority being highly connected (e.g. rabid domestic dogs (Hampson et al. 2009), lions (*Panthera leo*) with canine distemper (Craft et al. 2011), and Japanese macaques (*Macaca fuscata*) infected with nematodes (Macintosh et al. 2012)). This variation in contact frequency can be captured by a negative binomial distribution, as has also been seen in other wildlife-parasite systems, for example, in elk (*Cervus canadensis*) (Cross et al. 2013), European rabbits (*Oryctolagus cuniculus*) (Marsh et al. 2011), and yellow-necked mice (*Apodemus flavicollis*) (Perkins et al. 2009).

There is considerable interest in developing disease control efforts to identify and target highly connected individuals, because models suggest such control methods are likely to be considerably more effective at preventing and containing outbreaks than random control (Woolhouse et al. 1997; Lloyd-Smith et al. 2005). One study addressing this was carried out by Clay and colleagues (2009), who constructed social networks of deer mice (*Peromyscus maniculatus*) with respect to Sin Nombre virus, a hantavirus transmitted in wild mice by aggressive interactions. Clay et al. (2009) quantified aggressive encounters as contacts between individuals, finding that 20% of hosts accounted for 80% of contacts. The 20% most connected individuals were identified to be large animals that were more likely be infected with hantavirus, thereby giving an identifiable proportion of the population at which disease control efforts could be targeted or from which disease risk maps could be produced.

One of the first studies of animals to apply formal social network analysis to questions of disease transmission was that of Corner et al. (2003), who investigated the spread of tuberculosis (TB: *Mycobacterium bovis* infection) among captive brushtail possums (*Trichosurus vulpecula*). Possums that became infected were found to have greater closeness and flow-betweenness scores, indicating these animals to be more central to the network, than those that remained free of infection (Figure 10.1). As such, in this example, it was the position of these individuals in the network that rendered them more susceptible to acquiring infection, and so they could be considered for disease control based on this criterion.

It should be noted that skewed contact distributions (i.e. the negative binomial distribution) and superspreaders are not always present in a host-parasite system; instead network structure can reveal other patterns that are informative for disease control. Detailed investigation of a transmission network of Tasmanian devils (*Sarcophilus harrisii*), currently threatened by an infectious cancer revealed a 'giant component', an interacting population whereby all individuals were connected to all others (Hamede et al. 2009). Similarly, lions in the Serengeti were found to be sufficiently well connected within prides that persistence of canine distemper virus between prides relied on spillover from other reservoir species (Craft et al. 2011). The consequence of these well-mixed network structures is that targeted treatment or culling of infectious individuals would prove fruitless (Beeton and McCallum 2011; Hamede et al. 2012).

Collecting social network data to study disease transmission

A problem that is pervasive throughout ecology is 'what sample size is sufficient?' This is an issue especially pertinent in animal transmission networks, as the parasite may have a low prevalence, the host species studied may be rare, difficult to

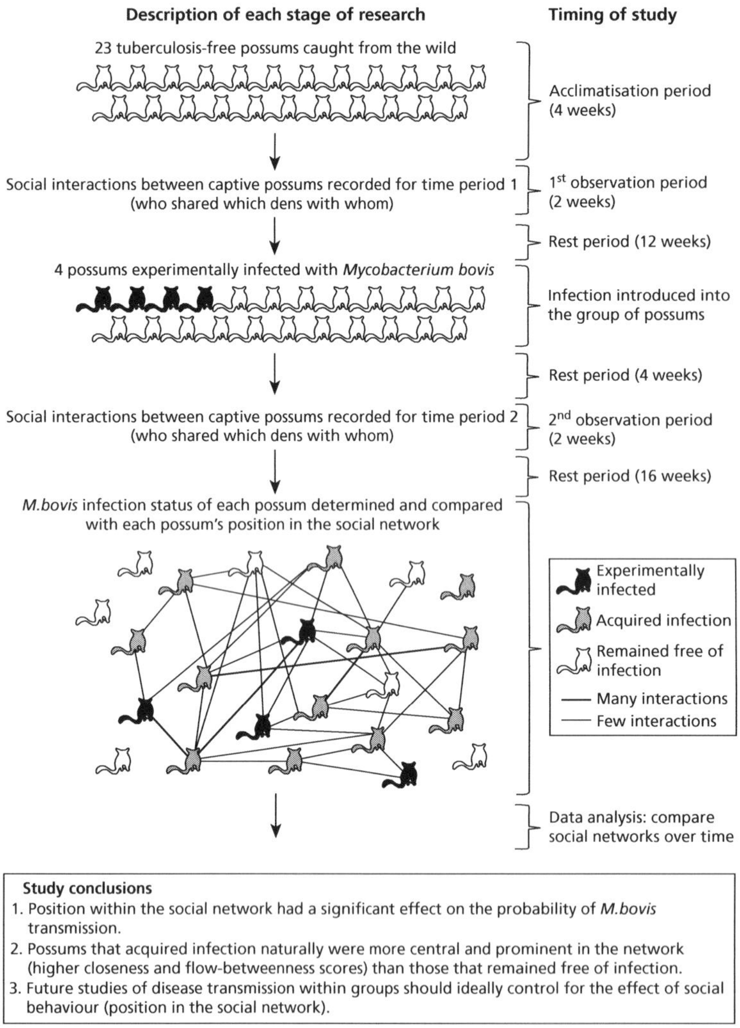

Figure 10.1 Application of social network analysis to help understand TB transmission in captive brushtail possums. This simplified example is based on data from Corner et al. (2003), who conducted the study on multiple groups and concluded that social networks based on den sharing provide a useful model for explaining TB transmission within captive possum groups.

observe, or capture and/or cost required for data collection is a limiting factor and so a trade-off with sample size exists. Sample size in this instance encompasses both host and parasite and so should consider the number of animals or groups (i.e. the nodes) included in the study and the temporal resolution over which data are collected, as well as the parasite(s) in question.

How many host–parasite associations should be included in a transmission network?

To ensure that the proportion of animals and interactions in a population that are included in a study of social networks is sufficient to represent a transmission network, both the nodes (individuals or subgroups) and edges (contacts between nodes) need to be considered. While a small sample size of animals will exclude both nodes and edges, too low a temporal resolution of data will lead both to loss of edges and to under-representation of the population. For studies of infectious disease, edges are exceptionally important, as they represent the transmission process, while the node is a host/subgroup that may be either susceptible or infected and could represent key individuals such as superspreaders. Host–parasite distributions tend to follow a 20:80 pattern, such that a minority of individuals (20%)

harbour the majority of parasites (80%) (e.g. Woolhouse et al. 1997; Perkins et al. 2003) and, given that 20% of hosts may account for 80% of the contacts (e.g. Clay et al. 2009), ensuring that parasite dynamics on animal social networks are well sampled presents a 'double whammy', as both highly connected and infected individuals may be rare events and the chances of 'sampling' them decreases with decreasing sample size (Figure 10.2).

The effect of incomplete sampling has been studied in biological and social networks where post-hoc methods have been used to sub-sample the network to assess whether the network is adequately sampled (e.g. Costenbader and Valente 2003). Incompletely sampled networks result in predictable errors, with some network measures more stable than others; for example, mean degree (contacts) is always lower, and in-degree is robust to under-sampling (Costenbader and Valente 2003; Han et al. 2005; Perreault 2010; Stumpf et al. 2005) (Figure 10.2). Omissions of key nodes, such as superspreaders, may have profound effects on the interpreted disease dynamics of the system. Empirical studies comparing incomplete sampling of

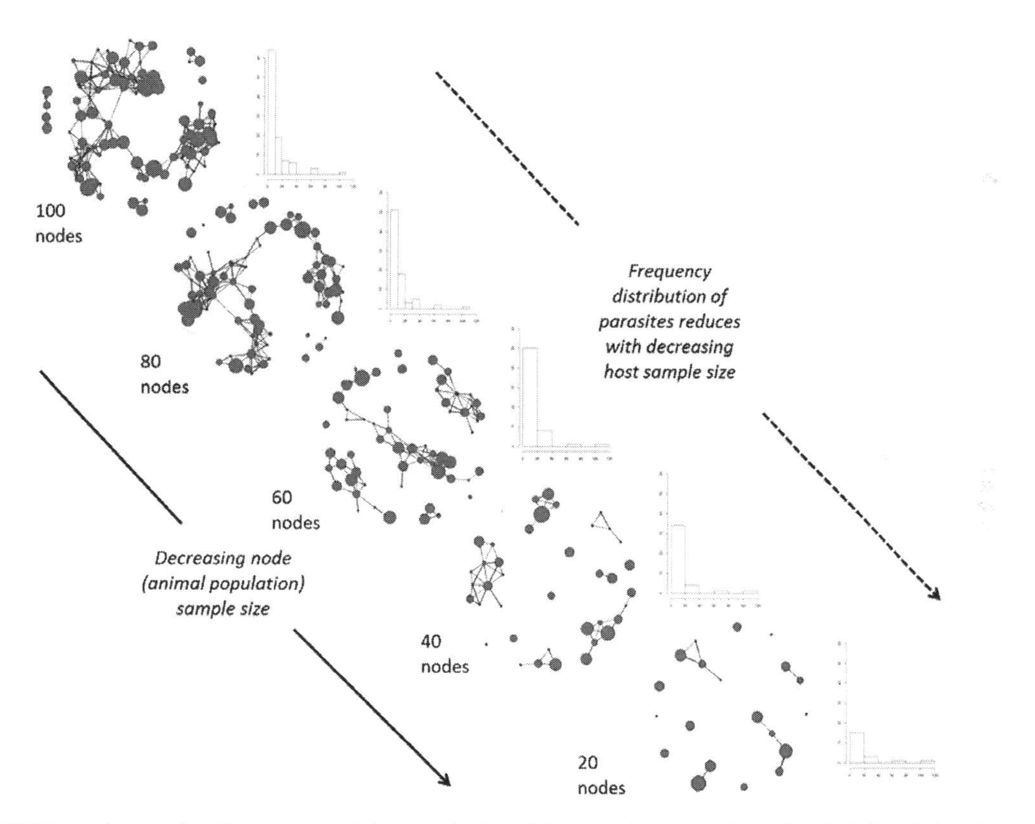

Figure 10.2 Incomplete sampling effects on a transmission network. Transmission networks are shown from a hypothetical population of 100 animals, with subsets of that population randomly sampled with sample sizes of 80, 60, 40, and 20 animals (nodes). Individuals are represented as nodes, and the edges (links between nodes) represent transmission-relevant contacts. The contacts between individuals within the population are assumed to follow a negative binomial distribution such that the majority of individuals have a few connections, and a minority have a high number of connections. Individual nodes are scaled according to their parasite load. The frequency distribution of parasites within this population was simulated such that it follows a typical host–parasite distribution, the negative binomial distribution, whereby the majority of the population harbour very low or no parasites, and a small minority of hosts are heavily infected. The parasites were distributed randomly among nodes, although it should be noted that some studies find that the most connected are the most infectious while others find no relationship (see main text). The frequency distributions of the contacts (edge distribution) are given adjacent to each network. As animal sample size decreases, clear differences in network structure can be seen, and the frequency distribution of parasites becomes truncated.

networks are rare in the literature (but see Perkins et al. 2009), while others note that conclusions on their comparisons should be interpreted broadly because transmission networks were not evaluated (Costenbader and Valente 2003). Where studied, systematic biases have been found in sub-sampled networks, such that the risk for of establishment and persistence of infection in a population (in this case HIV in humans) is likely to be underestimated (Ghani et al. 1998).

Cross et al. (2012) proposed a method for assessing network metrics that is of direct relevance for transmission networks. By using a generalized linear mixed model, their approach examines the variation that individuals and the environment contribute to different network statistics, thereby allowing an assessment of what processes underlie connections and so directing disease control towards either. In summary, the effects of incomplete networks on disease inference may be substantial but are currently poorly understood. This is an area that is ripe for further investigation.

Sampling considerations and the boundary effect

Time and financial constraints are such that most studies of animal social networks do not include all the animals in a population but rather collect data from a subset (a sample). Social network data collected from a sample is likely to be biased, because the chance of each individual interacting with other 'known' individuals is not constant. This bias may profoundly (and incorrectly) influence the conclusions that are drawn about disease transmission, and therefore it needs to be accounted for (see Figure 10.3).

Weighted or unweighted networks: capturing transmission processes?

In terms of disease transmission, not only is the contact important but so is the duration of that contact, something that can be captured by the 'weight' of the edge. Edges in a transmission network may be weighted by the frequency (intensity) or duration of

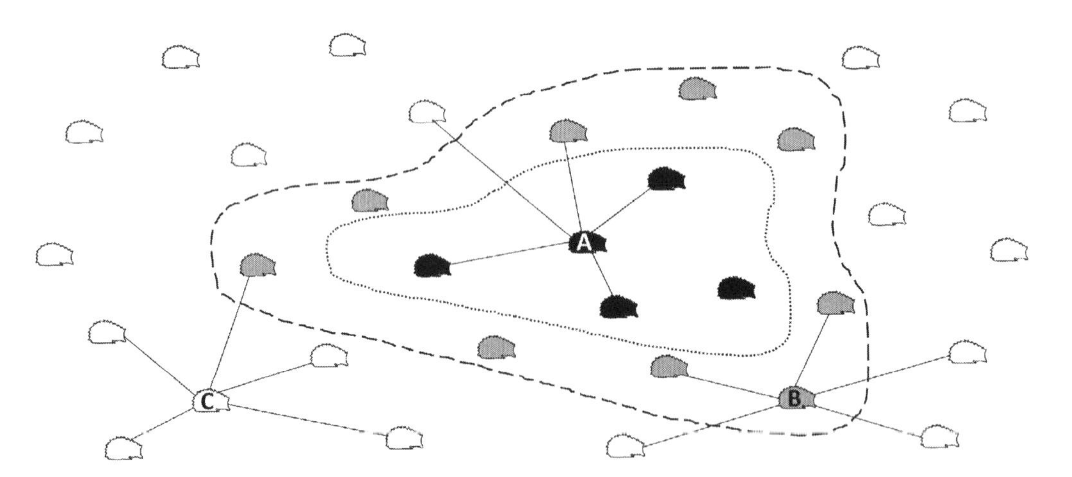

Figure 10.3 The boundary effect—and how to reduce its impact when sampling animal social networks. In this hypothetical example, the aim is to record the social interactions of all hedgehogs within a study site (denoted by the dashed line). Hedgehogs shaded white are outside the study site and are not studied. Hedgehogs that live at the periphery of the study site (shaded grey) are more likely to interact with unstudied animals than are hedgehogs living in the middle of the study site (shaded black). If this boundary effect is not accounted for, network parameters for the peripheral individuals may be underestimated, resulting in biased estimates. One way to deal with this is to use a constrained resampling procedure whereby an internal buffer zone (dotted line) is constructed, its size being proportional to the territory size of the study species (Godfrey et al. 2010). Network data are then calculated separately for animals living in the centre of the study area (e.g. hedgehog A, who in this example is most likely to contact other animals at the centre of the study area, moderately likely to contact animals in the buffer zone, and least likely to contact animals outside the study area) and for animals living within the buffer zone (e.g. hedgehog B, who is more likely to contact unstudied animals than hedgehog A). Note that unstudied animals (e.g. hedgehog C) may interact with study animals, but these interactions are likely to go unrecorded and therefore represent missing data.

interactions between nodes, or they may be left unweighted, in which case an edge is present if a minimum threshold of either frequency or duration of interactions is reached. Unweighted networks represent the system in binary form (an edge is either present or absent), and such networks therefore reflect who interacts with whom, but not how often or for how long. Networks weighted by frequency or duration of interactions incorporate a finer level of detail which may be very pertinent when considering disease transmission. A good example comes from the 2002–2003 pandemic of SARS: humans in an area of Canada who had at least 1 close contact (<1 m) with an infectious person of long duration (>30 min) were 20-fold more likely to become infected than were people whose contact with infectious patients was shorter or more distant (Rea et al. 2007).

Both the frequency and duration of social interactions may be important factors affecting the likelihood of transmission of infectious diseases, with the relative importance of each depending upon the nature of the disease under investigation. Frequency rather than duration of contacts may be more important for diseases that have frequency-dependent transmission (typically vector-borne and sexually transmitted diseases) (see McCallum et al. 2001). One such example is the Tasmanian devil and devil facial tumour disease (which is directly transmitted between individuals via bites), where the probability of a bite from infectious conspecifics is likely to increase with the frequency of bites (McCallum et al. 2009).

Importantly, whether weighted or unweighted, network skew (how dissimilar nodes are with respect to their network properties; a measure that might help identify superspreaders of disease) appears to vary depending on the social interaction measured. Centrality measures, for example, (which indicates the prominence of individuals or groups in a social network) for interactions between groups of wild meerkats (*Suricata suricatta*) were more skewed when data were weighted by frequency of contact (Drewe et al. 2009). Conversely, centrality measures for interactions between individuals within groups of the same meerkats were less skewed when data were weighted by frequency of contact (Madden et al. 2009). This indicates that the choice of whether or not to weight networks should be based on an exploration of the dataset, consideration of the type of interaction(s) being studied, and careful contemplation of the epidemiology of the disease under investigation.

Choice of time interval for constructing parasite transmission networks

A network constructed with a view to provide inference for disease dynamics is very different to that of a standard social network, due to one overarching issue: the timing and duration of the infectious period of the parasite in question. As such, an important but often overlooked consideration is the timescale over which a transmission network is created. There are four key issues which should be considered: these are described below.

Matching social network construction with the dynamics of disease

It is important that the network timeframe is epidemiologically meaningful (Box 10.1). If the time interval over which interaction data for social networks are collected does not match the dynamics of the disease being studied, then erroneous conclusions may be drawn. In a study of TB transmission between wild meerkats, a three-month time interval was chosen because this matched the stability of network structure over a range of timeframes with the chronic nature of the disease (Box 10.1 and Drewe 2010). For acute diseases such as rabies that may spread very quickly, much shorter time intervals, for example a few days, are required in order to match the social interaction information with the rapidity of disease dynamics (see Hampson et al. 2009). Where the infectious period of the parasite is not known, it may be informative to collect data on a very fine temporal scale and work backwards to determine the contact frequency that best describes the pattern of disease transmission seen.

There are many possible ways for collecting data on animal behaviour and disease status (Table 10.1). The method(s) should be selected following careful consideration of the dynamics of disease and the most appropriate technique to capture biologically and epidemiologically meaningful behaviour. Table 10.1 provides a summary of pros and cons of the most commonly used techniques for collecting data applicable for studying disease transmission networks.

Box 10.1 Social networks and TB transmission in wild meerkats

Meerkats (*Suricata suricatta*) are small (<1 kg) social mongooses from the arid regions of southern Africa. A population of wild meerkats living in the southern Kalahari Desert has been the focus of detailed behavioural ecology studies since 1993 (Clutton-Brock et al. 1998). TB due to infection with pathogenic mycobacteria is endemic within this study population (Drewe 2010). Habituation of these meerkats offers a unique opportunity to study the role of specific social interactions in the transmission of TB in a wild animal population.

The social interactions of 250 wild meerkats were observed and recorded on at least three days each week for two years (Drewe 2010). Interaction networks were constructed by dividing the data into blocks of different length: 1 day, 1 week, 1 month, and 3 months, for specific interactions between individuals within groups (intragroup); and 3 months, 12 months, and 24 months, for specific interactions between groups of meerkats (intergroup). Permutation tests were used to search for correlations between networks, with the assumption being that networks of similar structure would be positively correlated, indicating that contact rates between any pair of individuals (or groups) remained approximately constant.

Intragroup networks constructed from interactions between the same meerkats and which occurred over one day, one week, one month, and three months were found to be significantly correlated with each other across all four time intervals (Drewe 2010). In addition, intergroup networks generated from data collected over 3 months correlated with networks formed from datasets containing 12 and 24 months' data (Drewe et al. 2009). Thus, data collected over the relatively short time window of 3 months was considered representative of both shorter time intervals (down to 1 week for intragroup interactions) and longer time intervals (up to 24 months for intergroup interactions), because meerkat social structure was stable over these time intervals. Networks constructed over three-month intervals therefore allowed valid extrapolation and inferences to be drawn for meerkat social interactions over a wider range of time periods.

Importantly, the use of three-month time intervals for social network formation matched the chronic nature of TB (individuals could become infected within a three-month window, but pathological studies showed they were unlikely to go on to infect others within the same three-month period). Thus, meerkats were sampled every three months to ascertain their infection status, and these data were integrated with information on their social interactions over the same time periods (see Figure 10.4) to answer questions about disease transmission within a wild mammal population (Drewe 2010).

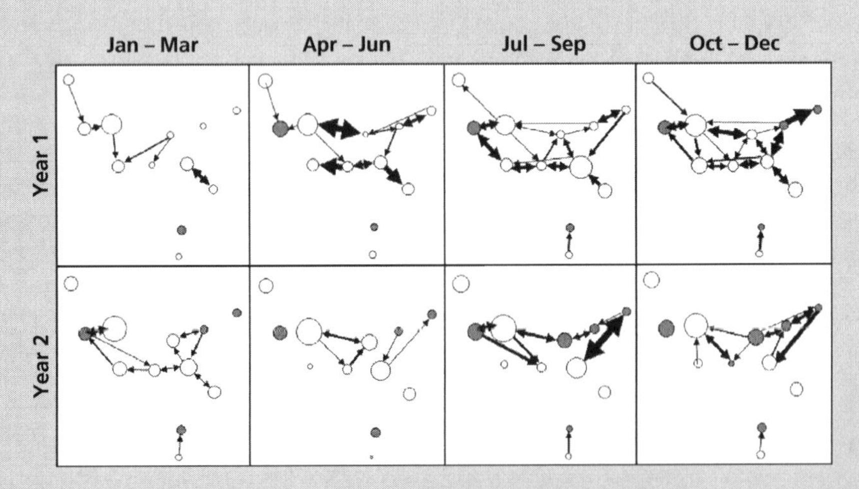

Figure 10.4 Seasonality in meerkat social network structure, and TB transmission over two consecutive years. Male meerkat intergroup movements (arrows) between 12 social groups (circles) are shown. Network structure (who interacted with whom) remained reasonably constant over time, although interaction rates were higher in the breeding season (September to December) than at other times. Arrow size is proportional to interaction rate. Node size is proportional to the number of meerkats in the group at each time point. Nodes are shaded grey when infection was detected, and white if not detected. Figure adapted from Drewe et al. 2009, with kind permission from Springer Science and Business Media.

Table 10.1 Summary of data collection methods that have been used to study disease transmission networks.

Technique	Advantages	Limitations	Types of disease, species, or research for which the technique is likely to be particularly suited	Examples of animal disease studies that have used this technique
Direct observations (may be matched with capture for disease sampling)	• Can provide very detailed records of 'complete' interaction rates • The exact types of social interaction can be recorded • Animal does not need to be captured	• Labour intensive • Not suitable for species that are difficult to observe • May require animals to be individually identifiable • Presence of observer might alter animal behaviour	• Widely suitable for many types of disease (acute and chronic)	• Ectoparasites/haemoparasites in tuataras (Godfrey et al. 2010) • Distemper virus in lions (Craft et al. 2011) • TB in meerkats (Drewe 2010)
Radio tracking	• Relatively cheap • High temporal resolution of data • Potential to be automated	• Requires animals to be captured to fit the devices (and possibly also to remove them, although auto-drop off collars are available) • Labour intensive, and sample size likely to be small • Spatial resolution of contact information may be too coarse for fine-scale transmission studies	• Identification of subgroups or clusters of individuals within which disease transmission may be more likely • Examination of specific individuals in the population (e.g. suspected superspreaders)	• TB in badgers (Böhm et al. 2008) • Encounternet (behaviour only: Mennill et al. 2012)
Live trapping (for capture–mark–recapture and/or disease sampling)	• Allows detailed information— including infection status—on each individual to be recorded at regular intervals	• Requires animals to be captured • Labour intensive • Coarse spatial scale	• Species which are easy to trap (e.g. small mammals) • Diseases for which accurate diagnostic tests exist	• Faecal–oral helminth transmission in wild rodents (Perkins et al. 2009)
Powder marking	• Gives good contact data if interactions involve direct physical contact • Very cheap	• Number of animals that can be monitored at one time point is limited by the number of powder colours • Requires animals to be captured	• Potentially useful for quantifying contacts for sexually transmitted infections	• Hantavirus in deer mice (Clay et al. 2009)
Proximity loggers (automatically record when animals fitted with loggers come within a preset distance of one another or a static base station)	• Time efficient • The frequency and duration of contacts are recorded • Complete temporal data • Loggers can be placed on animals and in static base stations in the environment	• Requires animals to be captured to fit the devices (and possibly also to remove them) • Ideally need to saturate the population, requiring all animals to be caught • Data requires careful manipulation to ensure it reflects real life patterns of contact • Does not tell you anything about the type of contact • Expensive	• Diseases in discrete populations or populations of limited size • More commonly used on medium- to large-sized animals (due to size and weight of equipment)	• Devil facial tumour disease in Tasmanian devils (Hamede et al. 2009) • TB in badgers and cattle (Drewe et al. 2013)

continued

Table 10.1 *Continued*

Technique	Advantages	Limitations	Types of disease, species, or research for which the technique is likely to be particularly suited	Examples of animal disease studies that have used this technique
GPS (collar, harness, or other form of attachment)	• Allows movement patterns of cryptic and long-range species to be recorded • Cheap versions are available	• Requires animals to be captured to fit the devices (and possibly also to remove them) • May not have sufficient precision or accuracy • Battery life is limited for devices, particularly those fitted to smaller animals • May not work in species living in dense forests or underground • Currently limited to larger animals due to unit size	• Diseases in wide-ranging populations (e.g. migrating animals or those that move long distances between countries/continents) • Suitable for marine animals	• Modelling of various diseases in feral/free-roaming sheep and goats (Mayberry et al. 2010)
Bait marking (markers in food are traced to the location of faeces)	• Does not require animals to be captured	• Requires extensive surveys to locate deposited bait	• Studies to establish space use by animals (e.g. by identifying latrines on territory boundaries)	• TB in badgers (Vicente et al. 2007)
Video cameras	• Data collection may be automated in some cases • Does not require animals to be captured	• Expensive • May require animals to be individually identifiable • Not suited to animals whose movement patterns are unpredictable unless camera is attached to animal and combined with a GPS	• 'Staking out' of likely contact points (e.g. areas of feed/water) for static cameras • Interaction patterns of a few individuals (cameras fitted to animals)	• Movement patterns and behaviours of birds (no specific diseases mentioned) (Bluff and Rutz 2008)
Biologgers (miniaturized, animal-attached tags that record data on an animal's movements, behaviour, physiology, and/or environment)	• May be used to collect physiological data related to infection (e.g. fever) • Multiple physiological parameters (e.g. body temperature, oxygen utilization, heart rate) as well as environmental variables can be collected at once	• Methods to infer animal behaviour from tag data are not fully developed • Battery life is limited • Data recovery may not be straightforward • Due consideration should be given to the ethics of such devices, particularly those requiring invasive implantation	• Most suited to easily captured and handled individuals • Diseases with specific clinical signs or quantifiable physiological responses • Useful for aquatic species	• Overview of loggers and their potential applications (no specific diseases mentioned) (Bograd et al. 2010)

Consistency of social networks over different timescales

The consistency of social networks over a variety of time intervals should be determined. For example, how similar in structure are networks constructed from interaction data collected over a day, a week, a month, or a year (Drewe 2010)? Determining this will help avoid choosing timescales that are too short (resulting in small datasets that are not representative of the true situation) or too long (which may result in several cycles of disease transmission being lumped together in one block and hence make it difficult to infer the dynamics of the disease).

Stability of social networks between different points in time

Network stability between different time points, for example seasons, times of the year or different stages of an animals' life ought to be considered. In a study of wild mice, the number of contacts reduced between time intervals from the start to the end of a breeding season (Perkins et al. 2009). In this case, the ecology of the system played a role in determining these patterns, with a natural reduction in population size over the course of the seasons likely reducing the contact patterns between individuals, although host factors such as testosterone may play a role in maintaining contacts and rates of infection (see Grear et al. 2009). The seasonal dynamics of a host's contact patterns may therefore drive the seasonality observed in infection, for example, with devil facial tumour disease (Hamede et al. 2009) and bovine TB in African buffalo (*Syncerus caffer*) (Cross et al. 2004).

Network structure in relation to different types of social interaction

Finally, consideration should be given to the epidemiological significance of the type of interactions observed. For example, the structures of networks of grooming interactions are different from those of networks of aggressive interactions in meerkats (Drewe 2010). This observation is important because different parasites are more readily transmitted via some behaviours than others. Grooming rather than aggression appears to be more relevant for transmission of TB in meerkats and individuals that initiate grooming are at highest risk of infection (Drewe et al. 2011). Conversely, aggressive behaviours, specifically biting, drive hantavirus transmission in wild mice (Clay et al. 2009). Differences in the structure of networks between different types of interaction mean that extrapolating between networks of dissimilar types of interaction may lead to invalid conclusions about disease transmission.

Of course, if we are interested in the detail of disease transmission, such as who infects whom, then the precise order in which interactions occur is important. A diseased individual who contacts a healthy individual may pass on the infection, but if the same two individuals had met before the first had become infected, then disease transmission would not have occurred. This illustrates the advantage of using time-ordered networks over the traditional static network approach (Blonder et al. 2012). Time-ordered networks represent data observed for a set of interactions that occur at certain times, thereby retaining complete information on the ordering, duration, and timing of events. A benefit of time-ordered networks is that they can be flexibly decomposed into multiple time-aggregated networks based on selected time windows (which should be determined to match the dynamics of disease being studied, see above), which can be analysed with standard methods (Blonder et al. 2012). This explicit consideration of time makes it possible to directly investigate how each time window impacts inferences about disease transmission through the network (Yeung et al. 2011).

Data analysis: which network measures are relevant to disease transmission?

Consideration should be given as to how the data will be analysed and how this will provide insight into the infectious disease process. Ideally, researchers should assess which network metrics are relevant to their question prior to data collection, so that enough data of the correct type, for example, inclusive of 'enough' infectious animals (see Figure 10.2), can be collected. A range of network measures, statistics, and analytical methods have been used to help understand the epidemiology of infectious disease transmission within social networks. In this section, we briefly review some of the more frequently used methods.

When considering infectious disease dynamics, we can divide the process into several steps: disease emergence, persistence, and control, for which different sets of network statistics come into play. When an epidemic emerges, a focus may be to ask, how quickly will disease spread through the population, or how resilient is the network to 'attack' by disease? In this instance, a first step in examining network statistics for disease ecology could be to examine the degree (contact) distribution. This is because the basic reproductive number (R_0) of a pathogen depends directly on the coefficient of variation of the degree distribution (May 2006). R_0 is the number of secondary cases caused by a single infectious host. It is immediately intuitive that an epidemic that is initiated in a highly connected individual will spread more rapidly than one that emerges in a weakly connected host. Indeed, contact distributions that follow a scale-free (or negative binomial) pattern are highly vulnerable to 'attack' by infections (Albert et al. 2000).

Understanding parasite persistence in a population presents questions that require different network measures to answer to that of emergence. Here, one may be more interested in asking, how does the parasite alter the network as the epidemic progresses? For example, what is the role of host behaviour in parasite persistence, and vice versa? For disease control, we may be interested in determining the characteristics of individuals that are pivotal in spreading disease (the so-called superspreaders) or looking for certain network structures, for example, 'fire-breaks' (individuals that link components), so that we can target our control on those key hosts. For example, a study on the social structure of badgers found that TB-positive hosts were socially isolated from their own groups but were more important for flow, potentially of infection, between social groups (Weber et al. 2013). Thus, efforts to control disease spread may focus on more egocentric (individual- or group-level) metrics than whole populations (see 'Networks and disease management in animal societies').

Network centrality and disease transmission

One network statistic that is used pervasively for understanding aspects of disease emergence, persistence, and control is that of centrality. Network centrality is a suite of measures reflecting the position of any given individual or group within a social network. As an epidemic spreads through a network, animals that are socially well connected—and therefore central to the social network—are usually considered or observed to be at higher risk of acquiring infection (Corner et al. 2003; Christley et al. 2005; Vicente et al. 2007; Godfrey et al. 2010) and are often assumed to perpetuate the infection.

The most socially connected individuals, however, are not always at highest risk of becoming infected (Drewe 2010). In this instance, teasing apart whether the most connected animals are a source or a sink of infection could determine how important these individuals are in parasite persistence. One approach to this is to quantify the ratio of in-degrees (contacts to) and out-degrees (contact arising from) a given individual. This method was adopted by Godfrey et al. (2010), who found that lizards with a high in-degree were most infected; in this case, individuals with high centrality were acting as sinks (receivers) of infection. Macintosh et al. (2012) used a similar approach to show that the out-degree of heavily infected female Japanese macaques was positively associated with host infectiousness; in this case, the most connected were the most infectious, and individuals with high centrality acted as sources of infection.

Calculation of network centrality offers a way to identify which individuals or groups within an at-risk population may be more likely to be involved in disease transmission (Corner et al. 2003), particularly when weighted data are used (Drewe et al. 2009). Definitions and formulae for network statistical terms can be found in Wasserman and Faust (1994). Also see Dube et al. (2011) for an explanation of network terms that may be useful when investigating disease transmission in social networks.

Of the centrality measures, degree centrality appears to be particularly useful in wild animal network analysis, especially where data may be incomplete, for several reasons. First, degree centrality is the simplest and most readily measured centrality measurement. Second, degree centrality has been found to be at least as good as other network parameters in predicting flow of information (in this case, risk of infection) when network analysis was used to identify high-risk individuals in a simulation study of disease spread (Christley et al. 2005). Third, it is far less sensitive to error than, for example, betweenness centrality—a measure of how often a node lies in the shortest path between all other pairs of nodes and thus how much control a node has on flow of information through the network—which can change dramatically if only one or two observations are missing or were wrongly recorded (Krause et al. 2007).

This can be illustrated by a case study in which two different networks were simultaneously constructed from the same population: contacts between many individuals were quantified at a coarse temporal scale (capture–mark–recapture), and a subset of this population was followed at a fine temporal scale (radio tracking) to produce networks using the same criteria (Perkins et al. 2009). The two networks had matching degree distributions, reflecting congruence in degree centrality, but estimates of betweenness differed between the two networks, a function of contacts 'missed' by using the different data collection methods. This study highlights how dramatically network statistics can change if data are missed—an important consideration when thinking also about sample size (see 'How many host–parasite associations should be included in a transmission network?').

Despite issues in sample size, betweenness could, in theory, provide insight into infectious disease dynamics. Conceptually, betweenness measures the flow of a parasite through the network, and an individual with high betweenness may link disparate parts of a network and so may be a 'hub' for rapidly spreading a parasite or could be targeted as a 'firebreak' for disease control. Betweenness, therefore, may be a useful statistic to apply to closed populations (for example, captive animals) where all individuals can be included in the network and, in theory, all contacts captured. In a captive population of brushtail possums, infected animals were more likely to have a high betweenness value than uninfected possums were, suggesting that this network statistic could help identify important individuals in parasite persistence (Corner et al. 2003).

While centrality measurements tell us something about the role of individuals, the spread of disease within localized areas of social networks can be explored by examining host clustering properties. Clustering coefficients measure the extent to which two neighbours of a focal animal are themselves neighbours. High clustering coefficients indicate that, on average, focal individuals are surrounded by other individuals that are well connected to each other (Wasserman and Faust 1994). The impact of this on disease transmission was investigated by Croft, Edenbrow, et al. (2011), who found that network clustering declined in guppies (*Poecilia reticulata*) following the introduction of an individual infected with gyrodactylid ectoparasites. In the control group, where no infection was present, the clustering coefficient showed a concomitant increase. This illustrates that presence of disease can alter host behaviour in a manner that may be protective, because a reduction in the amount of contact between hosts is likely to reduce the chances of transmission (Croft, Edenbrow, et al. 2011).

Relationships between network measures and host attributes

Examining for relationships between network measures and host attributes may reveal further implications for parasite transmission. For example, networks of social interactions between wild meerkats became less dense as group size increased, suggesting that individuals were limited in the number of interactions in which they could participate (Drewe et al. 2011). A simultaneous increase in network clustering (for both grooming and aggressive interactions—both known to be related to parasite transmission) implied that, as group size increased, individuals became more likely to interact locally with a subset of others rather than trying to maintain group-wide interactions. Clustering has been shown to be the dominant effect controlling the growth rate of epidemics (Keeling 1999), with increased clustering reducing R_0, the basic reproductive ratio (Cross et al. 2007). This has implications for the transmission of infectious diseases such as TB within meerkat networks, because such diseases may spread locally within clusters of interacting individuals but be limited from infecting all members of large groups by an apparent threshold in connections between clusters (Drewe et al. 2011).

Disease transmission network dynamics

Host and parasite-driven parameters in transmission networks

Host attributes may play a role in network structure and disease dynamics. One factor altering mixing patterns in social animals is behavioural heterogeneity, which is likely to be a function of an individuals' position in a dominance hierarchy (Ezenwa 2004).

Host attributes of size, sex, and age have consistently been shown to alter parasite loads (Enoksson 1988); for example, both the prevalence and intensity of parasitic infection is commonly observed to be male biased in mammals (Enoksson 1988; Schalk and Forbes 1997; Moore and Wilson 2002) and birds (Poulin 1996). These patterns have been observed on transmission networks where large males have been found to be most infected and connected (e.g. Clay et al. 2009; Zohdy et al. 2012).

From a parasite perspective, some species—for example, the rabies virus—can lead to an increased sociability of their hosts in an effort to pass on infection (Hampson et al. 2009). Mice infected with *Toxoplasma gondii* lose their neophobic behaviour towards cats—the definitive host for this trophically transmitted parasite (Webster 2007)—and so increase their chance of being eaten and passing on the infection. In addition to parasite-induced behavioural changes, it is increasingly recognized that animals have 'personalities'; a behavioural phenotype which is consistent and predictable over time and/or across contextually different situations (Reale et al. 2007). Pre-existing personality differences in animals have been found to influence individual susceptibility to infection, where the most active individuals were found to be least infected (Koprivnikar et al. 2011). This is an important and growing research area to be explored in the context of the interplay between personality, parasitism, and transmission networks (see Chapter 6).

Effects of infection on networks

Animal behaviours are inherently dynamic: they switch according to the ecological and biological pressures exerted on the system, altering susceptibility to infection and evolving behavioural strategies to avoid parasitism (Hughes, Border, et al. 2012). Although immune responses are mounted by individual members of a population, there is evidence to suggest that others may also gain benefit; termed social immunity (Cotter and Kilner 2010). Sick behaviours, such as reduction of activity and social interactions, can result in a form of social immunity via a reduction in contact rates that subsequently reduces transmission of an infection through the population (Lopes et al. 2012; Ugelvig and Cremer 2012). In a social network context, personal immunity may contribute towards herd immunity by preventing or reducing transmission via reduced contacts. This is expected to be the case if the social network structure is altered in response to infection. Additionally, infectious individuals may be actively avoided by susceptible individuals also leading to a reduction in contact frequency (Hughes, Border, et al. 2012).

There is substantial evidence to show that parasitism may result in behavioural avoidance—especially in the mate choice literature, where uninfected females have been observed to be preferential in selecting parasite-free or resistant males (e.g. Hamilton and Zuk 1982; Kennedy et al. 1987; Borgia and Collis 1989; Hillgarth 1990; Houde and Torio 1992; Wedekind 1992). Detailed studies using laboratory mice (*Mus domesticus*), have determined that females can discriminate between infected and uninfected males on the basis of odour and display a reduced interest in the urine of male mice infected with a range of parasites: influenza virus, the protozoan (coccidia) *Eimeria vermiformis*, and the common nematode parasite, *Heligmosomoides polygyrus* (Kavaliers and Colwell 1995; Kavaliers et al. 1997, 1998; Penn et al. 1998; Ehman and Scott 2001; Kavaliers et al. 2006). During an epidemic, therefore, we may expect either one of these behavioural responses (self-isolation or avoidance) to occur and be apparent in the social network geometry. To date, however, the effect of infection on network topology has not been examined empirically.

A fascinating insight into the role of pathogen-induced changes in network topology was revealed in a theoretical model by Ferrari et al. (2006). The authors performed stochastic simulations of the spread of disease on network structures commonly found in animal populations, by investigating the effects of the loss of individuals (and their edges) in the population through random immunization. Ferrari and colleagues found that, as an epidemic moves through a population, the degree distribution of the population changes from one of a negative binomial—where most individuals have few contacts and a minority have high number of contacts—to that of a normal distribution. During the simulated epidemic, the pathogen effectively 'wipes out' the most connected individuals (potentially the superspreaders), and the host population self-protects itself from further pathogen attack. It should be noted that this model stimulates loss of edges in the population as nodes become immune, but in reality the change of an individuals' status from infected to immune may result in behavioural changes that will rewire the network.

Tantalizing evidence that this might be the case come from 'natural experiments' involving the European badger and bovine TB. In the UK, badgers are a wildlife reservoir for TB and have been the subject of control programmes in the UK since the 1960s in an effort to reduce transmission of this pathogen to domestic cattle. Badgers have defined social groups, and culling has been found to alter the pattern of social contacts of badgers that remain in and around the cull area. Such badgers subsequently appear to have more frequent and wider-ranging movements that potentially lead to increased contact rates between infectious and susceptible individuals and hence increased disease transmission (Carter et al. 2007; Vicente et al. 2007). Further work on how the dynamics of the network itself change with respect to disease in response to perturbation—whether that be the disease itself or an intervention such as selective removal, vaccination, or natural immunity of nodes—is much needed.

Disease transmission in animal social networks: conclusions and future directions

In this chapter, we have highlighted some of the challenges and evaluated possible solutions associated with the use of transmission networks to study infectious disease processes. A particularly important consideration is the matter of which data collection methods are best to use (Table 10.1), an issue that is closely linked to that of establishing how well sampled the network is. Both nodes and edges may be under-sampled if sufficient knowledge of both host and parasite are not incorporated into the network methods.

Researchers should have ecological knowledge of their host–parasite system, along with behavioural knowledge of the host, and epidemiological insight into the parasite. By only looking at one factor in isolation, it may be that we ignore the role of others. For example, we might reasonably hypothesize that the most socially dominant individuals will be most important in perpetuating infection. However, observations of contact networks may yield surprising results, and the most socially dominant many not necessary be most infected (Drewe 2010). In addition, it should be appreciated that there is a duality in the host–parasite association such that perturbation of a systems (e.g. disease treatment) may cause the network to rewire.

Much of the work in this subject area to date has been carried out at the population level, but infectious disease processes operate across a range of scales, from within-host processes—for example interacting with host immunity—to links between populations and multi host–parasite systems (e.g. Craft et al. 2011). Network use to examine disease transmission has started to address some of these cross-scale issues. Bogich et al. (2013) have examined spillover of emerging zoonoses to predict emergence by classifying disease outbreak reports as nodes, and determining edges to be shared symptoms, seasonality, or case fatality—anything that might suggest a linkage. The method's success was rather parasite dependent but worked especially well for tracing outbreaks of Nipah virus, a zoonotic pathogen highly virulent to humans.

A similar approach has been successfully used by Teunis et al. (2013), who connected disease outbreaks using probability theory to carry out contact tracing for norovirus in humans. Within-host dynamics also offer fruitful opportunities for network theory, especially when combined with immunity studies, due to the complex nature of immune interactions. This has been addressed by Thakar et al. (2012), who used network model simulations to demonstrate how host immunity is altered by single and co-infection.

Other exciting areas of development that may reveal further insights into the epidemiology of infectious diseases are reviewed by Godfrey (2013). These include better incorporation of the dynamic aspects of the transmission process into network analysis; investigations into the effects of variations in ecological conditions on network properties and behaviour; understanding how animal disease networks respond to anthropogenic perturbations; and progressively increasing the complexity of the processes being modelled to include variations in individual infectiousness, co-infection with multiple pathogens, and immune-competence (Godfrey 2013). Such developments would be facilitated by the careful recording of detailed data—on behaviour, life history, infection status, environmental conditions, and other relevant parameters—of which the accurate collection of good quality and sufficient data from wild animals and ecosystems is likely to remain one of the biggest challenges.

Finally, it should be appreciated that social network analysis may not be the most appropriate method for helping to understand all disease processes in every situation. Our aim in this chapter has been to provide many examples of where it can be useful, how it may contribute to our knowledge of epidemiology, and to highlight some of the potential pitfalls that are commonly encountered in its use. The choice of whether or not to use social network analysis in the context of infectious disease epidemiology should be based on a careful consideration of the various factors reviewed in this chapter.

CHAPTER 11

Social networks and animal welfare

Brianne A. Beisner and Brenda McCowan

Introduction to the use of social network analysis in animal welfare

The use of social network analysis for the detection and description of the patterns and quality of interactions among animals has significant potential to improve animal welfare. Animals in a variety of managed settings face similar welfare issues, including injury, disease, and social stress, and social network analysis is uniquely suited for detecting patterns of interaction associated with these, as well as individual risk given observed patterns (Makagon et al. 2012). In fact, most species of food production animals (e.g. chickens (*Gallus domesticus*), cows (*Bos taurus*), sheep (*Ovis aries*), pigs (*Sus scrofa*)) and laboratory animals (e.g. mice (*Mus musculus*), rats (*Rattus norvegicus*), and non-human primates (e.g. *Macaca* spp.)) are social animals and typically housed in groups, making social network analysis particularly applicable.

Social network analysis involves the quantification of patterns of relationships (both direct and indirect), associations, or interactions among entities. A primary advantage of social network analysis over other analytical methods is the ability to quantify indirect relationships or associations, which allows for the detection of emergent network patterns and structures (as opposed to simple group-level rates of behaviours) as well as the identification of the different social roles in the group. As such, social network analysis is particularly useful for those living in moderate to large groups, where traditional focus on direct interactions yields limited information.

The detection of specific network patterns/structures and social roles can improve animal welfare through the identification of (1) patterns/structure which increase the risk of poor welfare, such as risk of infection; (2) individuals who either contribute to the maintenance of good welfare or facilitate the appearance or continuance of poor welfare; (3) the social dynamics underlying good versus poor welfare, such as structural features of group stability; and (4) avenues for improving welfare, such as social buffering of tension or stress, or prevention of poor welfare, such as node-targeted disease prevention. For example, at the group level, network measures can be used to track changes in social network structures over time (Blonder et al. 2012), evaluate the effect of the environment on social network structure, or compare social structures across groups, populations, or species. At the individual level, social network measures allow the heterogeneity of social experience within groups to be quantified. Furthermore, network measures can be used to test hypotheses about the relationship between an individual's social role and social stability or information flow throughout the network (Flack et al. 2006; Lusseau 2007a; Beisner et al. 2011b; McCowan et al. 2011).

Managed populations face a variety of welfare issues. Social network analysis is suited to address issues of a social nature, such as social aggression (and the pain/injury which stems from it), social stress and distress, and social transmission of disease. In addition, social network analysis may also be useful for addressing non-social issues, such as detecting illness, which can change an individual's social interactions or associations, or disease

Animal Social Networks. Edited by Jens Krause, Richard James, Daniel W. Franks and Darren P. Croft.

transmission that occurs via the use of common sites (rather than social transmission).

For the sake of simplicity, we divide animal welfare issues into two broad categories: (1) physical well-being and (2) psychological and social well-being, although this division in no way implies an actual separation between these types of welfare. In fact, it is well known that poor psychological well-being is often associated with poor physical health, and vice versa. The complex interrelationship between physical and psychological welfare goes beyond their potential influence on each other. The interrelationship between psychological and physical health stems from gene–environmental interactions as well as social behaviour and network structure. Let's consider, for example, how social aggression influences animal welfare.

First, gene × environmental interactions affect social behaviour and network structure (Figure 11.1): personality/temperament influences whether high-ranking individuals perform critical conflict management behaviours in macaques, which maintain network cohesion and stability by preventing the escalation of social aggression (McCowan et al. 2011), while group composition (i.e. sex ratio, number of high-ranking juvenile males, etc.) influences overall rates of social aggression as high-ranking juvenile males initiate a disproportionate amount of severe aggression (Beisner et al. 2011a). Second, network structure influences welfare: when key conflict managers are removed from the network, social aggression increases, and social niches contract and become more conservative (Flack et al. 2006), which simultaneously increases the likelihood of receiving severe social aggression and physical injury, and constructs a small, cohesive affiliative support system for coping with the social aggression. Not only that, but gene × environmental interactions can have effects here as well—personality will affect

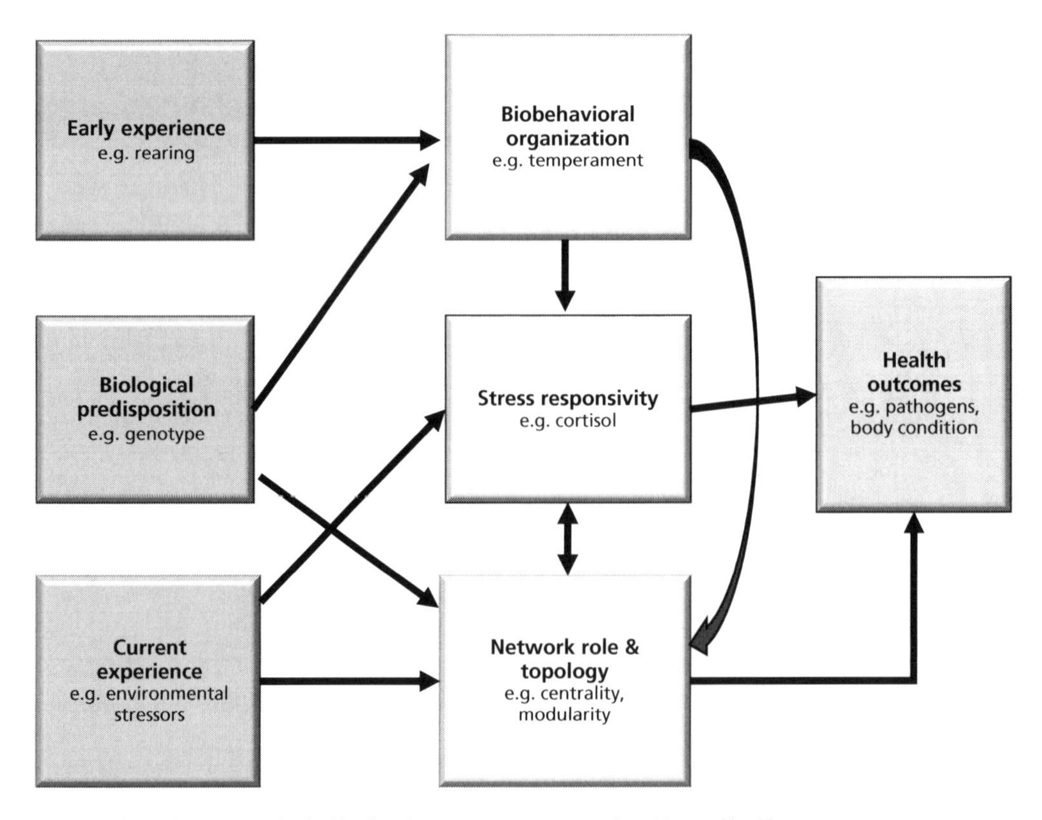

Figure 11.1 Interrelationships among individual-level attributes, environment, network position, and health outcomes.

how well (or how poorly) an individual responds to social aggression and stress.

Similar relationships between temperament, group composition, and social network structure and stability are likely present in many social species, as are other gene × environmental interactions affecting social behaviour and network structure. Thus, a multitude of factors interact across the level of the individual, the community, and the group to describe how social aggression affects welfare. In this chapter, we outline a number of social network measures and analytical approaches that can be used to address animal welfare issues, including examples of how social network analysis has already been used to investigate animal welfare as well as suggestions for future applications.

Physical health in animal social networks

Disease transmission is perhaps the most important social aspect of physical health. Infectious diseases represent a major threat to animal health and welfare (Thorne and Williams 2005; Tomley and Shirley 2009). For example, the US Department of Agriculture reported that in 2010 over 1.7 million cattle (1.85% of the 2010 population) in the United States died from respiratory (60.6%), digestive (29.0%), or other diseases (10.3%), accounting for 43.6% of all cattle deaths in the United States (National Agricultural Statistics Service, 2011). In addition, zoonotic disease transmission at the human–wildlife interface is also a growing concern. In rural Africa, for example, herds of livestock come into close proximity to herds of other livestock and wild ungulates at shared grazing and watering locations. Recent network analyses show that social proximity poses a greater threat of disease transmission than common use of water locations or sharing of home ranges (Waret-Szkuta et al. 2011; VanderWaal et al. 2014a,b). Across south and south-east Asia, free-ranging macaques live alongside humans in urban areas, and the prevalence of diseases such as TB in the macaque populations appears to match the prevalence in local human populations (A. Wilbur et al. 2012), suggesting zoonotic transmission. Social network analysis is well suited to address these welfare issues.

Disease transmission in animal networks

The transmission of infectious disease is a problem for all types of managed animal populations, including farm animals, laboratory animals, zoo animals, and wildlife. Infectious diseases present not only a welfare problem to the animals, but often also a management/production problem. Sick animals have higher rates of morbidity and mortality, which reduces production output in agricultural industries (Hassall et al. 1993; Radostits et al. 1994), reduces the number of study subjects as well as the quality and external validity of data in research facilities (National Research Council 2011), and may present a bad public image for zoos and sanctuaries.

Social network analysis can be used to investigate multiple aspects of disease transmission, such as calculating risk of infection based upon individual attributes, predicting the potential size of a disease epidemic and its rate of movement through a population, and identifying optimal target nodes (individuals or locations) for preventing or reducing the spread of disease within an animal population. We begin with calculation of individual-level risk of infection using node-based measures (see Chapter 10 in this volume for additional discussion regarding networks and disease).

Although many models assume homogenous interaction among the agents of a system, this is an inaccurate assumption for animal (or human) social interactions. Some individuals in a population are socially well connected and interact with others in a variety of ways (proximity, affiliation, aggression) while other individuals are somewhat socially isolated and interact with few others and at infrequent intervals. For example, in primate society, high-ranking males are typically well connected—they receive lots of grooming and initiate more fights than subordinates—whereas low-ranking males are less well connected. This variation in social connectedness, as well as variation in interaction strength and direction, can all be represented and analysed using node-based network measures.

Both theoretical and empirical evidence indicate that network centrality is associated with infection risk. Although most of this evidence comes from modelled systems and simulations, there is an increasing body of empirical evidence from real populations. Simulations of disease transmission show that the most central individuals in the network (those with high degree centrality and high betweenness centrality) have the highest risk of infection (Christley et al. 2005). Empirical studies of disease transmission confirm the simulation results. An experimental study of *Mycobacterium bovis* transmission in brushtail possums (*Trichosurus vulpecula*) (group size range 23—29) showed that experimental infection of highly connected animals resulted in 30–63% transmission to other group members as compared to 9% transmission under random selection for initial infection (Corner et al. 2002, 2003). Furthermore, secondary infection was more likely among group members with higher closeness and flow betweenness values (Corner et al. 2003).

Identification of more central individuals in populations may be used to inform surveillance and infection control strategies. However, it is not entirely clear which measures of connectedness or centrality are the best indicators of infection risk. In simulations, degree centrality was as good as other centrality measures which required more extensive calculation (and knowledge of the entire network), such as betweenness and farness, at predicting infection risk (Christley et al. 2005). It is important to keep in mind, however, that the importance of centrality measures may depend on disease biology—viruses that are only transmitted during an active shedding phase, or parasites that must complete a phase of their life cycle before attaching to the next individual may not follow this pattern, as centrality during specific time periods may be irrelevant. Regardless, the importance of network centrality has significant implications for monitoring the transmission of many diseases.

If degree centrality is sufficient for estimating risk of infection and time to infection, then all one needs to do is observe individual behaviour—degree distributions can be plotted by randomly sampling individuals in the population. Assessment of the entire network to obtain other network measures (e.g. betweenness) would then be unnecessary. However, in simulated networks, degree is often positively correlated with betweenness and farness (Christley et al. 2005), suggesting that degree centrality is only useful in predicting risk of infection in networks where these centrality measures are correlated, and many real populations may not be structured in that way. Empirical evidence from livestock movement networks in the UK (in connection with the 2001 foot and mouth disease epidemic) indicate that, at least, simple degree distribution is not sufficient and that direction must also be investigated (Kiss et al. 2006). In this study, livestock-holding premises (such as farms and markets) were the nodes, and edges were drawn if sheep were shipped between these premises. They found that a high correlation between in-degree and out-degree facilitated disease transmission (Kiss et al. 2006)—this was because those premises (nodes) that had livestock coming in frequently and shipping out frequently had the greatest potential contact between different animals, whereas premises that served only as shipping sources (high out-degree, no in-degree) did not.

The reason that more central individuals are at higher risk of infection may be connected to individual attributes if increased interaction is associated with sex, age, rank, or another individual attribute. This is exemplified in a study of the transmission of *Cryptosporidium* in wild populations of ground squirrels (*Spermophilus beldingi*) (VanderWaal et al. 2013): juvenile males, due to their greater exploratory behaviour, are more at risk of becoming infected with *Cryptosporidium*. Furthermore, the behaviour of juvenile males alters the structure of the network, which places other individuals at higher risk of infection due to the decrease in clustering that occurs when more juvenile males are present in the population (Figure 11.2). Thus, targeted surveillance and infection control strategies could be directed towards a single age–sex class of individuals, thus simplifying intervention strategies aiming at preventing or abating further disease transmission.

Network structure is equally useful for investigating disease transmission and provides information about global patterns that node-based measures cannot. In simulations, for example, network structure (e.g. modularity) has been found to be a predictor of the rate and extent of disease transmission in a population and might therefore be useful to monitor the spread of disease and predict the size of an epidemic (Salathe and Jones 2010). First, we describe the influence of clustering and modularity on disease transmission.

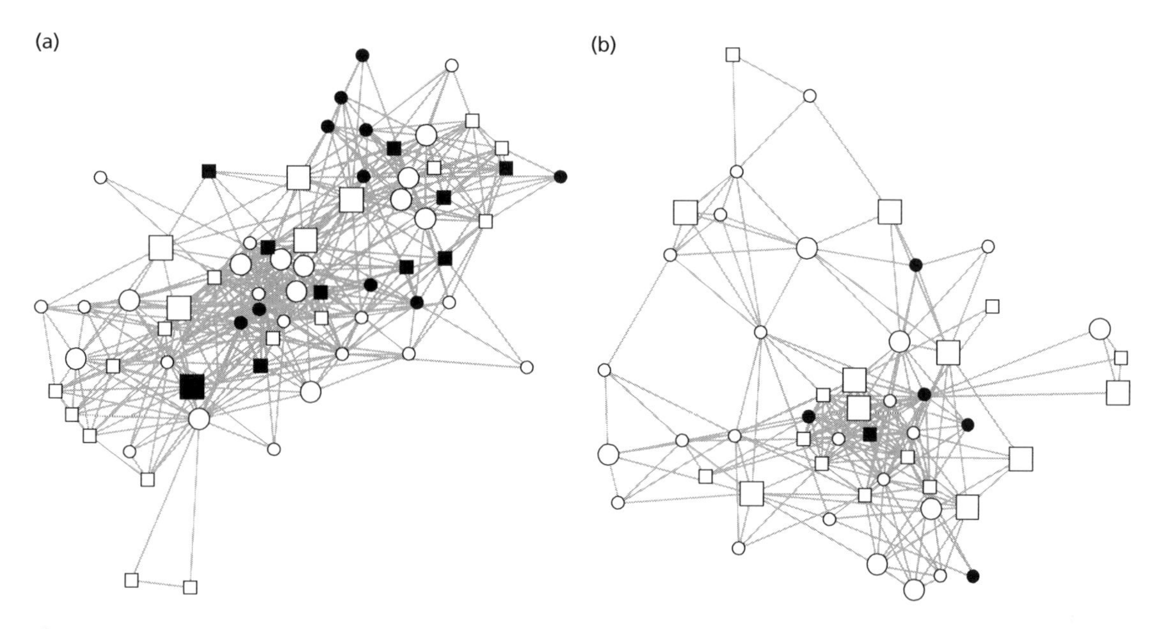

Figure 11.2 Contact network of a wild population of ground squirrels for (a) the first two-week period of observation ($n = 51$) and (b) the second two-week period of observation ($n = 62$). Squares and circles represent males and females, respectively. Large and small nodes denote adults and juveniles, respectively. White nodes are those that were infected during the contact period. Reproduced with kind permission from Springer Science and Business Media.

Clustering and modularity are network measures which describe a high degree of local connectedness. In the ground squirrel network described above, juvenile male ground squirrels facilitated the transmission of *Cryptosporidium* throughout the population because their high levels of exploratory behaviour (which brought them into greater contact with other squirrels) changed the overall network structure (VanderWaal et al. 2013). More specifically, exploratory behaviour by juvenile males reduced network-wide clustering, which facilitated the spread of *Cryptosporidium*. Similarly, community structure, or network modularity, is also associated with disease transmission. Modules are local subsets of highly connected nodes which are not well connected to nodes outside the module. A multi-species study of non-human primates has shown that populations with high levels of modularity have lower richness of socially transmitted parasites (Griffin and Nunn 2012). The biological reason behind this relationship between network structure and disease transmission is simple: socially transmitted diseases are less likely to become established in highly modular networks because infections tend to spread quickly within modules but die out before spreading to other modules (Salathe and Jones 2010).

The influence of network modularity on disease transmission suggests that managers of captive animal populations might be able to reduce disease transmission and prevalence if they artificially introduce modularity into the social network. This might be especially possible for housing and transport of livestock, as these domesticated animals typically live as social aggregates rather than socially dependent and cohesive units like many wild social species, such as primates and canids, kept in laboratory facilities, zoos, or sanctuaries. For example, several studies of the outbreak of foot and mouth disease in the UK have used livestock movement networks to understand how the pattern of movements between farms, markets, and other holding facilities contributed to the transmission of the disease (Kao et al. 2006; Kiss et al. 2006). Connections between livestock premises influenced the spread of disease. While the precise patterns associated with greater disease transmission are important, also important is the fact that shipping between premises can easily be changed to create a modular livestock movement network which reduces the transmission of disease.

A second pattern in network structure that influences disease transmission is the presence of a strongly connected component (SCC). In an SCC, all individuals in the component are reachable from all others and are connected in a cyclic manner (Sedgewick 2002). A network may have a single SCC, multiple SCCs, or perhaps none. A directed network may also have a weakly connected component (WCC) if, upon replacing directed edges with undirected edges, all nodes in the component become reachable from all others (Sedgewick 2002). The size of the largest SCC and the largest WCC in a network are estimates of the lower and upper bounds of the maximum epidemic size (Kao et al. 2006), because disease transmission is facilitated by the high degree of connectedness among nodes in these connected components. As with network modularity, increased connectivity among individuals facilitates the spread of disease, and transmission beyond the SCC or WCC into the periphery is much less likely due to their lack of connectedness.

Psychological and social health in animal networks

Social network analysis can also be used for promoting social and psychological well-being in social animals. Network structure may facilitate the maintenance of cohesive, stable patterns of social relationships within a group, as well as reduce the occurrence of negative social welfare such as deleterious social aggression or social stress. Although promotion of positive social and psychological welfare is a primary goal of animal managers, this may often be addressed as a reduction of negative welfare.

Deleterious social aggression, for example, is a real concern in breeding colonies at multiple national primate research centres, where approximately 20% of the rhesus macaque (*Macaca mulatta*) breeding population is hospitalized for trauma resulting from social aggression (McCowan unpublished data). Injuries from social aggression are also common in groups of agricultural animals, including pigs (Schmolke et al. 2004), chickens (Estevez et al. 2003), and horses (Hartmann et al. 2009) Social aggression is a natural part of social interaction and therefore cannot be eliminated. However, physical injury from serious social aggression can be kept to a minimum to maintain good welfare.

The negative impacts of social stress are also of concern. Chronic stress is associated with a wide variety of negative health impacts, including metabolic disorders (Tamashiro et al. 2011), gastrointestinal problems such as diarrhoea (Julio-Pieper et al. 2012), and impaired immune function (Cole et al. 2009). In fact, an animal's position in the social hierarchy is one predictor of social stress (Sapolsky 2005), suggesting that social network analysis is indeed appropriate for investigating the underlying causes of chronic stress as well as uncovering social coping strategies. Here, we outline several social network measures and approaches for both promoting positive social welfare and reducing negative welfare, such as identifying individual risk of poor social welfare, key individuals whose behaviour either reduces or exacerbates group-wide social aggression, group stability, or social stress, and network patterns and structure associated with decreased rates of social aggression or stress, such as social buffering via close-knit affiliative clustering. We begin with social aggression.

Social aggression in animal social networks

Social aggression arises in groups of animals for a variety of reasons. Aggressive interactions may simply be an intrinsic part of normal social dynamics. Aggression may arise when group membership changes, such as during regrouping of agricultural animals (Schmolke et al. 2004; Von Keyserlingk et al. 2008) or after removal of key conflict managers (Flack et al. 2005b). Variance in group size and density (Erwin and Erwin 1976; Sosnowka-Czajka et al. 2007), as well as the size and distribution of resources (Isbell et al. 1999), have also been found to influence aggressive interactions. Nearly all social animals use aggressive behaviour to establish dominance relationships, including domesticated animals such as pigs (Ewbank et al. 1974), cows (Beilharz and Mylrea 1963; Arave and Albright 1976), and chickens (Pagel and Dawkins 1997), laboratory animals such as non-human primates

(Kawamura 1958; Sade 1967), and zoo animals such as ungulates (Wirtu et al. 2004). Since dominance hierarchies develop to prevent unnecessary and costly fighting, investigation of hierarchical structure via social network measures is a logical step towards maintenance of healthy competitive relationships (see Chapter 7 for further discussion about networks and dominance).

Cohesion in the dominance network is likely to be critical, because group stability is maintained when a large proportion of the group is in agreement over the social hierarchy (McCowan et al. 2008; Oates-O'Brien et al. 2010). Dominance cohesion can be measured in a number of ways, depending upon the precise behavioural structure one is looking for. Fragmentation, a measure of the proportion of mutually reachable nodes as each node is unconnected/removed from the network (Borgatti 2003), is a good way of estimating redundancy in dominance interactions. A dominance network in which multiple pathways exist between any two nodes is both well connected and highly redundant, meaning that dominance information is communicated frequently and suggesting that most dominance relationships are clear and settled. Indeed, we find that less fragmentation in rhesus displacement networks (approach–avoid interactions) is associated with less contact aggression and wounding (McCowan et al. 2008).

Transitivity is another good measure of cohesion and stability in a dominance network. A high degree of transitivity in a dominance network indicates that most dominance relationships fall into a linear pattern and few relationships run contrary to that linear pattern (Shev et al. 2012). Thus, networks with low transitivity may have higher levels of deleterious social aggression due to a large proportion of group members actively competing over rank, whereas networks with high transitivity may have lower levels of social aggression (this topic is treated in more detail in Chapter 7 of this volume).

Centrality measures, such as aggression out-degree, in-degree, and betweenness, can be used to identify instigators of aggression, individuals most at risk of physical injury from social aggression, or individuals whose involvement in aggression may spread to encompass more individuals. For example, network centrality was important in understanding how feed restriction influenced social structure and fin biting in Atlantic salmon (*Salmo salar*) (Cañon Jones et al. 2010). In feed-restricted groups, fin damage was associated with aggression and fin biting, and social network analysis showed that key instigators and targets of aggression were identified by high out-degree and high in-degree, respectively (Cañon Jones et al. 2010). This study highlights how network measures may be especially useful when comparing before versus after certain management procedures (e.g. food restriction, such as during transport) to evaluate the impact of management actions on the social dynamics and social aggression within the social group (Blonder et al. 2012).

When trying to reduce the levels of intrinsic aggression that is not caused by management practices, identification of aggression targets does not necessarily solve the problem. Removal of aggression targets is likely to be unsuccessful, as former targets of aggression will quickly be replaced by new targets in the network. A better way to reduce the risk of physical injury from social aggression is to use network analysis to identify evolved mechanisms of conflict management or social coping. Below we discuss network patterns associated with conflict management, social aggression, and wounding in non-human primates and discuss applications to other species.

Intervention behaviour to control conflict, also known as policing, appears to be an important robustness mechanism in primate social groups (McCowan et al. 2011; Von Rohr et al. 2012), and policing individuals are best identified using node-based network measures. Subordination in-degree and strength best characterize individuals with policing ability (Flack et al. 2005a; McCowan et al. 2008). Subordination is a peaceful signal (silent-bared-teeth display) given by a subordinate to a dominant (Flack and de Waal 2007), and individuals with the highest in-degree and strength have the highest group consensus of their social power, thereby allowing these policers to intervene successfully and at relatively low cost. Not only are node-level measures of these subordination signals important for identifying key policers, but also the transitivity and complexity of the power structure created by network-wide subordination signals appear be associated with group stability and outbreaks of deleterious aggression in captive macaque groups (Fushing et al. 2013). Since dominance relationships underlie the power structure found in subordination networks in macaques, it is likely that many animal societies possess an optimal dominance-driven power structure that maximizes group stability. Social network analysis is uniquely suited for detecting this sort of power structure (for a detailed discussion of the use of signals for constructing social networks, see Chapter 9).

The comparison of network structure from two different time points is a powerful analytical tool for detecting more complex changes in the society as a result of environmental or social change. Longitudinal network comparisons may be applicable to understanding the social stability and welfare consequences of a variety of environmental changes that commonly occur in managed settings, including changes in management procedure, changes in housing conditions, or social introductions. For example, a comparison of network structure before and during experimental removal of policers demonstrates exactly how these key individuals influence patterns of behaviour and group stability (see Blonder et al. 2012 for a discussion of methods for the analysis of temporal dynamics in social networks). Flack and colleagues (Flack et al. 2005b) conducted a temporary experimental knockout of key policers in a pig-tailed macaque (*Macaca nemestrina*) society. They then analysed several affiliation networks, including proximity, grooming, and social contact, for structural changes. Results show that removal of the top three policers destabilizes social niches—groups had lower mean degree (smaller ego networks), greater local clustering (and therefore less group-wide cohesion), and higher assortativity (lower nearest neighbour diversity) when policers were absent (Flack et al. 2006).

Similarly, the removal of key individuals from leaf-roosting bat networks affects group stability. Chaverri (2010) found that, unlike most forest-dwelling bats, which live in fission–fusion societies in which individual bats switch roosts regularly (Kerth 2008), Spix's disc-winged bats (*Thyroptera tricolor*) form stable communities (i.e. roosting groups) of individuals who associate more closely with each other than with individuals from other roosting groups. However, in one study group, emigration of key individuals caused a loss of clustering structure over time, suggesting that the stability of these social networks is susceptible to the removal of key individuals (Chaverri 2010). If, in a captive setting, such key individuals needed to be removed for reasons of individual health, regrouping, or other management procedure, social network analyses could be used to identify whether such individuals held key network roles that could impact group stability.

A social network need not represent the entire social group. In fact, network construction for particular subgroups within the social group can reveal important information that is inaccessible at a group network level. Kinship groups within a larger social group or age–sex class represent one such example. Kinship is an important organizing principle in many animal societies (Packer 1986; Rubenstein 1986; Emlen 1994), which suggests that cohesive relationships among kin may be important in maintaining group stability. Age–sex class is also often an important organizing principle, and in populations such as livestock, where management practices such as regrouping are based upon age–sex class, analysis of cohesion within these subgroups may also be important in achieving group stability.

Lower modularity (i.e. higher network-wide cohesion) in kin groom networks is associated with lower rates of severe aggression and wounding among kin in captive groups of rhesus macaques (Beisner et al. 2011b). Behaviourally, this suggests that more cohesive kin relationships reduces serious fighting among family members over relative ranks, as kin form alliances to establish and maintain both individual and family rank (Sade 1967). Interestingly, cohesion in kin grooming networks in these captive macaques was mediated by genetic relatedness—higher average genetic relatedness within a matriline was associated with fewer grooming modules and less fighting among kin (Beisner et al. 2011b). Importantly for welfare, genetic fragmentation can be managed by preventing the formation of genetic fragments (i.e. unconnected branches within the pedigree), and thereby minimize intra-familial fighting and risk of injury. Similarly, kinship structure promotes network cohesion in bottlenose dolphins, as network association strength is correlated with genetic relatedness (Wiszniewski et al. 2010), which suggests that preservation of kin structure may also improve welfare in dolphins.

The dynamics underlying social stability, or instability, are complex, and no single network can represent this complexity. We can consider social behaviour as a multidimensional network composed of many individual behavioural networks, such as spatial proximity and social contact. Joint modelling is a new powerful network analysis tool for detecting inter-behavioural patterns by jointly modelling multiple behavioural networks. Developed by our colleagues (Chan et al. 2013), joint modelling involves the empirical construction of multiple social networks, such as affiliation and aggression, that allow us to synthesize complex information and reveal collective behaviours which arise from several interconnected social realms. The basic idea behind jointly modelling two networks is to prescribe the probability of a link in one network being associated a link in the other network.

Joint modelling is an ideal analytical tool for perturbation experiments because the relational pattern between multiple networks from before and after the perturbation can be compared and thus reveal key features of how these behaviours are interdependent. For example, the global relationships among spatial association, groom, and aggression networks in baboons changed after the death of a high-ranking female, such that the joint entropy across the multiple networks decreased, indicating greater social conservatism as a coping strategy (Barrett et al. 2012). Joint modelling has also been used to detect group instability using data from a stable time period and an unstable period (identified as unstable post-hoc). Joint modelling of four behaviours (aggression, alliance, groom, and status) at these two time points shows that the status network becomes independent of the other behavioural networks during the unstable time point (Chan et al. 2013). Independence between behavioural networks means that, for example, knowing the nature of a dyad's aggressive relationship tells us nothing about the nature of their status signalling relationship. Joint modelling therefore identified a single behavioural network (status) whose loss of interdependence with other behaviours was associated with social instability, which suggests that behavioural dependence and interconnectedness may be an important feature of healthy social complexity.

The presence or absence of this interdependent nature of social complexity in any managed population can be monitored using social network analysis to compare networks from two different time points. The ability to detect emergent global patterns of relationships across multiple networks could be useful in for a variety of welfare issues. The direction of disease transmission (which would be network 1) could be jointly modelled with many other potential social behaviours (which would determine network 2) such as aggression, affiliative contact, proximity, or even kinship to find out if disease transmission overlaps with agonistic or friendly relationships, with spatial proximity, or with family relationships. The interrelationship between grooming and aggression networks could be used to determine whether affiliative relationships buffer individuals' physiological or behavioural response to social aggression.

Social stress and health in animal social networks

Chronic stress can cause a variety of health problems, including gastrointestinal issues (like diarrhoea) (Julio-Pieper et al. 2012), metabolic disorders (Tamashiro et al. 2011), and impaired immune function (Cole et al. 2009). Chronic stress that is socially based can be investigated using social network analysis. Like social aggression, the emergence of chronic social stress and stress-related health problems is highly complex, indicating that social network analysis may be more useful than other analytical methods for identifying the underlying social causes of these health problems.

Network analysis can be used to identify which individuals are experiencing the greatest amount of stress and why, as well as reveal avenues for ameliorating stress. Node centrality measures can be compared with biological markers of stress, such as glucocorticoid levels, chronic diarrhoea, hair loss, or weight loss, to test, for example, whether individuals in particular social positions are also more likely to suffer from or be buffered against chronic stress and its associated health problems. The network structure and measures predictive of severe social aggression (discussed in 'Social aggression in animal social networks') are likely to also be predictive of social stress, as groups with higher levels of severe aggression and wounding are likely to be more stressful to live in. For example, in baboons (*Papio anubis*) (Sapolsky and Share 2004), being subordinate is more stressful when the highest ranking males are more aggressive in nature. Bonacich power may also be a useful measure to test whether socially isolated individuals (e.g. those on the periphery of an affiliation network) are more likely to suffer from chronic stress than well-integrated individuals, as social isolation is likely to have a negative impact on the health of highly social species (Bonacich 1987).

Another potential source of stress for managed animal populations is regrouping (Von Keyserlingk et al. 2008; Correa et al. 2010). Individual members of livestock groups (e.g. cows) are regrouped with new, unfamiliar individuals for a variety of purposes, such as regrouping according to age, nutrient requirements, and lactation period to enhance productivity (Raussi et al. 2005). Regrouping is also common in zoo settings, where individuals from one zoo are often shipped to another zoo to be combined with existing group members. Regrouping can cause social stress (Correa et al. 2010), as individuals are not only cut off from their previous social ties but also are required to establish new social ties (e.g. new dominance hierarchy) (Schmolke et al. 2004). Reduction of social stress can improve productivity in agricultural animals (e.g. milk production (Von Keyserlingk et al. 2008)) and may improve reproductive capacity in socially housed zoo animals whose reproduction is impaired by chronic stress (Rivier and Rivest 1991). Maintenance of reproductive capacity is particularly important for vulnerable/endangered species in zoos, where loss of reproductive opportunity represents a setback in conservation effort. Measures of subgrouping, such as modularity and clustering coefficient, may be quite useful for identifying cohesive affiliative subgroups of individuals that can be moved together, which may reduce the stress of these management processes and permit social coping via affiliative ties.

A related topic to regrouping is simply the decision of what age or sex classes of individuals to house together. Some zoo animals, such as bats, may be housed in single-sex groups (Ingrid Russell-White, San Francisco Zoo, personal communication), and their social structure could affect the success of captive housing decisions. For example, females are highly central in social networks of big-eared bats (*Corynorhinus rafinesquii*), whereas males consistently hold the least central positions, most likely because males benefit from less social roosting (J. Johnson et al. 2012). This suggests that female-only groups may be more successful that male-only groups, particularly if forced socialization is stressful on males. Social segregation by sex, and therefore network segregation by sex, is common in many species (Ruckstuhl and Neuhaus 2005), and this may influence how these animals are housed and managed. For example, clear segregation by sex has been found in the networks of spider monkeys (*Ateles geoffroyi*) (Ramos-Fernandez et al. 2009).

Previous work in non-human primates suggests that network modularity and clustering are relevant to understanding social stress. When key policers were temporarily removed from a pig-tailed macaque group (Flack et al. 2005b), the mean clustering coefficient in the proximity network increased, indicating that individuals became more socially conservative in order to cope with this stressful situation. Similarly, when an alpha male was ousted from a baboon troop by a new alpha male, the females contracted their grooming circles (i.e. groom degree centrality decreased) and those who already had a small number of groom partners had the smallest rise in glucocorticoids (Wittig et al. 2008).

Grooming is a coping mechanism for dealing with social stress, as social grooming reduces social tension by lowering heart rate and cortisol level in participants (Shutt et al. 2007). Therefore, these results indicate that coping mechanisms may need to be studied in conjunction with aggression networks in order to predict which individuals are most likely to experience negative health effects from social stress. As mentioned in 'Social aggression in animal social networks', joint modelling may be particularly appropriate for understanding the joint relationship between social aggression (network 1—the potential cause of social stress) and grooming (network 2—the potential coping mechanism for social stress), which will allow us to better understand the emergence of stress-related health issues. Joint modelling of groom and aggression networks can address whether dyads with two-way aggression (a sign of active competition for rank, which may be stressful) also show grooming more often than dyads with one-way aggression, as a coping mechanism.

Social network analysis in animal welfare: conclusions and future directions

Social interactions underlie many of the most complex aspects of animal health and welfare. There is a growing realization that this complexity lies at the heart of many health and welfare issues and that standard measures of rates, means, and local relationships cannot account for this complexity. As such, social network analysis has the potential to be a highly useful tool in monitoring the health and welfare of a wide variety of managed animal populations. Social network measures are amenable to measuring both physical and social–psychological aspects of health and welfare. Global patterns in the spread of disease, the ability to cope with psychosocial stress, represent emergent properties of interaction networks, and social network analysis is uniquely suited to identify and investigate the emergence of such properties.

We have reviewed a variety of individual-level and population-level network measures that have already been successfully applied to the study of animal health/welfare, as well as some measures which we suspect will be useful in future applications. For example, the risk of disease transmission can be monitored from the individual level using node-based attributes such as network centrality (e.g. degree, closeness, and betweenness) as well as at the population-level using network modularity and the detection of SCC or WCC. Similarly, the status of animals' social health can be monitored at the individual level using centrality measures to identify instigators or targets of social aggression, at the group level using measures such as cohesion and transitivity to estimate the stability and robustness of dominance networks, and even across time, using joint modelling to identify changes in network structure that parallel management actions, such as regrouping. We believe that social network analysis will become an increasingly useful method for managing the health and welfare of animal population, particularly as the community of people involved in animal management move towards a more holistic understanding of animal health and welfare. Therefore, this discussion of network applications to problems in animal health and welfare represents only the beginning of a new generation of analytical approaches and perspectives.

Acknowledgements

We would like to thank Dr Rob Atwill for helpful discussions and perspectives on disease transmission, and Dr Hsieh Fushing for guidance and collaborative research in the application of social network theory in our work on rhesus monkeys. We thank Dr Jessica Vandeleest for helping design the flowchart in Figure 11.1. We also thank the editor Dr Darren Croft and two anonymous reviewers for their many helpful suggestions and comments on the chapter.

SECTION 3

Taxonomic Overviews of Animal Social Networks

Jens Krause

The third section of this book contains chapters that are specific to particular taxonomic groups and study systems where biological questions are often entwined with methodological problems and particular research traditions. Here, we review how the network approach has contributed to a better understanding of particular taxa, what challenges have to be met in this context (for example in terms of identifying individual animals), and what the perspectives are for future studies in this field. By assessing the impact that the network approach has had on the study of specific taxa, and not only on our understanding of conceptually oriented topics such as information transmission and sexual selection, we follow the tradition of most research journals, where article authors report how our understanding of a general topic and a study system have progressed at the same time.

The selection of the chapters (non-human primates, cetaceans, fish, insects, birds, terrestrial ungulates, and reptiles) reflects where most of the activity on animal social networks studies has focussed. However, there are several species that have been studied in some detail and which are not explicitly covered in this section because they did not slot easily into larger taxonomic groups. In other cases, we were not convinced that a sufficient body of work had been accumulated for a particular taxon to warrant a chapter. Undoubtedly, these are subjective choices to a certain extent and we have tried to make amends by providing a short coverage in this introduction to some of those species and taxonomic groups that do not have entire chapters devoted to them.

Among the carnivores, the meerkats (*Suricata suricatta*) take a special place because they have been investigated in considerable detail with regards to intergroup interactions, intragroup contact patterns, and individual network positions (Drewe et al. 2009; Madden et al. 2009; Croft, Madden, et al. 2011). Networks were constructed from interaction data on grooming, dominance, and foraging competition for multiple meerkat groups. The authors concluded that the pattern of interactions between group members is not consistent between groups but instead depends on general attributes of the group, the influence of specific individuals within the group, and ecological factors acting on group members. Furthermore, differences in an individual's attributes did not consistently influence association patterns across different interaction network types, whereas, within network types, some trends were detected, such as negative assortativity by age and mass in grooming networks (Madden et al. 2011). Other work on carnivores includes a study by J. B. W. Wolf et al. (2007) on the social structure in Galapagos sea lions (*Zalophus wollebaeki*); this study not only detected the existence of multiple communities but also the substructuring of these communities into cliques which require further investigation. There are also recent network studies on lions (*Panthera leo*) (Abell et al. 2013), hyenas (*Crocuta crocuta*) (Holekamp et al. 2012), and racoons (*Procyon lotor*) (Hirsch et al. 2013).

In rodents, the work by Blumstein and co-workers on the social organization of yellow-bellied marmots (*Marmota flaviventris*) stands out for its depth and detail. The authors of these studies used the social network approach to investigate the social organization of marmots and found (among other things) that juveniles played a more important role in maintaining social cohesion in marmot colonies than had been expected (Wey and Blumstein 2010, 2012). This work is also notable for making a connection between social attributes and several performance measures such as annual reproductive success, parasite infection, and basal stress (Wey and Blumstein 2012) and provides a nice example of using agonistic as well as affiliative relationships to understand the connections between sociality and fitness (Lea et al. 2010). Other studies on rodents include work on disease transmission in Belding's ground squirrels (*Spermophilus beldingi*) and yellow-necked mice (*Apodemus flavicollis*) (also see Chapter 10).

Research on bats has produced a number of interesting network studies on different species in recent years. One investigation focused on the fission–fusion dynamics of Bechstein's bats (*Myotis bechsteinii*) and used network analysis to look at community structure within populations. Links between communities were maintained primarily by older individuals (Kerth et al. 2011). Furthermore, changes in community structure were observed over time as a result of population decline that resulted in the fusion of subunits (Baigger et al. 2013). The network approach was also used to look at parental care in northern long-eared bats (*Myotis septentrionalis*) (Patriquin et al. 2010) and to characterize and compare patterns of sociality between three populations of the leaf-roosting Spix's disc-winged bat (*Thyroptera tricolour*) (Chaverri 2010), to pick out just two additional studies from the growing amount of network literature on bats. Marsupials have also attracted some interest in the context of social network analysis. Disease transmission in brushtail possum (*Trichosurus vulpecula*) (Corner et al. 2003; Porphyre et al. 2011) and Tasmanian devils (*Sarcophilus harrisii*) (Hamede et al. 2009) has been investigated, as well as the social organization of eastern grey kangaroos (*Macropus giganteus*) (E. Best et al. 2013).

A surprising absence from the network literature concerns the amphibians. This might reflect (a) the research emphasis in this field, (b) the fact that group living is not very widespread in amphibian species, and (c) the difficulty with marking individuals more permanently. Nevertheless, topics such as chorusing in male frogs (Boatright-Horowitz et al. 2000; Jones et al. 2014) could be very promising to investigate from a network perspective linking communication networks and social networks (see Chapter 9). In addition, at the larval stage, some amphibians are social, and association patterns of tadpoles could be another potential topic for network applications (Leu, Bashford, et al. 2010).

A brief overview makes clear that the vertebrates have much greater representation than invertebrates. This is unsurprising given that the social network approach originated in humans and therefore transferred more easily to species and taxonomic groups where individual recognition has a long research tradition. Nevertheless, with the growing emphasis on the role of the individual in insect colonies and an increasing number of studies on behaviour types (i.e. animal personalities) in insects, we might expect to see a corresponding increase in social network studies in this taxon.

CHAPTER 12

Primate social networks

Sally Macdonald and Bernhard Voelkl

Introduction to social network analysis in primatology

Why is social network analysis useful for primatologists?

The primate order is incredibly diverse. There are currently over 370 recognized primate species (Groves 2005), and individual species differ from each other enormously. In terms of body mass, primates can range from around 30 g (the Madame Berthe's mouse lemur (*Microcebus berthae*)) to over 200 kg (male eastern and western gorillas (*Gorilla beringei* and *Gorilla gorilla*, respectively)). They make use of extremely varied habitats, from dense rain forests and open savannas to snow-covered mountains and arid semi-deserts, and exploit a huge array of food sources, including leaves, fruits, insects, meat, gum, nuts, and roots, among others (Napier and Napier 1997).

But perhaps of greatest interest here is the extreme variety that exists in primate social systems, ranging from the almost completely solitary aye-aye (*Daubentonia madagascariensis*) (Sterling 1993), to the complex, multilevel communities of up to 1000 individuals found in gelada baboons (*Theropithecus gelada*) (Snyder-Mackler et al. 2012). Analysis and comparisons of this social variation is challenging, even when dealing with a smaller subset of the wider range of this variety (Aureli et al. 2008). The social network analysis approach provides primatology with a new perspective and a new set of tools for approaching these problems.

In applying social network analysis to primate social behaviour, we are taking an approach developed to study the social lives of a very specific primate, *Homo sapiens*, and broadening it to study the social lives of their closest relatives. Unfortunately, we cannot confront non-human primates—or any other animals for that matter—with questionnaires to provide us with the data we need, so we must rely on behavioural observations. Thankfully, primates display many different social behaviours which are suitable for network analysis, for example, grooming (Sade 1965; Flack, Krakauer, et al. 2005; L. Lehmann and Boesch 2009; MacIntosh et al. 2012), copulations (Cheney 1978a; J. Lehmann and Ross 2011), displays of aggression and submission (J. Lehmann and Ross 2011; Flack 2012; Brent, Heilbronner, et al. 2013), play (Cheney 1978b; Flack et al. 2006), co-feeding (King et al. 2011), etc. While the collection of such data is certainly more time consuming (direct observations of behaviour taking much longer to gather than self-report questionnaires), most of these behaviours are easily observed, and it is possible to avoid some of the disadvantages associated with self-report data—so much so that, as an interesting side-note, we are beginning to see a trend in human studies towards automated behavioural sampling (dubbed 'reality mining') which brings the methods of human and animal network research closer together (Krause et al. 2013). We may never manage to match social scientists in terms of the precision with which they can fine-tune their analysis of human social behaviour but, within animal network research, primatologists are fortunate in having a larger range of behaviours to work with.

Researchers working on other species generally have much more limited options. More importantly, these options often differ not only in quantity but in quality, as many researchers are often forced to

Animal Social Networks. Edited by Jens Krause, Richard James, Daniel W. Franks and Darren P. Croft.

base their analysis purely on association data alone (e.g. for studies on ungulates (Wittemyer et al. 2005; Sundaresan, Fischhoff, Dushoff, et al. 2007; Stanley and Dunbar 2013) or cetaceans (R. Williams and Lusseau 2006; Lusseau 2007a)). From these associations, social relations between animals are then inferred, the argument being that, if two individuals spend significantly more time in close proximity to each other than to others, then one can assume some sort of social bond between them. This reasoning may be correct, but it is ultimately only indirect evidence of a relation.

Primatologists, on the other hand, can build their networks upon observations of direct interactions (e.g. grooming, copulation, aggression, etc.). These interactions directly signify a relation between individuals: if two individuals groomed each other, then we can state that they have a 'grooming relationship'. Thus, these networks are relationship networks based on direct evidence.

With these networks in hand primatologist can then begin to explore the wide range of analysis social network analysis has to offer. One of the most widely lauded benefits of social network analysis is the ability to look beyond the prospect of a single dyad and consider the interactions of individuals (or even subgroups of individuals) within the context of their entire group. This could help primatologists address questions about the role of matrilines in determining the social structure of a group, how dominance relates to the integration of individuals into a groups social structure, or whether an individual's position in the social structure puts them at a higher risk for diseases.

Another commonly praised benefit of social network analysis is its ability to provide quantitative replacements for verbal classifications of social style and structure. Among primatologists, this could help to identify subtle subgroupings with larger groups, or valuable measures of more subtle gradients in species apparently displaying very similar social structures. Social network analysis also has the potential to provide primatologists with an interesting new perspective for phylogenetic comparisons via a range of measures which summarize overall group structure in a variety of ways. These same measures can also be used to make comparisons within species, for example, to examine how ecological factors affect the social interactions and structure of the same species at different sites.

A brief history of social network analysis in primatology

For many primatologists, this apparently new approach of social network analysis may seem rather familiar. There is a feeling that they have seen a lot of what social network analysis has to offer before, and to a certain extent this feeling is justified. Social network analysis, or at least some of the elements of social network analysis, has been around in primatology for a long time. However, a recent surge in computational power and production of ready-to-use software has now made social network analysis a much more interesting and accessible tool. In this section, we provide a brief history of social network analysis in primatology in the hope of providing a useful insight into the roots of social network analysis in primatology and also help primatologists to identify recent changes and developments which might make social network analysis useful for them today.

During the 1960s to the mid-1970s, it became clear that the social structures of non-human primates were much more complex than had previously been suspected. As a result, researchers began to consider whether adapting the concepts and methodologies of human social science would be both appropriate and necessary to fully understand these non-human primate 'societies' (Sade 1972; Hinde 1976). Social network analysis, a thriving field in the human social sciences at this time, appeared to provide the perfect template from which to work (Brent, Lehmann, et al. 2011).

One of the first and most widely adopted social network analysis methods used was the sociogram. This useful diagram allowed primatologists to visualize their observations of the social interactions between individuals within a group. In particular, sociograms allowed primatologists to clearly display specific patterns in interactions that might otherwise have been hard to spot.

Perhaps the first student of animal behaviour to make use of these diagrams was Hans Kummer (1957), in his detailed study of the social organization of hamadryas baboons (*Papio hamadryas*). Kummer made extensive use of sociograms (Figure 12.1)

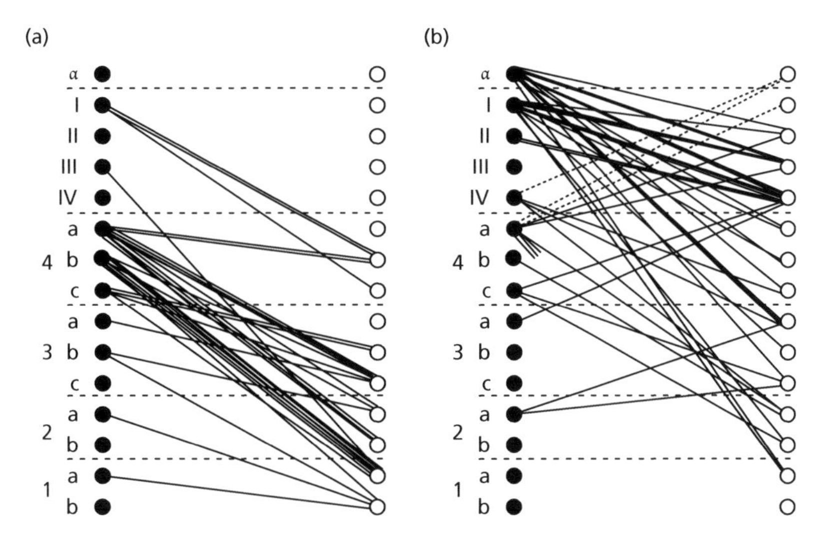

Figure 12.1 Sociograms for a group of hamadryas baboons for (a) 'embracing' and (b) 'brow-lifting'. Each line represents one individual: the left side of each panel (black circles) shows the individuals as actors, and the right side (white circles) as receivers of the behaviour. Reproduced from Kummer 1957: Soziales Verhalten einer Mantelpavianen-Gruppe. *Schweizerische Zeitschrift fuer Psychologie* Beiheft 33: pp. 23 and 40, with permission from Huber Verlag.

throughout this and subsequent studies (e.g. Kummer 1968) to document the structural development of the one-male units, as well as changes in the interactions between males and their females during various phases of their reproductive cycle. He would later describe with enthusiasm how he first came to use this tool:

'At first I was most fascinated with the "social structure" of the group. But what that actually was, and how it could be studied, neither my zoology teachers nor the ethological textbooks could tell me. I worried for weeks about this question. The one thing that was clear was that primates—monkeys and apes—treat one another as individuals and not as interchangeable members of a class, as happens among ants and termites, in schools of fish, or in enormous bird colonies. Eventually I thought I had found a solution: by diagramming how often each group member directed a particular signal at each other member, and by doing that for every signal, whether aggressive, caring, sexual, or playful, I must ultimately obtain a picture of the network of social relationships in the group. The complete diagram would then show, for instance, whether Pasha had a sexual preference for one female and whether lower-ranking mothers really punish their children more often than higher-ranking ones. Friendships and enmities would surely be revealed in the graphic network. So that is what I did.' (Kummer 1995, p. 37)

Unfortunately Kummer published his first report in German, in a supplement to the Swiss *Journal of Psychology*—a journal perhaps not closely monitored by primatologists or animal behavioural researchers.

In 1965 Donald Sade also began to use sociograms, seemingly independent of Kummer, to visualize the grooming interactions among a free-ranging group of rhesus macaques (*Macaca mulatta*). He used his diagrams to illustrate how grooming relations in the group were focused mainly among related individuals, and in particular how the grooming relationships between a mother and her offspring remained strong even once the offspring was fully mature—both novel insights at the time (Sade 1965). It is not clear at which point Sade got to know Kummer's earlier work—there is no reference to Kummer in Sade's 1965 article, but in a later study (Sade 1972), Kummer's work is cited. After these initial publications and through the influential books by E. Wilson (1975) and Hinde (1983), sociograms became an increasingly popular way to display all kinds of social interaction data (e.g. Ploog et al. 1963; Soczka 1974; Cheney 1978a,b) before becoming gradually less common in the early 1990s.

Another social network analysis tool (although not always recognized as such) which became popular among primatologists in the 1980s and early 1990s was cluster analysis. Like sociograms, cluster analysis provided useful diagrams, such as dendrograms, which could be used to display subgroupings (i.e. clusters), based on a specific behaviour of interest, within a group of individuals. However, in addition to being a means of displaying grouping patterns, clustering analysis is also a technique for identifying clusters of more closely connected individuals within a social group. Cluster analysis was mostly used to look for patterns in association data (C. Chapman 1990; Corradino 1990; Yeager 1990) and grooming data (Byrne et al. 1989; Chepko-Sade et al. 1989) and proved useful for identifying subtle patterns of subgrouping which otherwise might have been missed. However, there remains a certain arbitrariness in the technique when it comes to determining the boundaries of these clusters. As a result, different algorithms can produce very different groupings (Scott 2000; Whitehead 2008a), and this in turn has led to cluster analysis falling in and out of fashion over time.

For most primatologists, these two methods were the only brushes with social network analysis at this time; the one major exception was Donald Sade and his colleagues. After being one of the first among primatologists to make use of sociograms, Sade went on to author a series of four interconnected papers which investigated the utility of several social network analysis measures for the study of primate social structure (Sade 1972; Sade et al. 1988; Chepko-Sade et al. 1989; Sade 1989). His work focussed specifically on the social structure and grooming behaviour of a colony of free-ranging rhesus macaques living on the island of Cayo Santiago. He made use of a range of different social network analysis methods and measures, including sociograms, *n*-cliques (Sade 1972) and clustering analysis (Chepko-Sade et al. 1989); however, the measure which seemed to interest him most was a measure of centrality referred to as '*n*-path' centrality (Sade 1989).Like modern eigenvector centrality, *n*-path centrality calculates an individual's centrality based not only on their own direct connections within the network but also their indirect connection (i.e. the connections of the individuals to whom other individuals are connected) (see Chapter 2 for a discussion of the different centrality measures). Sade described the measure as follows:

> 'Consider the two hypothetical sociograms in table II [here reproduced as Figure 12.2]. Using grooming statues based on one-step links . . . it would appear that monkeys b and d have equal status, since each is groomed by one individual. However, the monkey who grooms b is himself groomed by 3 monkeys, but the monkey who grooms d is only groomed by one monkey. Therefore monkey b should be assigned higher grooming status than monkey d because b is chosen by a monkey with higher status than the monkey who chooses d.' (Sade 1972)

In this quote, Sade describes '2-path' centrality, as there are two sets of links or 'paths', the links that

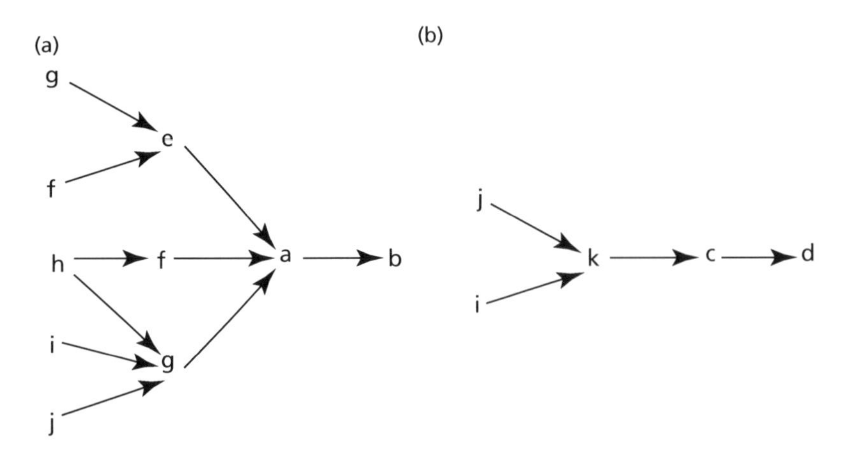

Figure 12.2 Graphical representation of grooming networks as introduced by Sade (1972).

exist between the first individual and other group members and the links between those individual(s) and the rest of the group. Sade went on to investigate the centrality of individuals based on 1 to 3 paths (Sade 1972), later extending this up to 15 paths (Sade 1989), arguing that, by increasing the length of the path used to calculate centrality scores, one could increase the variation in the scores and therefore better distinguish between the different statuses of individual monkeys. However, his analysis in fact showed that, while there was a clear increase in the variation of score using two paths and three paths compared to one path, there was little increase in the variation of scores using path lengths longer than this. He later went on to use 3-path centrality scores to show what many researcher might intuitively have expected, that, in rhesus macaques, higher-ranking individuals are more central (i.e. more integrated rather than physically central) in their social grooming networks (Sade 1972).

Overall Sade had hoped that these new measures would help primatologists move away from what he believed to be an intuition-based analysis of social structure to more objective, comparable methods. However, he also admitted that these new methods could be computationally burdensome to perform (at the time) (Sade 1989). And unfortunately, despite successfully attracting interest from those studying human social networks, in the form of a dedicated session on non-human primate networks at the 1988 Sunbelt Social Networks Conference and a subsequent special issue of the journal *Social Networks*, social network analysis methods would play only a very marginal role in the analysis of primate social behaviour until the beginning of the twenty-first century.

In 2006, social network analysis returned to primatology with something of a bang, in the form of a now widely cited study by Jessica Flack and colleagues (Flack et al. 2006). In this study, Flack and her colleagues used four different network measures (degree, reach, assortativity, and mean clustering coefficient) to examine the effect of removing the three highest-ranking males from a group of captive pig-tailed macaques (*Macaca nemestrina*) on social networks of grooming, play, contact sitting, and proximity. They were able to show that the removal of these high-ranking males affected the social networks more severely than predicted by simulated removals. From this point onwards, a new surge of studies embracing a social network analysis approach began to appear. While many of these studies focused on what could be considered the traditional topics addressed by social network analysis (i.e. describing social relationships/structures and their potential functions in primate groups (Sueur and Petit 2008; Henzi et al. 2009; Kasper and Voelkl 2009; J. Lehmann and Dunbar 2009; L. Lehmann and Boesch 2009; Ramos-Fernández et al. 2009)), many others went on to make use of the network approach to address a broader range of less conventional topics, including social learning and information flow (Voelkl and Noë 2008; Franz and Nunn 2009; Franz and Nunn 2010; Hoppitt, Kandler, et al. 2010; Kendal, Custance, et al. 2010; Voelkl and Noë 2010), the influence of social structure on cooperation (Voelkl and Kasper 2009; Voelkl 2010), and the welfare of captive groups of primates (McCowan et al. 2008; Beisner, Jackson, et al. 2011a; Clark 2011; Dufour et al. 2011).

Levels of primate social network analysis

Social network analysis can be split up into three broad levels of analysis: the individual (nodal) level, the subgroup level, and the group level (see also Chapter 2). The question of which level of analysis to use usually goes hand in hand with the question of its purpose. In this section, we will take a brief look at what analysis can be done at each of these three levels and how it can be of use to primatologists in particular.

Primate social network analysis at the individual level

The individual level, sometimes also referred to as the nodal level, is where social network analysis is used to ascribe certain 'nodal properties' to an individual animal. As such, this is the level of analysis that allows you to consider an individual's social position within a group beyond the level of its dyadic interactions, by incorporating its interactions with the group as a whole. This can be done using a single network metric such as a centrality measure or a clustering coefficient, or with a list of several different nodal network measures. In the latter case, it has been suggested that these measures could even be summarized to form a personality profile for each individual (Crofoot et al. 2011). The purpose of such an individual-based approach is usually to correlate these nodal network measures with other biological characteristics of the individuals such as sex, age, social rank, and 'personality scores' (Sade 1972; McCowan et al. 2011; Tiddi et al. 2011), reproductive success (Brent, Heilbronner, et al. 2013), parasite load (MacIntosh et al. 2012), or physiological measurements such as faecal glucocorticoid levels (Brent, Semple, et al. 2011).

Sade (1972) was one of the first to use nodal measures in this way to show that higher-ranking female rhesus macaques were more integrated into their grooming network than lower-ranking females. This finding has recently been replicated in a broader study of macaque social structure (Sueur et al. 2011). The study used eigenvector centrality scores to investigate how female rank was related to female integration into a social network based on body contact (i.e. grooming and contact sitting combined) in 12 groups of macaques from 4 different species. Their results showed that in socially 'intolerant' species of macaques, such as rhesus macaques and Japanese macaques (*Macaca fuscata*), high-ranking females were significantly more central (i.e. more integrated) in the social network than lower-ranking females. However, the same pattern was not found in two more socially 'tolerant' species of macaques, Tonkean macaques (*Macaca tonkeana*) and crested macaques (*Macaca nigra*), suggesting a potentially interesting link between social style and network structure in this genus.

Nodal network measures have also been used to investigate the heritability of social traits in primates. In a seminal study, Brent, Heilbronner, and colleagues (2013) showed that betweenness and eigenvector centrality of the grooming network, outstrength of the aggression network, and eigenvector centrality of the proximity network of the rhesus macaques at Cayo Santiago Island all appeared to have a significant additive genetic component. In addition, a quantitative genetic analysis in the study suggested a correlation between eigenvector centrality for grooming and the genetic variation at two loci involved in serotonergic signalling; hence, this is the first study to link genes with nodal network measures.

Primate social network analysis at the subgroup level

The purpose of analysis at the subgroup level is usually to better understand how a given group is structured, by identifying clusters or subgroups within it, and to identify how these subgroups are connected to each other. This identification process is often referred to as subgroup or community detection (Dow and deWaal 1989; Matsuda et al. 2012). A nice example of this level of analysis in primatology comes from Snyder-Mackler et al. (2012), who studied the multilevel societies of gelada baboons (*Theropithecus gelada*), which consist of two or more nested levels of organization: one-male units, teams, bands, and communities. While the multiple levels of gelada society were described decades ago and are generally accepted, no operational definition existed for the higher levels (i.e. levels above the one-male unit). Making use of clustering analysis methods, the authors were able to find a significant discontinuity at the 50% association level in their study groups, indicating a sharp distinction between members of the same band and members of the same community. This finding allowed them to provide a meaningful delineation between these two levels of gelada society.

Similarly, using hierarchical cluster analysis, Zhang and colleagues (2012) nicely demonstrated the multilevel character of snub-nosed monkey bands (*Rhinopithecus roxellana*). Based on a simple ratio association index using proximity data, they were able to identify 9 discrete one-male units, in a band of 58 individuals. All members of a one-male unit were frequently seen within one metre of each other. Females could also occasionally be seen in close proximity to females from other units; however, males were virtually never seen in close proximity to any member of another males' unit. Based on this and sex differences in eigenvector and betweenness centralities, Zhang et al (2012) argued that lactating adult females play more important social roles than males in connecting the one-male units. In contrast to these results, spider monkeys (*Ateles geoffroyi*) studied by Ramos-Fernandez and colleagues

(2009) form groups where usually young males have high centrality scores and, as such, appear to act as brokers tying subcommunities together.

Another useful measure for primatologists at the subgroup level is community modularity. This measure gives an indication of how fragmented a given group is by examining the frequency at which interactions occur within or between subgroups. For example, Sueur, Petit, et al. (2011) used community modularity to examine how frequently matrilines within different species of macaques interact. They were able to show that 'intolerant' species of macaques had significantly higher modularity scores than more 'tolerate' species, indicating that intolerant species of macaques concentrate their interactions more within their matrilines than more tolerant macaques.

Likewise, Beisner and colleagues (2011b) made use of community modularity as a measure of the degree of subgrouping within matrilines of rhesus macaques. Each matriline was partitioned into all potential arrangements of subgroupings, and community modularity scores were calculated for each subgrouping arrangement. The arrangements which lead to the highest modularity scores (i.e. the strongest within subgroup interactions and the weakest between subgroup interactions) were chosen. Based on these arrangements, the authors were then able to show that the number of subgroups and the degree of modularity found within matrilines were significantly related to relatedness within those matrilines. That is, as relatedness within a matriline decreased, the number of subgroups found within the matriline increased, as did the modularity score of the matriline, indicating that these matrilines as a whole were becoming less cohesive.

The subgroup level of analysis can also prove useful for researchers interested in the flow of information, parasites, or disease within a group. For example, Voelkl and Noë (2008, 2010) used simulations of social learning to show that the predicted flow of socially transmitted information can be significantly altered by the social structure of a group. They found that the expected average path length of a transmission process is clearly determined by the community modularity of a group; that is, the greater the modularity of a given group, the longer the average path length. This clearly shows the importance of incorporating social structure into any predictions of the spread of social information. A similar simulation approach was used by Griffin and Nunn (2012; Nunn 2012) to show that a stronger subdivision of larger social groups should slow the spread of infectious diseases.

Primate social network analysis at the group level

The purpose of group-level analysis—sometimes referred to as population level analysis—is normally to either compare the properties of different groups or the properties of the same group over different periods of time. The measures used for this sort of analysis are usually either graph-wide measures which describe a property of the whole network (e.g. graph density), or moments of the distribution of nodal measures like the skewness of the degree distribution or edge-weight disparity. Group-level analyses have been made for different purposes: for descriptively summarizing the structural characteristics of a specific study group (e.g. Sade 1965); for detecting temporal changes—either seasonal long-term changes (McCowan et al. 2008; Henzi et al. 2009; Ramos-Fernández et al. 2009; Brent, MacLarnon, et al. 2013; Macdonald et al. 2014) or short-term changes due to a specific event, such as individual migration events (Flack, Krakauer, et al. 2005; L. Lehmann and Boesch 2009; Lusseau, Barrett, et al. 2012) or habitat changes (Dufour et al. 2011); for comparing different study groups of the same species (Beisner, Jackson, et al. 2011b; Crofoot et al. 2011), or for comparing the group structure of many different species (usually in meta-analyses, e.g. Kasper and Voelkl 2009; J. Lehmann and Dunbar 2009; Lehmann et al. 2010; Sueur et al. 2011; Matsuda et al. 2012).

The group level of social network analysis opens up new and interesting perspectives for phylogenetic comparisons. Primates are a well-studied group, with a considerable number of species currently being studied, in some cases at multiple sites and for several years or even decades. As a result, it is entirely possible to construct networks based on the same behavioural categories (e.g. grooming, associations, aggression) for many different species.

So far, only a few broad comparisons have been attempted: one covering a wide range of primate species (70 groups from 30 different species (Kasper and Voelkl 2009)) and another which focussed more specifically on macaques (12 groups from 4 different species (Sueur et al. 2011)). In another study, J. Lehmann and Dunbar (2009) examined the relationship between neocortex size and social network structure across 11 different primate species. They found that neocortex size appears to be negatively correlated with the density and fragmentation of social networks in these species (i.e. the larger the neocortices, the sparser and more fragmented these networks appear to become). These have all been rather rough initial comparisons with relatively modest sample sizes which may not allow for sensible estimates of within-species variance of network measures, and as such there remains a lot of potentially interesting work to be done.

In group-level analysis, the unit of measurement is the group; thus, for studies focussing on a single group, the sample size is one. Significance tests are therefore often made with randomization procedures, comparing the observed network metric against a distribution of metrics from randomized networks with the same number of vertices and edges. Yet, the question as to whether a random network is really a plausible null assumption remains open for now (Croft, Madden, et al. 2011). Intraspecific comparisons of several groups still suffer from relatively modest sample sizes, while large-scale interspecific comparisons come with the usual problems of meta-analyses (mainly questions of how comparable methods and study designs were between studies). Nevertheless, group-level analyses are among the most common network analyses in primatology and, even in studies that focus on the individual or subgroup level, researchers will usually also give some group-level summaries of their study groups.

Potential pitfalls and limitations in primate social networks

Compared with other animal social network research, the problems facing studies of primate social networks are in no way special or distinct. There are, however, a few of these general problems which are especially prominent in primatology.

Group size in primate networks

Most primate groups are relatively small, which can be a serious limiting factor for certain network measures. While there are a few primate species where group sizes can reach, or even exceed 100 individuals (e.g. macaques, baboons, mandrills, or snub-nosed monkeys), a comparative study based on published socio-matrices from 70 primate groups from 30 different species has shown that most primate groups consist of just a dozen or so animals (Kasper and Voelkl 2009). Group sizes were found to range from 4 to 35, with a median group size of 9 and an inter-quartile range of 6–16 individuals. These numbers were close to those found during a rough literature survey of 184 primate species, with a median group size for free-living primates of 9 animals and an inter-quartile range of 4–20. Such small group sizes can severely limit the scope of a social network analysis, as several metrics show little variation for small groups (Perreault 2010).

For example, community modularity (Newman and Girvan 2004) is a measure defined between 0 and 1; but if we take 1000 random graphs with $n = 9$ and a density of 0.73 (the observed average density for primate grooming networks), we get community modularity scores ranging only between 0.03 and 0.04. Likewise clustering coefficient, reach, or characteristic path length will also vary only little for small networks. This problem can be amplified by the dense nature of most primate social networks. In particular, networks built on association data are prone to high densities (if one just waits long enough, one might see even the biggest rivals close together for a short time), to the point that most group members are directly connected with each other, although some perhaps just very weakly. This reduces variation even further, especially for binary network measures.

The problem can be less severe for certain interaction networks (like grooming), although a median density of 0.75 for socio-positive interactions (Kasper and Voelkl 2009) or 0.73 for undirected grooming networks and 0.51 for directed grooming networks (n = 77, Kasper and Voelkl, unpublished data) can still be considered as relatively dense in comparison to the sparse networks reported elsewhere. The recent surge in the development of weighted networks measure goes some way to help combat these problems. However, researchers should still take time to consider, even when using weighted measures, whether the network measure they wish to use makes sense given the size of their network.

Observation frequency in primate social networks

Students of animal behaviour usually infer relationships between individuals based on observed interactions between those individuals. As they repeatedly observe the same animals, they can take these repeated observations to quantify the strength of the relationship (i.e. the weights of the edges in such networks). This gives rise to a different problem, the reliable estimation of the edge weights (for discussion of edge weights in this context see e.g. Whitehead 2008b; Kasper and Voelkl 2009; Sundaresan et al. 2009; Perreault 2010; Voelkl et al. 2011). With interdependent edge weights and no reasonable assumptions about their underlying distribution, the best response to this estimation problem is a bootstrap re-sampling plan (Lusseau et al. 2008; Franks, James, et al. 2009; Franks, Ruxton, et al. 2009).

It is of course extremely difficult to judge how many observations are needed in order to have an accurate picture of the relationships in a group. It could also be argued that, with the small size of primate groups and the range of easily observable primate social behaviours, primatologists are unlikely to face serious problems in terms of quantity of observational data. However, many primate social behaviours, such as grooming, copulations, aggression, etc., are relatively rare. Dunbar (1991) summarized that primates spend on average only 5.2% of their time grooming, while a cross-species comparison of agonistic interactions reported a mean rate of 0.61 ± 0.09 agonistic interactions per individual per hour (Wheeler et al. 2013) (focussed on female agonism). Therefore, long hours of detailed observations are still needed to get an accurate picture of these social behaviours.

For example, in a study of chacma baboons (*Papio cynocephalus ursinus*), Henzi and colleagues (2003) collected 278 h of focal observations in one 17-month study period, during which they observed 263 bouts of grooming between adult females. As the group contained 12 adult females, this gives an average of only 2 grooming bouts per dyad in the female grooming network. In such a case, it would be treacherous to estimate individual edge weights, as many estimates would be based on either just a few or even just a single observation. Furthermore, edge-weight distributions based on grooming data are usually skewed (Kasper and Voelkl 2009), which means that the better part of grooming events will fall on just a few dyads, while estimates for the majority of existing edges will be based on very few observations (e.g. Tiddi et al. 2011; Brent, Heilbronner, et al. 2013). This is definitely a fact one should worry about and, as such, we would encourage researchers to give serious consideration to the volume of data available to them before attempting social network analysis.

One way to help judge whether enough observations have been collected is to consider the number of isolates in the network (i.e. individuals with no connections at all to any other individual in the network). The observation of large numbers of isolated individuals is generally a clear warning that, rather than individuals not interacting, one simply has not seen these individuals interact. There are of course situations where the presence of isolates makes sense. For example, in a copulation network, certain adult males may be isolated completely from the network as a result of being excluded from virtually all mating by more powerful higher-ranking males or, in a social play network, one might not be surprised to see a large number of isolated adults. However, in networks based on more widespread behaviours such as grooming, the presence of even a few isolates can be a strong indication that more data is needed.

Not all network measures are equally sensitive to small samples; thus, how much observational data are required to reliably answer a certain question can also depend on the type of network measure used. Voelkl and colleagues (2011) investigated how network measures for networks of 44 primate groups changed under re-sampling and simulated error-attack. By re-sampling at lower rates from the original datasets, they asked how stable measures were if researchers had collected fewer data. Unsurprisingly, they found that, in general, networks based on few observations (less than 100) were relatively sensitive to re-sampling, corroborating the general wisdom that larger samples give more robust results. However, the authors could also show that network measures differed in their sensitivity towards re-sampling. Some basic measures like density, degree variance, or edge-weight disparity proved to be relatively robust, while other more complex measures like clustering coefficient, eigenvector centrality, or vertex strength variance were much less so.

In summary, the small group sizes and the high density of most primate networks makes it more difficult to find strong, or meaningful, contrasts in network measures between groups. This is especially true when one considers binary network measures. There is hope that weighted network measures can give a more accentuated picture; however, researchers must ensure they have large quantities of reliable data in order to obtain good estimates of edge weights, especially if they wish to make use of the more complex network measures.

Specificity in primate social networks

In 'Introduction to social network analysis in primatology', we highlighted the fact that primates are one of those few taxons where researchers are able to build networks based on a variety of different behaviours. This should—in principle—enable them to fine-tune the measure to the specific question. Yet, more often than not, the choice of the measure for primate networks seems to be based on practical considerations (i.e. how easily and reliably the behaviour can be observed) than on principal considerations (i.e. which type of network would be most relevant for the specific question). When it comes to the transmission of social information, a proximity network might be of interest, as one could argue that individuals that spend much time in close proximity are more likely to learn from each other, while an aggression or a copulation network, for example, might be less appropriate. On the other hand, copulation networks might be very informative when one is interested in disease transmission.

That the type of network should fit to the question might sound obvious, yet errors in this area have repeatedly slipped into primatological research. For example, the membership to a primate group has, in some cases, been exclusively defined by a single behaviour, such as grooming. Disruptions of this network—either observed or simulated—are then interpreted as a group split, neglecting the fact that while these subsets of individuals may not groom each other, they may still continue to roam together, forage together, defend a territory together, and sleep together at the same site. Speaking in such a case of a 'group split' would, in our opinion, be a clear misinterpretation.

Thus, whenever a social network is built upon a single behaviour, one should keep its specificity in mind. The importance of considering multiple behavioural measures to examine sociality in primates has long been emphasized (Kummer 1968; Hinde 1976), and social network analysis itself provides primatologists with an excellent means of comparing and integrating multiple social behaviours for this purpose (J. Lehmann and Ross 2011; Barrett et al. 2012; Brent, Heilbronner, et al. 2013; Brent, MacLarnon, et al. 2013; Chan et al. 2013; Macdonald et al. 2014). In addition, given that network structures for different social behaviours can vary markedly—particularly proximity or associations networks (Barrett et al. 2012; Macdonald et al. 2014)—a reliance on single behaviours in general seems unadvisable if multiple social behaviours could be used to give a more rounded and reliable picture.

Intraspecific variability in primate social networks

Finally, for many primate species we find a very high intraspecific variability in social organization and group composition (e.g. Smuts et al. 1987; Fleagle et al. 1999). This is expected to be reflected in an equally high variability in network metrics for different groups of the same species, making it difficult to make species-level generalizations without data from a large number of groups. This argument appears to be supported by a phylogenetic comparison of network metrics within the order of primates, as it found no evidence for systematic differences (Kasper and Voelkl 2009).

To illustrate the extent of intraspecific variation of group-level network measures, we took a dataset of 77 primate groups from 35 different species (Kasper and Voelkl, unpublished data) and focussed on two species for which data from several study groups were available: vervet monkeys (*Chlorocebus aethiops*) ($n = 7$), and Japanese macaques (*Macaca fuscata*) ($n = 9$). The results for four network measures—the community modularity, edge-weight disparity, group-closeness centrality index, and weighted clustering coefficient—are depicted in Figure 12.3.

A non-parametric means comparison found significant differences for the first two measures (Mann–Whitney U test, U = 7.5, p = 0.009, and U = 10, p = 0.023, respectively), but none for the latter two measures (U = 31, p = 0.95, and U = 47, p = 0.10, respectively). The results of this comparison suggest that Japanese macaques have a stronger modular structure and that the distribution of link-strengths—representing the quality of the relationship—is more skewed in the macaques. Both results fit well with our understanding of macaques being organized in matrilines with strict hierarchies, and groups of vervet monkeys being less structured and relatively more egalitarian. The black bar at the bottom of each panel in Figure 12.3 gives the inter-quartile range for the respective measure for the whole dataset. It shows that Japanese macaques and vervet monkeys do not only differ in these measures relative to each other but are in both cases quite at the other ends of the distribution of the whole primate sample.

For the other two measures, however, the picture is a different one. In the case of the closeness centrality score we cannot find a pronounced difference between the species (the median value for vervet monkeys is 0.0161 and for Japanese macaques is 0.0165), although looking at the inter-quartile range for the complete primate sample, we

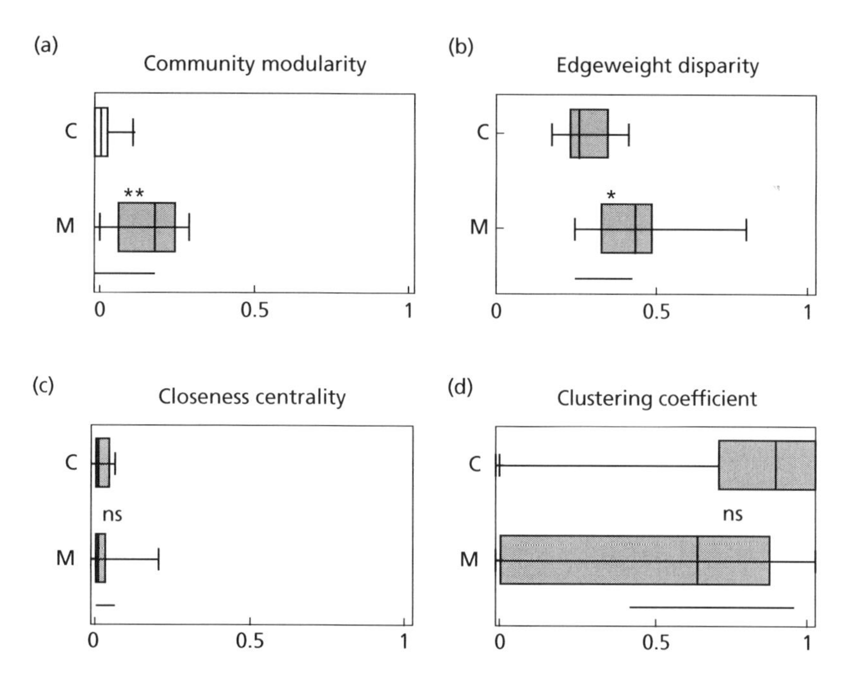

Figure 12.3 Group-level community modularity, edge–weight disparity, closeness centrality, and weighted clustering coefficient for seven groups of vervet monkeys (*Chlorocebus aethiops*) (C), and nine groups of Japanese macaques (*Macaca fuscata*) (M). Networks are built upon data from Hasegawa and Hiraiwa 1980; Seyfarth 1980; Furuichi 1985; Mehlman and Chapais 1988; Oi 1988; Cheney and Seyfarth 1990; Chapais and Mignault 1991; Takahashi and Furuichi 1998; Scheid, unpublished data; Ventura et al. 2006; Fruteau et al. 2009). Boxes indicate median and inter-quartile ranges, and whiskers indicate the full ranges. The grey bar at the bottom gives the inter-quartile range for a sample of 77 primate groups from 35 species; $^{**}p < 0.01$; $^{*}p < 0.05$; ns, not significant (see text for details).

see that this, too, is quite small. For the weighted clustering coefficient, on the other hand, we find a pronounced difference in the median values (0.62 for vervet monkeys and 0.88 for the Japanese macaques) although, due to the enormous intraspecific variation, this difference is not significant.

Social network analysis in primatology: conclusions and future directions

In this final section, we take a look at some of the most promising directions for social network analysis in primatology and sketch out a picture of how the future of primate network analysis might look. This is, of course, quite a speculative endeavour. Our motivation is not to test our magical abilities of foretelling the future, but rather to provide some suggestions of what might be the most fruitful applications of social network analysis to primate social behaviour in the future.

Our first, almost trivial, prediction would be that grooming will continue playing a prominent role in the construction of primate social networks. As we have mentioned several times, primatologists have a distinct advantage over other researchers of animal behaviour when applying social network analysis, in that their study species show a considerable range of social behaviours that seem to be relatively universal and uniform in their functions. Among these 'universal' social behaviours, grooming is by far the most prominent. It is found in nearly all primate species, is easily recognizable, unambiguous, frequent, and frequently recorded by field researchers. Regarded as a directed sociopositive interaction, grooming seems predestined for the construction of social networks and, being widespread within primates, it positively invites itself as a basis for interspecies comparisons (Kasper and Voelkl 2009; Sueur et al. 2011).

Our second prediction would be an increase in the development and use of multilayered social networks. Early discussions of primate sociality often emphasized the importance of incorporating the influences of multiple social behaviours in order to gain a well-rounded view of primate social relationships and social structures (Kummer 1968; Hinde 1976); however, the reality of incorporating measures of multiple behaviours has proved challenging. Social network analysis provides primatologists with new opportunities in this regard. Until recently, primate social networks have always been constructed based on a single behavioural measure. If researchers were interested in more than one behavioural dimension, then they constructed separate networks for each of them.

Making comparisons between networks based on exactly the same set of individuals is relatively straightforward (Croft et al. 2008) and can allow primatologist to examine and directly contrast subtle differences in the overall structure and the different positions individuals might take in each network (Flack et al. 2006; J. Lehmann and Ross 2011; Brent, Heilbronner, et al. 2013; Brent, MacLarnon, et al. 2013; Macdonald et al. 2014). Alternatively it is possible to construct a single score which integrates multiple social behaviours such as, for example, the composite sociality index promoted by Silk and colleagues (2006), and base a single network on this measure (Macdonald et al. 2014). However, it has been argued that we could gain a much more comprehensive picture of primate sociality by combining separate behavioural networks to form a single multilayered network (Barrett al. 2012).

Such a multidimensional graph object has the potential to give a much more accurate and detailed picture of an animals' social niche. Barrett and colleagues constructed their multilayered network based on aggressive and grooming behaviour as well as spatial proximity data for all adult females from a troop of baboons in the De Hope Nature Reserve in South Africa. The resulting three-layered multidimensional network was characterized in terms of its information entropy. Using both natural and simulated 'knock-outs', this case study suggested that the overall entropy of the network changes more after the disappearance of a high-ranking individual than after the disappearance of a low-ranking female (Lusseau, Barrett, et al. 2011).

Another method to integrate information from multiple single-behaviour networks was recently introduced by Chan and colleagues (Chan et al. 2013), who used a multidimensional framework based on joint probabilities of edge co-occurrences from four different behavioural networks to extract patterns of relationships (via constraint functions).

Using this approach, they compared the social structure of a rhesus macaque group during a period of rank stability and during a period leading up to a major rank disruption. They were able to show that during the period leading up to the rank disruption, subtle structural properties, linked to a stable social structure, appeared to have disintegrated, resulting in a simpler, yet arguably less stable, social structure. Such multilayered approaches are still in an early stage of development; however, they appear to be very promising and given the range of easily observable types of behavioural interactions, primates seem to be a well-suited taxonomic group for developing such an approach further.

Another interesting topic that we predict will receive more attention is the question of how animal social networks change over time. Here, again, primates would be a well-suited model system, as primates are relatively long lived and tend to form stable social groups that often persist over several generations. At some study sites, groups of chimpanzees, macaques, or baboons have been continuously monitored for up to five decades, and while not all the historic data will be suitable for a detailed reconstruction of the animals' social networks, the recent interest in networks may convince researchers to change their observational protocols in order to collect network-fit data.

There have been some interesting initial studies that have focused on temporal changes in primate social networks. The first, published by Henzi and colleagues (2009) showed a clear cyclicity in the degree to which female baboons maintained differentiated associations. Similarly, a recent study of free-ranging rhesus macaques showed that female social network structure varied around the species breeding cycle, being more centralized and more clustered in the mating season than in the birth season (Brent, MacLarnon, et al. 2013). This contrasts with a study by Macdonald et al (2014) which shows relative stability in the social relationships of females Assamese macaques (*Macaca assamensis*) despite significant changes in their social and ecological environment. Undoubtedly, these will not be the last studies to focus on temporal changes in primate networks.

In a recent review, Brent, Lehmann, et al. (2011) envisage that social network analysis may allow a new quantitative approach to social role theory that might be especially useful for non-human primate research. Role theory was originally developed by social psychologists but later also adopted by some primatologists (Bernstein 1964; V. Reynolds 1970; Fedigan 1972). It largely fell out of favour in the 1980s (for a review see Roney and Maestripieri 2003) due ultimately, according to Brent and colleagues, to three reasons: first, role theory was criticized for producing little more than descriptions of activity patterns of members of given age or sex classes; second, its implications were partly misunderstood; and third, it was criticized for lacking clear operational definitions. As some of these misconceptions—especially concerning the link between role theory and group selection—could now be cleared out (Roney and Maestripieri 2003), and social network measures offer a way to define roles operationally via nodal properties of the individuals, Brent, Lehmann, et al. (2011) predict a revival of role theory in primatology. Time will tell to what extend this will be the case, but we would predict that primatologists and, more generally, behavioural ecologists will begin to embrace network analysis as a tool for quantitatively describing the 'social niche' of an individual or defining 'socio-behavioural phenotypes'.

Another recent trend which is currently seen in behavioural ecology and which we predict is likely to also affect primatology is the increased use of automated recording techniques, where individuals are fitted with either geolocation loggers or with radio frequency identification tags (Rutz et al. 2012). These devices record the geographical position and sometimes motion parameters (direction and acceleration) or further physiological parameters of the animals with a high spatial and temporal resolution, generating huge amounts of data. For example, by using passive integrated transponder tags on a population of great tits, and recording the resulting signals with an array of antennas sampling at a rate of more than 1 Hz, Garroway and colleagues were able to record several million observations and more than 10,000 co-occurrences of birds during a 5-month period (Chapter 16). This is clearly something that would not be possible with the classical boots–binocular–pen-and-paper approach still practised by many primatologists. With such enormous datasets, it is now possible to construct accurate weighted networks; however, this technique is, of course, limited to addressing questions based on association data. Despite this drawback, this technique has the potential to be an extremely useful tool for primatologists by helping them to address important topics where traditional data collection methods often struggle, such as large-scale group movements/ranging patterns, intergroup encounters, ranging among fission–fusion societies, dispersal patterns, etc.

Finally, researchers have started, with the help of social network analysis, to begin to investigate the genetic basis of social behaviours. As it can be assumed that behavioural traits have a highly polygenetic background, one needs two things to tackle this topic: first, high-resolution sequencing techniques that allow the construction of detailed quantitative trait locus maps and, second, good quantitative descriptors for the social behaviour of an individual. For the latter, social network analysis has the potential to offer valuable contributions. Having nodal network measures for single individuals describing their particular social niche allows one to then relate these measures to the genetic data.

Such an approach has been recently attempted by Brent, Heilbronner, and colleagues (2013). The authors appear to find both additive genetic variance for several network measures as well as a link between centrality in the grooming network and a gene complex in the serotonin pathway. This is the first study of its kind to take this direction. Given this study's reception, no magical skills are required to predict that more studies along these lines are to come and that this will be a hot topic for the near future.

Despite several enthusiastic reviews directly aimed at primatologists (Brent, Lehmann, et al. 2011; Jacobs and Petit 2011; Sueur, Jacobs, et al. 2011) the social network analysis approach has been slow to develop in this field. This is likely in part due to the barrage of new jargon that accompanies it, and in part due to the somewhat unfamiliar statistical techniques required to adequately deal with the extremely dependent data. For some primatologists, there is also a sense that this approach is not actually providing them with anything very new. We hope that by giving a realistic overview of what social network analysis of primate groups can and cannot achieve, and by clarifying what is and is not new in the approach, this chapter will encourage more primatologists to embrace social network analysis in their research and help establish social network analysis as a useful tool for investigating questions of primate social behaviour.

Acknowledgements

We thank the editors of this book and one anonymous reviewer for very useful comments on the manuscript.

CHAPTER 13

Oceanic societies: studying cetaceans with a social networks approach

Shane Gero and Luke Rendell

Introduction to network analysis of cetacean societies

Cetacean behaviour has in recent decades attracted increasing levels of scientific attention, largely because this group represents an evolutionary peak in terms of their cognitive capacities (Marino et al. 2007), their complex communication systems (Janik and Slater 1997; Tyack and Sayigh 1997), and their multilevelled societies (Connor et al. 1998; Mann et al. 2000). This peak is fascinating because it has arisen quite independently, in an environment radically different from that which has given rise to similar peaks in terrestrial taxa. It thus has the potential to richly inform comparative analyses of a range of behavioural characteristics.

Cetacean ecology is typically very different from that experienced on land, in several respects. Compared to terrestrial habitats, variation in marine environments is substantially weighted towards lower frequencies—that is, over long timescales (Steele 1985; Cyr and Cyr 2003; Vasseur and Yodzis 2004)—resulting in high levels of variation in resources across time and space. Furthermore, it is difficult to defend those resources in a three-dimensional habitat. Finally, travel costs are typically much lower (T. Williams 1999); as a result, cetaceans show little evidence of territoriality (Connor et al. 1998; Connor 2000) and operate on relatively larger spatial (Stevick et al. 2011) and temporal scales (George et al. 1999). Combining these features with the fact that these species spend that vast majority of their lives underwater and out of the view of observers, the study of cetacean social networks has its challenges. Nonetheless, a remarkable amount is being learned about the social complexity in this order, through a combination of long-term studies, advances in data-gathering technologies, and conceptual advances in areas such as social network analysis.

While knowledge of cetaceans has historically lagged behind that of their terrestrial mammalian counterparts, they have been at the heart of the recent rise in the application of network analysis methods to the study of animal behaviour since it began (e.g. Lusseau 2003; Lusseau and Newman 2004). Here, we outline how this approach has aided the current understanding of cetacean societies and what these studies have brought in terms of a broader understanding of animal social dynamics. We briefly cover the methodological challenges encountered when studying social interactions between wild cetaceans at sea and conclude with perspectives on future avenues for inquiry.

Oceanic social networks

Before discussing the application of social network analysis as it is meant in the context of this book, that is, the interdisciplinary use of methods and tools originating from the study of quite different types of networks, we begin by noting that the study of social structure in cetaceans was well advanced before this cross-fertilization occurred. The social complexity of cetacean societies covers a wide continuum, so before describing recent applications and methodological developments originating from cetacean research, we will briefly outline some of the more interesting features found

Animal Social Networks. Edited by Jens Krause, Richard James, Daniel W. Franks and Darren P. Croft.

in cetacean societies, to provide some context as to why they have proven fertile ground for the social network approach. Most cetaceans are highly social, living most of their lives in some form of close association with conspecifics. Even among those that apparently do not, we may be missing quite a lot of their social interactions, which can potentially take place at great ranges and are mediated by acoustic signals that can propagate for tens and sometimes hundreds of kilometres (Payne and Webb 1971; Širović et al. 2007). For social species, their position within the social network can apparently have major impacts on their survival and reproduction (Frère et al. 2010; Stanton and Mann 2012), so the shape and structure of cetacean social networks are evolutionarily significant.

Starting with the baleen whales (*Mysticeti*), we find in general plenty of associations, some of which, such as surface-active groups of right whales (*Eubalaena* spp.; Kraus and Hatch 2001), or cooperative foraging groups (Whitehead and Carlson 1988), are highly social, but long-term bonds beyond the mother–calf pair appear to be rather rare (Weinrich 1991; Weinrich and Kuhlberg 1991; Clapham 2000). Thus, baleen whale social networks, where studied, are dense in that many individuals associate, but relatively homogeneous, reflecting a tendency for multiple short-term associations (Figure 13.1). Not all individuals are equally connected, however, as they vary in both degree and betweenness, two measures of how central an individual is in a network (see Chapter 2). Why such variation exists, and what the implications are for the animals, are just two of the questions with which social network analysis has enriched the study of cetacean sociality.

In contrast, the toothed whales (*Odontoceti*) exhibit more variation in social structure, including within-species differences across habitats. The best studied small odontocete is the bottlenose dolphin (*Tursiops* spp.), whose communities often inhabit coastal waters that are relatively easy for researchers to access, such as those in Shark Bay, Australia, or Sarasota Bay, Florida. Within these communities, there are often sex differences in social structure. Females form flexible networks of associates, while males form stable alliances, which are necessary to compete with other males for access to females (Connor et al. 2000). In some cases, this competition has led to complex and dynamic networks of alliances and alliances of alliances (termed second-order alliances). For example, a 'super-alliance' containing 14 males which were already in first-order alliances has been documented in Shark Bay (Connor et al. 1999, 2001, 2006; Connor 2007, 2010). While these are striking features to find in non-human societies, they are not universal. For example, in the bottlenose dolphin community of Doubtful Sound, New Zealand, males and females are both integrated into a dynamic network of fission–fusion associations (Figure 13.2). In the deep ocean, dolphin groups are much larger (Gowans et al. 2007), sometimes in the thousands, probably as a response to the elevated predation threats in pelagic waters, but we know virtually nothing about how they might be structured on a finer scale.

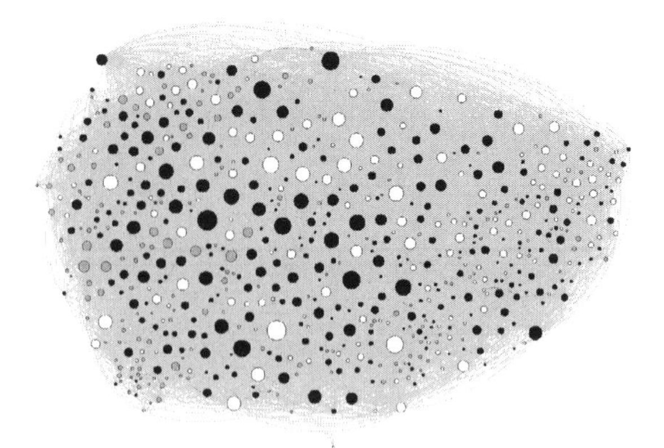

Figure 13.1 The social network of 653 humpback whales which summer in the Gulf of Maine. Data are from Allen et al. (2013). Edges are weighted based the half-weight association index, with association defined as displaying coordinated behavioural patterns while less than two body lengths apart. Nodes are sized relative to their measure of degree and shaded according to sex (males in white, females in black, and nodes of unknown sex in grey). The large cloud of weak edges and an absence of clear clustering are typical of this species' social structure. All networks in this chapter were plotted using Gephi 0.8.2 beta (<https://www.gephi.org>) and a Force Atlas 2 layout algorithm (Jacomy et al. 2012). This is a force-directed layout in which nodes repel each other whereas edges attract nodes they connect. In all cases, nodes are sized relative to their measure of degree (the number of edges connecting to the node) and coloured based on sex.

The larger odontocetes, however, have more highly structured, multilevelled societies which have at their base long-term groups that may associate with each other and, in some cases, occasionally

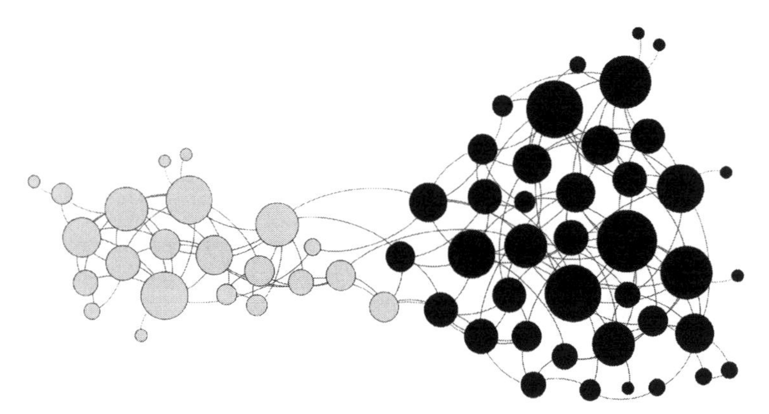

Figure 13.2 The binary social network of 62 bottlenose dolphins living in Doubtful Sound, New Zealand. Edges represent pairs which were observed in schools together more than expected by chance based on permutation tests. For more details see Lusseau (2003), but see also Whitehead et al. (2005) for caveats about representing associations in this way. Lusseau and Newman (2004) identified two communities within the population which are labelled as black and grey nodes, respectively. Nodes are sized relative to their measure of degree. No edge weighting or sex data were available.

swap members, but maintain their identity over the timescales for which they have been studied. Here the emphasis is on matrilineal groups (Whitehead 2003; R. Baird and Whitehead 2007). Perhaps most striking are the 'pods' of 'resident', fish-eating killer whales (*Orcinus orca*) in the waters of the north-east Pacific, off British Columbia, Canada, and Washington State. These pods provide a very rare, if not unique, mammalian example of natal social philopatry by both sexes—neither males nor females disperse from their natal pods but obtain mating opportunities during temporary interactions with other pods (Figure 13.3).

Finally, sperm whale social structure is based upon broadly matrilineal social 'units' containing females and their offspring, which often associate into 'groups' of two or three units over periods of hours to days. While group formation takes place in the Pacific, it appears to occur far less often in the Atlantic (Whitehead 2003; Whitehead et al. 2012; Gero, Milligan, et al. 2013). In both oceans, males disperse from their natal units in their early teens

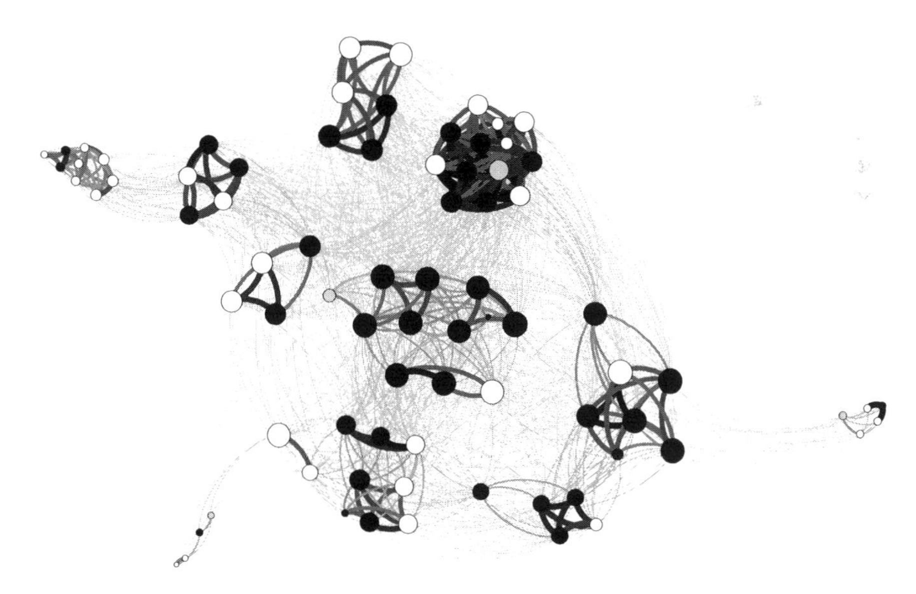

Figure 13.3 The social network of a portion of the 'northern resident' killer whale community off northeastern Vancouver Island, Canada. Edges are weighted based on observations of individuals associating within subgroups during 15 min scan samples across a daily sampling interval, collected from June to September between 1995 and 2002. Nodes are sized relative to their measure of degree and shaded according to sex (females in black, males in white, and unsexed in grey). Data are from R. Williams and Lusseau (2006). The clear modular structure of this network highlights the natal social philatropy by both sexes within pods. The less frequent associations between matrilineal pods link most of the individuals in an interconnected network that encompasses the entire community.

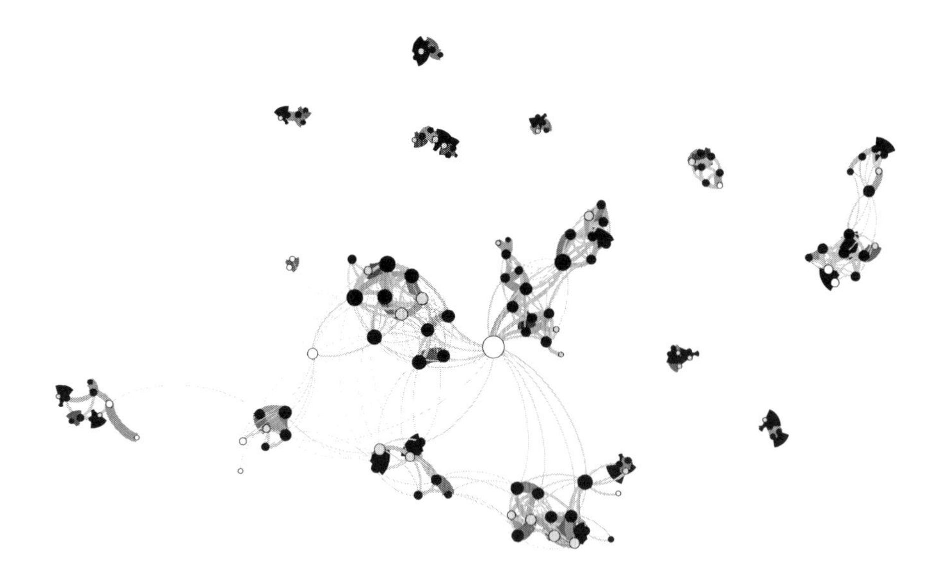

Figure 13.4 The social network of 17 sperm whale units from the community living in the Caribbean Sea. Edges are weighted based on the half-weight association index within clusters (<3 body lengths apart) at the surface within a 2 h sampling period across an 8 year study (see Gero, Milligan, et al. 2013). Nodes are sized relative to their measure of degree and shaded according to sex (females in black, calves in grey, and mature males in white). Note the modular structure in the network, with the stronger edge weight within units and the weaker edge weight between units. Many units are not connected and have spent no time together. Mature males are often the links between units by having consorted with different social units on different days. This is particularly evident for the one male in the centre of the plot, as he is associated with six different units.

(Gero, Gordon, et al. 2013), and very little is known of their subsequent social interactions, except that, when both sexually and socially mature, they rove between female social units in search of mating opportunities (Whitehead 2003). While sperm whale units in the Atlantic are smaller and may, like killer whales, only contain one matriline (Gero et al. 2008), in the Pacific, units can contain multiple, unrelated matrilines, which suggests a more dynamic social structure, albeit on timescales we find difficult to observe directly (Mesnick 2001). Nonetheless, sperm whale societies are tightly structured around the long-term bonds between females within social units (Figure 13.4).

These societies are also multilevelled, such that killer whale pods can be clustered into 'clans'. Clans were initially defined based on acoustic similarity rather than social patterns (Ford 1991)—pods in the same clan share at least one call—but this structure is also reflected in the patterns of relatedness between pods (Deecke et al. 2010). Similarly, sperm whale units in the Pacific can be clustered into 'vocal clans' and prefer to associate with other units that share the same vocal dialect (Rendell and Whitehead 2003). Again, this differs from the Atlantic, where sympatric clans have not been observed and repertoires appear to vary geographically (Antunes 2009; Whitehead et al. 2012). We will explore further this link between social networks and communication networks in 'Social networks and communication networks among cetaceans'.

Studying cetaceans using a social network approach

The focus by cetologists on the link between social structure and vocal communication has in part been driven by the relative ease with which the acoustic interactions between animals can be recorded while underwater, compared with directly identifying individuals or observing the behaviours involved in dyadic social interactions. The latter, while challenging, has been greatly helped by natural markings, which can be used to identify individuals visually in many species.

Identifying individuals in cetacean social networks

In most cases, individual cetaceans are not physically marked or banded for identification. Individuals are identified through photographic methods by natural marks on their dorsal fins or ridges (for the delphinids (e.g. Wursig and Wursig 1977)), or their tail flukes (e.g. Arnbom 1987). These methods rely on the individuals to surface in order to be identified.

Recently, however, new methods for acoustically identifying individuals have advanced our abilities to follow individuals while underwater. This is accomplished more readily in species like the bottlenose dolphin that have individual-specific signature whistles (S. King and Janik 2013; S. King et al. 2013), but innovative new technology has increase our ability to do so in species for which no individually specific vocal cues have been identified (Rutz and Hays 2009; Krause et al. 2013; Kühl and Burghardt 2013). Acoustic accelerometer tags such as the DTAG (M. P. Johnson and Tyack 2003) have made it possible to assign calls to the tagged whales after those whales have been identified by standard photographic methods at the surface during the deployment of the tag. In addition, these tags can facilitate relative localization of other individuals while submerged (Zimmer et al. 2005), allowing for interactions between whales at depth to be studied for the first time (e.g. Wiley et al. 2011). Even without the assistance of technology, studies with a thorough understanding of both how the vocalizations are produced, and the specific individuals under study, have been able to reliably identify individual whales during vocal interactions underwater (e.g. Schulz et al. 2008; Schulz et al. 2011). While these advances in underwater observation are exciting, currently the vast majority of social interactions are still quantified based on what is observable from the surface.

Interactions between individuals in cetacean social networks

Interactions between individuals form the basis for any social network analysis. In many cases however, it is difficult or impossible to directly observe or quantify directed interactions between individual whales or dolphins. As a result, social 'interactions' are often inferred by assuming that associations between individuals within specifically defined groups observed at the surface are representative of social interactions between the individuals that are associated. In most cases, these groups are defined with biologically informed spatial and temporal limits. The 'gambit of the group' assumption (Whitehead and Dufault 1999; Franks et al. 2010) is often made in nocturnal or aquatic species in which visibility is a methodological challenge or when group sizes are large and mobile, as in flocking birds (see Chapter 16). While this technique assumes that all social interactions occur during spatial and temporal proximity and that all members of groups interact equally, these assumptions can often be justified or supported through study.

In some species of cetacean, it has been shown that the majority of social interactions occur at or near the surface (Whitehead and Weilgart 1991; Watwood et al. 2006) and that individuals within spatio-temporally associated groups often do interact through vocal communication (Schulz et al. 2008; Quick and Janik 2012; S. King and Janik 2013). Furthermore, with tightly defined levels of spatial and temporal association, this assumption is likely upheld given that synchrony is costly (Conradt and Roper 2000). Therefore, in the absence of the ability to directly observe behavioural interactions due to the reality of these species ecology, the 'gambit of the group' should not be seen as inherently faulty if its assumptions can be justified.

Contributions from studies on cetaceans

Given these inherent methodological limitations, social network analysis has become the norm when describing cetacean societies. Basic network measures are now commonly included as a part of cetacean social analyses (e.g. Wiszniewski et al. 2009; Wiszniewski et al. 2010; Parra et al. 2011; Wiszniewski et al. 2012).

Methodological advances in studying cetacean social networks

Since Lusseau's (2003) paper on bottlenose dolphin social networks, further methodological and theoretical research has incorporated ideas derived from the study of cetacean societies. In particular, Lusseau and his colleagues have applied a number of novel methods, including those which enable researchers to try to identify individual social roles among cetaceans (Lusseau and Newman 2004) and quantify individual influence within cetacean networks (Lusseau et al. 2006). Others have shown that social network analysis can be used to study dynamical processes in networks with only a small number of individuals (Guimarães et al. 2007) and can also be adapted to account for the inherent uncertainty around social network measures (Lusseau et al. 2008). While subtle, these are critical points when studying cetaceans (and many other species), as many of the original network methodologies were developed for technological networks in which there is little uncertainty and a large number of nodes. Theoretical contributions stemming from studies on cetaceans include a framework on how the resulting emergent structures in social networks depend on the spatial and temporal scale of the study (Cantor et al. 2012), as well as a network formalization of the link between social learning, behavioural variation, and culture (Cantor and Whitehead 2013), and the modelling to support it (Whitehead and Lusseau 2012).

Cetacean networks and management: resilience and survival

Cetaceans can be highly reliant on their social associates for cooperative foraging (Hoelzel 1991; Gazda et al. 2005), defence against predators (Pitman et al. 2001), successful raising of young (Gero et al. 2009; Frère et al. 2010; Gero, Gordon, et al. 2013), and, for males, success in competition for mating opportunities (Connor et al. 2001). This reliance means that changes in social networks can have conservation and management consequences, and social network analysis has a role to play in understanding how cetacean social networks respond to external pressures, such as anthropogenic disturbance (Ansmann et al. 2012), disease outbreak (Guimarães et al. 2007), targeted removals (R. Williams and Lusseau 2006), and survival (Stanton and Mann 2012). Here, the term 'resilience' is used to describe the extent to which the structure of cetacean social networks is robust to such pressures.

By analysing the association patterns of bottlenose dolphins (*Tursiops aduncus*) both during and after closure of a commercial fishery in Moreton Bay, Australia, Ansmann et al. (2012) showed that the dolphins' social network radically changed structure. Prior to the closure of the fishery, 'trawler' dolphins and 'nontrawler' dolphins were socially segregated and associated only with those which shared their foraging specialization. In contrast, the post-fishery social network is more compact and less differentiated, with dolphins associating regardless of their previous foraging habits. This suggests that, while fisheries can dramatically affect social interactions, dolphin communities may be resilient to this type of anthropogenic disturbance.

Another study using bottlenose dolphins (Stanton and Mann 2012) used common social network measures to show that connectedness (in this case eigenvector centrality) was a good predictor of calf survival among bottlenose dolphins (*Tursiops* sp.). This study vividly illustrates how the network approach can produce real insights into the evolutionary drivers of sociality, even in species that are difficult to study. In contrast, Guimarães et al. (2007) studied the susceptibility of highly connected social networks among killer whales to disease outbreaks, suggesting that both the topology of the social network (high clustering and short path-length) and tie strength between individuals in this population makes them particularly vulnerable.

Lastly, R. Williams and Lusseau (2006) found in their study of killer whales that certain individuals, particularly juvenile females, had apparently central roles in their social networks, as defined by their high degree (i.e. high number of links). The authors interpreted these data as showing that juvenile females played central roles in maintaining cohesion of the social network of the community of fish-eating, 'resident' killer whales. Furthermore, when the authors simulated targeted removals of very few individuals, they found that the loss of one or two individuals might have dramatic effect on network stability. They concluded that management cannot consider all individuals in killer whale social networks as equal, as these individuals differ in the roles they play in maintaining the structure of the social network.

Social roles and decision making in cetacean social networks

Network analysis provides a useful set of tools for identifying the social roles of individuals in their families, communities, or populations that may not be overtly evident through traditional observational methods. Lusseau and Newman (2004) demonstrated that certain individuals, whom they called 'social brokers', are more important to the connectivity of their dolphin social network than all others, as they have much higher betweeness than average. The removal of these individuals, the authors suggested, could have a disproportionate effect on community cohesiveness as measured by network connectivity; a finding similar to that of R. Williams and Lusseau (2006). Interestingly though, Lusseau (2007) found that many of these same 'brokers' also performed aerial behaviours correlated with shifts in activity states, an observation which does imply a particular social role for these highly connected individuals. The performance of 'side-flops' by males was linked to the start of travelling bouts, while upside-down lobtails by females were correlated with the ends of such movements (Lusseau and Conradt 2009). Lusseau (2007) suggests that these individuals benefit from increased social knowledge of their potential competitors because of their position in the social network and as a result may be more reliable in the decision-making process.

Drivers of network structure in cetacean societies

Network analysis has also been used to examine behavioural, ecological, and cultural determinants of social structure among cetaceans. Recent studies have addressed each of these in different species. Among sperm whales, which regularly dive in excess of 300 m to forage, it has been hypothesized that the evolution of communal care for the calves was an important driver of social evolution, leading to the maintenance of long-term bonds within their stable units (P. Best 1979; Arnbom and Whitehead 1989; Whitehead 1996). Calves under 2 or 3 years old cannot, or will not, make deep dives alongside their mother while she forages (Whitehead 1996; Gero, Milligan, et al. 2013). As a result, the calf is left vulnerable at the surface for long periods. If other individuals are available to accompany the calf while at the surface, then this provides some defence from predation.

Gero, Gordon, et al. (2013) used a network approach to test if calves were in fact central nodes in the network of association within social units. By calculating common network measures across years and through changes in maternal status, as females became mothers and lost or weaned their calves, the authors determined that calves were significant nodes within units. Thus, by using network analysis, Gero, Gordon, et al. (2013) were able to provide quantitative support for this hypothesis about the evolution of social structure in sperm whales.

Ecological constraints are often drivers of emergent structures in social networks (Croft et al. 2008), as the content and quality of individual interactions often relate to the abundance of food or other resources (e.g. Tanner and Jackson 2012). E. Foster et al. (2012) used network measures to test for ecological drivers of sociality among fish-eating, 'resident' killer whales. They found a significant correlation between salmon abundance and social network structure: individuals are more interconnected in years of high salmon abundance, even when controlling for group size (E. Foster et al. 2012). While the connection between food abundance and group size has been shown previously, this study focussed on the stability of social interactions and was really the first to show the link between resource availability and the local and global connectedness of a free-ranging population's social network.

More broadly, Hill et al. (2008) suggested that ecological drivers for network structure may differ in the three-dimensional aquatic environment that cetaceans experience. In terrestrial multilevelled societies, including those among elephants, humans, and other primates, successive structures often increase in size following a scaling ratio of approximately 3 (Zhou et al. 2005; Hill et al. 2008); however, Hill et al. (2008) find that this ratio is closer to 4 in killer whales. Just how a three-dimensional environment might create differing selective pressures on network structure remains to be understood.

Culture can also influence social networks, in that behavioural preferences and social segregation between identifiable cultural groups can create population structure (Boyd and Richerson 1987; Nettle 1999; McElreath et al. 2003; Efferson et al. 2008). Several studies which used a network approach illustrate this among three different populations of bottlenose dolphins. In Shark Bay, Australia, bottlenose dolphins (*Tursiops* sp.) use marine sponges, apparently to protect their rostra when foraging in rocky habitat (Smolker et al. 1997; Krutzen et al. 2005; Mann et al. 2008). This is a cultural tradition which is primarily transmitted vertically, within matrilines, from mother to daughter (Bacher et al. 2010; Kopps and Sherwin 2012). Mann et al. (2012) showed using a network approach that 'spongers' preferentially associate with other spongers, providing evidence that behavioural traditions create structure in their social network.

Associations based on foraging specializations also appear to have structured another population of bottlenose dolphins off Laguna, Brazil, where dolphins cooperatively catch fish with human fisherman using mutually understood signals (Simões-Lopes et al. 1998; Daura-Jorge et al. 2012). In the Laguna population, dolphins that forage cooperatively with fishermen live sympatrically with those that do not, but Daura-Jorge et al. (2012) used network analysis to show that they form distinct modules within their social network. Lastly, Ansmann et al.'s (2012) study shows the result of a natural experiment on how dolphin communities can quickly respond to and become structured by cultural innovations in response to new ecological opportunities, in this case provided by the discards thrown overboard by prawn trawlers. Both the behaviour and the modularized network structure disappeared once the fishery ended. This provides strong evidence that it was the cultural behaviours which were driving the structure in the social network.

Cultural transmission and cetacean social networks

The shape and dynamics of a social network act as a substrate for cultural behavioural diversity (Whitehead and Lusseau 2012; Cantor and Whitehead 2013). Recent modelling suggests that increasing network structure, as measured by modularity, could promote behavioural diversity when social learning, of vocal dialects or foraging tactics for example, is frequent (Whitehead and Lusseau 2012). Cantor and Whitehead (2013) illustrate an emerging diversity of empirical evidence from several different species of cetacean in support of this notion. The variability of the marine environment creates a situation in which social learning is favoured over individual learning or genetic determination of behaviour (Whitehead 2007). Together with the fact that many cetaceans have modular social structures with stable social groups and prolonged parental care, including some species with natal philopatry in which both sexes remain in the family unit (Connor 2000), the opportunities for cultural transmission of behaviours are plentiful, with traits more likely to spread within groups than between them (Whitehead 1998; Whitehead and Lusseau 2012). As a result, several species exhibit high levels of behavioural variation between social groups, much of which is thought to be due to the interaction between cultural inheritance and network structure (Connor 2000; Connor et al. 2000; Connor 2001; Rendell and Whitehead 2001; Yurk et al. 2002; Sargeant and Mann 2009; Riesch et al. 2012; Whitehead et al. 2012).

This makes cetaceans a particularly interesting taxon for studying transmission of information or behaviour through social networks. This application of network analysis has been a major research topic for some time in the social sciences, where it is referred to as 'social influence theory' (Robins et al. 2001; Shoham et al. 2012). In animal behaviour, related methods have been collectively termed network-based diffusion analysis (see Chapter 5). This is a set of techniques that fit data on the time or order of acquisition of a behaviour to an adapted Cox proportional hazards model (Franz and Nunn 2009; Atton et al. 2012).

One recent study demonstrated the cultural spread of a foraging technique through a population of humpback whales using network-based diffusion analysis (Allen et al. 2013). These researchers were able to track the transmission of this novel behaviour from its apparent innovator, a single whale in 1980, and show that models with a social transmission component produced much better fits to the observed spread of the behaviour even when ecological and genetic factors were taken into account. Given what has previously been shown for song transmission in this species (Noad et al. 2000; Garland et al. 2011), this study showed that humpback populations can maintain multiple independently evolving behavioural traditions.

Social networks and communication networks among cetaceans

There is considerable overlap between social networks and communication networks (see Chapter 9), and cetaceans have contributed some very powerful comparative data to the study of this interaction. Across species of cetacean, variation in vocal communication is closely related to the complexity of social structures (Tyack 1986; Tyack and Sayigh 1997). Humpback whales produce hierarchically complex songs (Payne and McVay 1971; Suzuki et al. 2006) which change over time (Noad et al. 2000), but all of the males in a breeding population converge on the same song (Payne and Guinee 1983). By contrast bottlenose dolphins (*Tursiops* spp.) live in a dynamic fission–fusion system of constantly changing social partners, some of which are longer term associates and others rare encounters within their wider communities (Wells 1991; Connor et al. 2000).

Bottlenose dolphins have individually specific signature whistles that label individuals and are apparently used to facilitate interactions within these dynamic communities (Caldwell and Caldwell 1965; Quick and Janik 2012; S. King and Janik 2013). Furthermore, the similarity between the signature whistles of pairs of individuals relates to their type of association. Males, who generally disperse farther from their mothers than females, have signature whistles which are similar to their mothers', while female offspring develop novel whistles quite distinct from their mother and other close associates in their community (Sayigh et al. 1990; Fripp et al. 2005). Females will retain their distinct whistles (Sayigh et al. 1990), while males alter their whistles to be more similar to their partners' whistles when forming stable male–male alliances at sexual maturity (Smolker and Pepper 1999; Watwood et al. 2004). Sperm whales have a particularly interesting multilevelled society in which long-term, stable social units of adult females and their dependents appear to use socially learned dialects of vocalizations termed 'codas' to socially segregate among sympatric 'vocal clans' (Whitehead et al. 1991; Rendell and Whitehead 2003; Rendell et al. 2012; Whitehead et al. 2012). These clans may contain tens of thousands of animals and represent the largest mammalian cooperative groups outside of humans (Rendell and Whitehead 2003).

Finally, the co-evolution of diverse communication systems and complex social structure among cetaceans is best exemplified by the fish-eating 'resident' killer whales. They live in matrilines, within which both sexes remain their whole lives. Matrilines associate into 'pods' with other matrilines that share a similar socially learned traditional vocal dialect. Several genetically related 'pods', which share some or all of their call repertoires, are grouped into 'clans', and several clans make up vocally distinct communities that use distinct, although sometimes overlapping, geographic areas but never mix socially (Ford 1991; Ford 2002a,b; Ford and Ellis 2006; Deecke et al. 2010; Ivkovich et al. 2010). As a whole, this diversity of social and vocal systems means that the cetaceans are a potentially fruitful group in which to investigate comparative hypotheses about the relationship between the evolution of social structure and communication.

Current challenges and avenues for inquiry concerning cetacean social networks

Cetaceans have proven to be fertile soil for novel applications of network theory to social behaviour in animals, and we see no reason that this should not continue to be the case. Therefore, we conclude by highlighting certain areas where we feel further investigation could rapidly yield interesting findings.

Linking cetacean social networks with vocal complexity

The social intelligence hypothesis proposes that individuals living in complex social environments must not only deal with the challenges of the physical habitat but must also solve problems related to their social interactions (Byrne and Whiten 1988; Whiten and Byrne 1997). In particular, individuals need to be able to recognize and remember their interaction histories with other individuals in order to manage and express their behaviour towards each other. As a result, these individuals face selective pressures for more complex communicative signals to mediate this diversity of social interactions (Freeberg et al. 2012).

Cetaceans are well known for their vocal imitation and learning abilities (Janik and Slater 1997). Vocal learning may have evolved in response to selection for making advantageous social decisions by allowing cetaceans to recognize the signals of a wide number of conspecifics and neighbouring social groups encountered in their large ranges (Deecke et al. 2010). This link between large ranging patterns, encounters with less-known individuals across multiple levels of social structure, and vocal learning is also evident in the African elephant (*Loxodonta africana*). Elephants have the ability to learn and recognize the vocal signals of familiar and unfamiliar individuals and social groups, across multiple levels of social structure, and over large spatial and temporal scales (McComb et al. 2000; McComb and Semple 2005; Poole et al. 2005; Soltis et al. 2005).

The ocean is a good medium for sound transmission, making vocal markers for social structures even more likely among cetaceans by potentially allowing the marking signals to be effective over large distances (Tyack 1986; Connor et al. 1998). Given the cetaceans' diverse communication systems, these species provide an opportunity to study communication networks (see Chapter 9). To date, however, little has been tested, in part due to logistical difficulties in identifying callers and receivers underwater during detailed directional vocal interactions. With the development and increasing use of acoustic behavioural tags (M. P. Johnson and Tyack 2003), these types of dyadic vocal interactions will be easier to quantify and study using a network approach, and potentially be amenable to experimental investigation with playback studies in the wild (e.g. S. King and Janik 2013).

Cetacean social network analysis: beyond associations

Interpreting the outputs of network analyses is not always straightforward when assumptions about the nature of the interactions being measured by association indices can be hard to verify. This is not uniquely a problem of the 'gambit of the group'; but, as the 'gambit' features heavily in cetacean studies, we use it here to illustrate a more general point. Using the 'gambit of the group' in equating spatial and temporal synchrony as indexes of interaction means that the resultant social networks are agnostic to the nature of the actual interactions they are assumed to index. In effect, this lumps differing types of behavioural interactions rather than treating each as a different interaction network, as has been done in primates (Barrett et al. 2012).

Without supporting data on the nature and quality of the interactions being indexed by association, care is needed in interpreting social network analyses. So, for example, it may be that juvenile female killer whales are more fluid in their associations, giving them a high degree, compared to other age–sex classes, but it can be difficult to understand the importance of those associations, beyond providing network connectivity, without knowing what happens within them. Improved observation of fine-scale behaviour through technologies such as acoustic and motion recording tags will hopefully aid these interpretations by giving some indications of the types of interactions that are taking place during the observed associations. Careful observation of behavioural states can also help here, allowing researchers to construct interaction networks of associations involved in particular activities, such as socializing or foraging (e.g. Gero et al. 2005). The interpretations of roles suggested by network analyses, such as 'social brokers', will be stronger when we can describe what is being brokered.

Collective motion and decision making in cetacean societies

To stay together, cetacean social groups must synchronize movement within their three-dimensional habitat. While recent work has used a network approach to study collective motion (Bode et al. 2010;

Bode et al. 2011a,b; Bode, Franks, and Wood 2012; summarized in Chapter 8), this has not been applied to cetaceans. Given Lusseau and Conradt's (2009) finding that highly social individuals appear to lead the decision making related to travelling among bottlenose dolphin groups, or Whitehead's (1996) finding that sperm whales change their patterns of dive synchrony when calves are present, there is likely far more to be understood about how cetaceans coordinate movement, and there is much relevant theory waiting to be applied (e.g. Berdahl et al. 2013).

Network analysis of cetacean societies: conclusions and future directions

Cetaceans provide a much needed comparison with terrestrial species, particularly other big brained mammals, for the study of the evolution of social structures. The diversity of social systems and oceanic habitats found in this taxon offer a potentially rich resource for comparative research. While they are a challenge to study and are not the best subjects for experimental designs, cetaceans have provided a great deal of new insight into animal social dynamics and will likely do so in the future as novel network analysis techniques are applied to their oceanic societies.

Acknowledgements

Both S. G. and L. R. were supported by the Marine Alliance for Science and Technology for Scotland pooling initiative and their support is gratefully acknowledged. The Marine Alliance for Science and Technology is funded by the Scottish Funding Council (grant reference HR09011) and contributing institutions. We thank David Lusseau and Rob Williams for contributions of data towards the figures.

CHAPTER 14

The network approach in teleost fishes and elasmobranchs

Jens Krause, Darren P. Croft, and Alexander D. M. Wilson

Introduction to networks in teleost fishes and elasmobranchs

Teleost fishes are the most species-rich taxon among vertebrates, with over 25,000 marine and freshwater fishes around the world (Moyle and Cech 2004). They also hold, together with the amphibians, the record for the smallest vertebrates (genus *Paedocypris* (7–10 mm) and male anglerfish (*Photocorynus spiniceps*) (6–7 mm)) and range up to 11 m (oarfish (*Regalecus glesne*)). Given the great number of species and their wide geographic distribution, it may not be unexpected that teleost fishes show huge diversity in morphology and behaviour, and this in turn provides many research opportunities (Magnhagen et al. 2008; Rocha et al. 2008; Moyle and Cech 2004).

Many smaller fish species have been studied extensively and have become model species in ecology and evolution. For example, our understanding of speciation has been advanced in major ways by the study of threespine sticklebacks (*Gasterosteus aculeatus*) and African cichlids (Schluter 2000). Work on the guppy (*Poecilia reticulata*) which has become a major model organism in evolutionary ecology over the last 50 or so years, has shed light on questions regarding sexual selection, life-history evolution, and the evolution of group living. For example, transplants of guppy populations from high-predation to low-predation sites in Trinidad provided detailed information regarding changes in shoaling behaviour and other behavioural and morphological traits over a period of 30+ years (Magurran 2005).

Teleost fishes also show some of the greatest diversity in mating systems (monogamy, polygyny, polygynandry, and polyandry) and parental care (uniparental care, biparental care, and the complete absence of parental care) among vertebrates (Rocha et al. 2008). For example, some fish species are clonal (e.g. mangrove killifish (*Kryptolebias marmoratus*) and the Amazon molly (*Poecilia formosa*)), whereas the male of the above-mentioned anglerfish (*Photocorynus spiniceps*) fuses permanently with the body of the female. Pipefishes and seahorses provide excellent systems for the study of sex role reversal, and some gobies and wrasses undergo sex changes in response to environmental changes (Patzner 2008; Taylor and Knight 2008).

Some fish species are largely solitary (such as pike (*Esox lucius*)), whereas others form some of the largest aggregations of vertebrates on record (e.g. sardines in South Africa (Freon et al. 2010) and Atlantic cod (*Gadus morhua*) (DeBlois and Rose 1996)). Among those species living in groups, many form so-called fission–fusion systems, which are characterized by highly dynamic changes in group size and composition (see also Chapter 8). This social organization is typical of many fish species (particularly in pelagic environments). In contrast, we find many territorial species in coral reefs and also in tropical freshwater habitats that spend large parts of their lives within just a few square metres (e.g. damselfishes (Pomacentridae)). Identifying an approach that can handle this huge diversity in social organization is a challenge and, in recent years, the social network approach has become a front-runner for precisely that.

Animal Social Networks. Edited by Jens Krause, Richard James, Daniel W. Franks and Darren P. Croft.

The modern approach of using social networks to analyse animal behaviour started primarily with studies on teleosts and cetaceans (Ward et al. 2002; Lusseau 2003; Croft et al. 2004; Krause et al. 2007), although more conventional sociograms had long been in use in primatology (see Chapter 12). This might seem surprising: given that the network approach was first developed for use on human relationships (see Scott 2000 for a review), it should easily transfer to non-human primates, but aquatic organisms do not seem an obvious candidate for social network studies. However, methodological advances in statistical physics heightened the importance of quantitative analysis techniques and replication, which are much easier to obtain for small organisms such as guppies or sticklebacks since these species can be experimentally manipulated both in the laboratory and in the field (Ward et al. 2002; Croft et al. 2004, 2005). Furthermore, there was already a rich literature on assortative behaviour by factors such as body size, sex, and species for teleost fishes (see Krause et al. 2000 for a review) and it was only a small step to extend this approach to looking at associations between individuals in the context of social networks (Croft et al. 2004).

Teleost fishes provide important study systems in the fields of animal behaviour and behavioural ecology, including work on predator–prey relationships, familiarity, cognition, sexual selection, cooperation, collective behaviour, and decision making (Krause and Ruxton 2002). In this chapter, we review those topics where the network approach has made a particular contribution. Given that fish are increasingly used not only as model systems for pure research but also play an increasingly important role in aquaculture, we have included a section on welfare (see also Chapter 11). We also have included a section on methodological issues regarding biometry and biologging and discuss promising developments in this area and their implications for the study of fish behaviour. Given that many of the methodological issues and research questions regarding sharks and rays are similar to those in teleost fishes, we decided to include the literature on elasmobranchs in our review (Table 14.1).

Population structure of teleost fishes and elasmobranchs

It has long been recognized that fish can have preferences (or the reverse) for particular conspecifics regarding different attributes in the context of, for example, shoaling behaviour, mate choice, or predator inspection behaviour (Tinbergen 1951; Metcalfe and Thomson 1995; Krause et al. 2000; Candolin and Wong 2008). However, this information on social and sexual preferences had not been put into the context of social units larger than that of the dyad or the group (Krause et al. 2000). There are at least two reasons for this restriction. First, an analysis of preferences and association patterns of individuals of an entire population requires that all individuals be identified over the course of the study period so that, in some cases, sophisticated marking and tracking techniques are needed. Second, the study of individual association patterns required a new conceptual and analytical framework.

As mentioned in Chapter 1, the social network approach provides an excellent tool for studying social behaviour at different organizational levels (individual, dyad, group, community, population, meta-population, and species), and novel marking and tracking procedures (see 'Techniques for identifying individuals in teleost fishes and elasmobranches') helped address the other problem (see also Krause et al. 2013). It remains, however, a great methodological challenge to carry out monitoring programmes of individual association patterns within entire populations. As this continues to be a stumbling stone for many projects, we discuss it here in some detail.

Techniques for identifying individuals in teleost fishes and elasmobranchs

The network approach requires that individuals be recognized via unique marks and thus often requires the use of identification tags. However, identification tags are not necessary for animals such as whale sharks (*Rhincodon typus*) (Arzoumanian et al. 2005; Holmberg et al. 2009), eagle rays (*Aetobatus narinari*) (Krause, Mattner, et al. 2009),

Table 14.1 Overview of the teleost and elasmobranch species and subject areas which have been investigated using the network approach.

Species	Topic	Reference
Teleosts		
Guppy (*Poecilia reticulata*)	Social structure, cooperation, personality, sexual behaviour, environmental factors, fission–fusion, parasite transmission	Croft et al. 2004, 2005; Couzin et al. 2006; Croft et al. 2006; Morrell et al. 2008; Thomas et al. 2008; Croft et al. 2009; Darden et al. 2009; Edenbrow et al. 2011; Kelly et al. 2011; Croft et al. 2012
Threespine stickleback (*Gasterosteus aculeatus*)	Personality, partner preferences	Ward et al. 2002; Pike et al. 2008; Atton et al. 2012; Webster et al. (2013)
Zebrafish (*Danio rerio*)	Social role	Vital and Martins 2011
Atlantic salmon (*Salmo salar*)	Welfare, aggression	Cañon Jones et al. 2010, 2011, 2012
Sunbleak (*Leucaspius delineatus*)	Invasive species	Beyer et al. 2010
Cichlids (*Neolamprologus pulcher*)	Aggression, dominance	Schürch et al. (2010); Dey et al. (2013)
Elasmobranchs		
Lemon shark (*Negaprion brevirostris*)	Partner preferences, leadership	Guttridge et al. (2010, 2011)
Eagle ray (*Aetobatus narinari*)	Partner preferences	Krause et al. (2009)
Catshark (*Scyliorhinus canicula*)	Social structure	Jacoby et al. (2010); Jacoby, Croft, et al. (2012)
Blacktip reef shark (*Carcharhinus melanopterus*)	Social structure	Mourier et al. (2012)

and manta rays (*Manta alfredi*) (Kitchen-Wheeler 2010), which have natural variations in pattern that allow for the identification of individuals. In recent years, rapid advances have been made in animal biometrics, so that automated recognition of individuals on the basis of their phenotypes is now possible for a wide range of species (Kühl and Burghardt 2013). This has resulted in the creation of online databases such as the ECOCEAN Whale Shark Photo-identification Library (Holmberg et al. 2009) and is likely to become more widespread given technological developments in pattern recognition such as machine learning (Kühl and Burghardt 2013). In addition, recent work shows that mosquitofish can be individually recognized (when alone and even when in small groups) with high probability based on their movement patterns (Herbert-Read et al. 2013).

Different types of marking techniques exist depending on whether the fish are studied in the laboratory or the field, the size of the animals, and whether observations are done directly by a human observer or remotely by a machine (Krause et al. 2013). Remote sensing involves the attachment of an electronic device (often surgically implanted) such as a passive integrated transponder tag, an active transmitter for telemetry purposes, or a proximity logger. In the case of passive integrated transponder tags, data is collected when fish swim over an antenna which reads and records the code identifier together with a time stamp (Klefoth et al. 2012). The arrival sequence of individuals at the location where the antenna is positioned can provide information from which association patterns of individuals can be reconstructed (Psorakis et al. 2012). However, receivers often have problems with the registration of the simultaneous arrival of two or more individuals (Klefoth et al. 2012).

Hydroacoustic telemetry can be used to obtain the three-dimensional location of fish in entire lakes or marine bays, but the spatial resolution of these systems can be a limiting factor in defining associations (Jacoby, Brooks, et al. 2012; S. Cooke et al. 2013; Krause et al. 2013). Finally, proximity loggers can record the identity of other loggers (carried by fish) but do not provide information on where the encounter took place. The signal strength and frequency is usually a function of the distance between the loggers (Guttridge et al. 2010).

For direct observations in the laboratory and field, various wire tags and subcutaneous dyes can be used (Beukers et al. 1995). A popular method is the use of a fluorescent elastomer, which comes in different colours, can be injected subcutaneously, and is externally visible over periods of months or even longer (Croft et al. 2004). Other methods include applying tags externally to the fish: Webster (Webster and Laland 2009), for example, developed a method for sticklebacks whereby polyvinyl chloride discs are mounted on the dorsal spines. All marking or tagging procedures should be accompanied by controls which establish whether and to which degree the behaviour of the fish is affected.

Guppies and sticklebacks—a case study

Partly because of their small size but also because of the great research interest, guppies and threespine sticklebacks were the first fish for which social network studies were carried out (Ward et al. 2002; Croft et al. 2004). Given that they live in so-called fission–fusion systems (see 'Collective behaviour among teleost fishes') it seemed unlikely, if not outright impossible, 10–15 years ago that particular individuals would show a tendency to remain together over periods of days and weeks in the wild, where hundreds or potentially thousands of different fish encounter each other every day (Croft et al. 2003). Novel marking procedures in combination with the network approach greatly enhanced our understanding of the social substructure of some freshwater fish species such as guppies and sticklebacks. Social network analysis allowed the identification of population substructures, so-called communities, in which some individuals are more closely connected than with others in the population (Figure 14.1). It is known that community structure in humans has important implications for processes such as disease and information transmission (Granovetter 1973; Newman 2010). However, in fish populations this idea remains to be tested.

An assessment of the social networks of guppies revealed that individuals with many social contacts were found to interact more with each other than with other conspecifics—a phenomenon known as a 'positive degree distribution' (see Chapter 2). Positive degree distributions have been observed in many social networks (Newman 2010) and have implications for transmission processes of information and disease, affecting, for example, the speed with which a pathogen could spread in a population. Furthermore, guppy social networks were found to be assorted by personality traits, as individuals of similar attributes were more likely to be connected in the wild (Croft et al. 2009; A. Wilson et al. 2014).

In a study where two behaviour types, shoaling and predator inspection, were measured, a strong negative correlation was found between them. Fish were found to be highly assorted by a composite variable of the two behaviours, possibly indicative of the willingness of individuals to cooperate (Croft et al. 2009). Pike et al. (2008) reported in sticklebacks (*Gasterosteus aculeatus*) that bold and shy fish differed in their connectivity, with bold individuals having fewer interactions, which were more evenly distributed, and shy ones having more interactions, with stronger preferences for particular individuals. It has recently been shown that this variation in behaviour types within populations can have ecological consequences in the context of behaviourally mediated trophic cascades (Ioannou et al. 2008; Wolf and Weissing 2012).

Outlook for population applications of social network studies

The social network approach, when applied to entire populations, potentially provides interesting opportunities for studying the ability of some fish species to disperse and invade new areas. For example, guppies (*Poecilia reticulata*) and mosquitofish

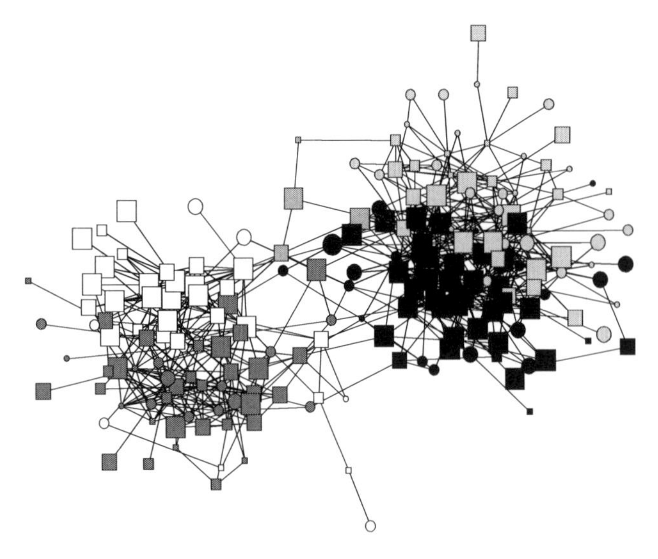

Figure 14.1 A social network of a guppy population in Trinidad. All guppies from two interconnected pools were marked and released. Over the next two weeks, approximately 20 shoals were captured daily, and fish that belonged to the same shoal were connected in the network. Over time, a completely connected network developed that comprises 197 fish. Each circle represents an individual male fish, and each square an individual female. The size of the symbol is indicative of the body length of the fish. Individuals interconnected by lines were found together at least twice. Five distinct communities (indicated by different shades of grey) were identified in the guppy network.

(*Gambusia holbrooki*) are two species that are known for their ability to invade and colonize new areas around the world. The question is whether their network structures can provide us with clues as to why these species are so successful. A study by Beyer et al. (2010) on sunbleak (*Leucaspius delineatus*) provides a first attempt in this interesting direction. The authors suggest that part of the invasive success of the sunbleak may be due to its ability to socially integrate with native fish species.

Another promising area with untapped potential for population applications of social network studies can be found in coral reefs, which are characterized by large numbers of fish (and invertebrate) species (in the adult stages) that show high degrees of site fidelity where they frequently interact with each other. The interconnected nature of life on reefs via mutualistic, aggressive, and other types of interactions provides the perfect setting for network studies to investigate the importance of particular species for the reef community (which can be experimentally tested in removal experiments). The fact that many isolated portions of reef exist also means that replication can be achieved.

Familiarity and site fidelity in teleost fish and elasmobranchs

Investigations of social networks usually require that researchers can identify individual fish (Croft et al. 2008) but they do not necessarily require that the fish can individually recognize each other. Social recognition mechanisms can operate at different levels, and individual recognition is just one of them. In fact, surprisingly few critical tests have been carried out on individual recognition in fish, given how important this ability is for the study of reciprocal altruism (Griffiths and Ward 2011). In contrast, numerous tests have been done on familiarity, identifying the ability of fish to recognize others which share the same habitat and therefore smell familiar (cue familiarity) (Ward et al. 2004).

Most of the work which has been done on social networks in fish has been carried out on individuals that were from the same local area or pool and were familiar with each other. However, this means that work on fish has largely missed out on a topic of particular interest in the social network literature: namely, how different communities are interconnected via weak ties by a few individuals which have links to more than one community (Granovetter 1973). Exceptions are the work on adjacent guppy pools where some individuals crossed between pools (Figure 14.1), and on eagle rays (*Aetobatus narinari*) and blacktip reef sharks (*Carcharhinus melanopterus*) roaming around islands (Croft et al. 2006; Krause, Lusseau, et al. 2009; Mourier et al. 2012). The study on eagle rays highlights the need to assess and control for spatial preferences (site fidelity) of individuals when testing for association patterns (Krause, Lusseau, et al. 2009). Moreover, studies on captive guppies (Darden et al. 2009) and catsharks (*Scyliorhinus canicula*) (Jacoby et al. 2010) showed that the presence of males which sexually harass females can disrupt social structure and prevent females from developing social familiarity. Thus, more work is needed to understand the influence of group composition on the emergent network patterns in the field.

Cooperation in teleost fishes

Work on threespine sticklebacks and guppies in the late 1980s and 1990s pushed the boundaries of our understanding of the evolution of reciprocal altruism, suggesting that some fish species have the ability to individually recognize conspecifics, to remember the outcome of social interactions, and to use this information to determine their future social interactions (Dugatkin 1997). Most of the early work on these species was done in the laboratory and thus begged the question as to whether the patterns of cooperation that were observed in captivity were also present in and relevant for wild populations of fish. The expectation (based on lab work) was that fish in the wild would have preferred social partners with whom they spend large amounts of time, particularly during those periods when cooperative behaviours are required (e.g. during predator inspection). On the other hand, fission–fusion models predicted a regular exchange of individuals between groups (Couzin et al. 2002; Couzin and Krause 2003), and empirical evidence showed that fish were not faithful to a particular shoal (Hoare et al. 2000) but had frequent encounters with other shoals and switched between them (Croft et al. 2003).

A number of studies combining both laboratory and field work on guppies indicate that individuals, particularly females, have preferred social partners, with which they spent more time than with others (Croft et al. 2004, 2006). Preference tests showed that fish recognize each other (Croft et al. 2006; Ward et al. 2009) and prefer those individuals which they are often seen associated with in the wild (Croft et al. 2006). It was observed that strong associations between pairs of fish were a good predictor of which individuals cooperated during predator inspection (of a potentially dangerous pike cichlid (*Crenicichla frenata*)) (Croft et al. 2006) and that female guppies formed cooperation networks of individuals that frequently cooperated with each other and avoided defectors (individuals which do not cooperate during risky predator inspection) (Croft et al. 2009). Importantly, studies on adult guppies demonstrated that kinship is unlikely to explain the social associations observed in networks (Russell et al. 2004; Croft et al. 2012), although it seems to play a role in juveniles (Piyapong et al. 2011).

The studies we describe above were all carried out on guppies and sticklebacks. However, it is clear that many other study systems that are focussed on the evolution of cooperation could benefit from the network approach. One of them is the cleaner–client system, which has produced fascinating insights into the mutual relationship between cleaner wrasses (*Labroides dimidiatus*), which clean a vast number of different fish species in tropical coral reefs around the world, and the fish they clean (Bshary 2003). Clients of cleaners could be marked or fit with tracking devices that allow long-term monitoring of visits to cleaning stations in response to the differences in cleaner strategies. Similarly, studies on the fish community response to the presence or absence of cleaner fish in a reef system could be monitored by studying changes in the intra- and interspecific interaction patterns.

In addition, in many cichlid species, dominance hierarchies are studied in connection with cooperative breeding. Social networks provide an ideal tool for quantifying aggressive interactions within breeding groups (Dey et al. 2013). Therefore, they might be useful for generating predictions regarding removal experiments.

Fish cognition and social learning

In the past, teleost fishes were looked down upon as being cognitively limited compared to other more 'highly' developed vertebrate groups such as mammals and birds. This has changed, and studies on fish cognition have produced evidence for an impressive range of abilities, from social learning, traditions, and culture to Machiavellian intelligence (C. Brown et al. 2011). Today, teleost fishes are frequently used for the study of collective cognition, providing excellent examples of how groups can overcome environmental challenges which are problematic for singletons (Ward et al. 2008, 2011). In studies on collective decision making in birds, Nagy et al. (2010) showed that the social network is an important predictor of who follows whom in a small flock of flying pigeons. However, despite the fact that variation in behaviour types is common in fish populations, little information exists linking social network structure to decision-making processes (alhtough see Chapter 8).

Empirical evidence on mate choice copying in guppies was provided by Dugatkin (1992) and has since been reported for a number of different fish species as well as species from other taxonomic groups. Who copies whom, in what context, and what pathway the information takes through a group or population has been the subject of many studies (Witte and Nöbel 2011). The social network approach provides many opportunities to study this topic experimentally in fish populations and should be an interesting area for future work (see also Chapter 5).

Collective behaviour and social networks in teleost fishes

Studies on teleost fishes have made great contributions to our understanding of animal collective behaviour and more specifically issues of leadership, decision making, and collective cognition (Huth and Wissel 1992; Couzin et al. 2002; Couzin and Krause 2003; Sumpter et al. 2008; Ward et al. 2008; Hartcourt et al. 2010; Herbert-Read et al. 2011; Katz et al. 2011; Ward et al. 2011; Ioannou et al. 2012). Methodological progress has been made with interactive fish robots that are recognized as conspecifics by live fish and can be used for investigating decision-making problems in fish schools (Faria et al. 2010; Krause, Winfield, et al. 2011). This development, combined with largely automated tracking of individual fish (Herbert-Read et al. 2011; Katz et al. 2011; Herbert-Read et al. 2013), opens up the potential for building up large databanks of association patterns. These developments create opportunities for addressing one of the big challenges in the field of network studies in vertebrates: closing the gap between group patterns and dynamics (Krause, Mattner, et al. 2009; Krause et al. 2013).

Usually, scan samples or transect data are recorded to provide information about social patterns of animals, but these techniques themselves do not shed much light on the underlying dynamics of such groups. However, A. Wilson et al. (2014) developed an approach which captures the fission–fusion dynamics of fish populations in the wild and demonstrates how the gap between pattern and dynamics may be closed. The authors used focal follows of individual fish whose nearest neighbour was recorded every 10 s over a period of 1.5 min, and thereby built up information on the fine-scale temporal association patterns of fish in the wild. From this information, they developed a model of the social dynamics of guppies (*Poecilia reticulata*) which allowed them to characterize the fission–fusion behaviour and social network structure of this species (Figure 14.2) (this approach could also be used to examine other behaviours such as foraging, aggression, mating, and cooperation).

Using this model, it was possible for the authors to captures the general social dynamics of the guppy, including individual preferences, (Figure 14.2). Furthermore, the authors used this information to explore what impact guppy social dynamics could potentially have on important transmission processes within the population, thereby changing the emphasis of the social network approach (as applied to animals) from generating purely descriptive, mechanistic studies to producing functional and predictive ones. Finally, the study by A. Wilson et al. (2014) showed that some levels of social dynamics are not individual dependent and might facilitate population or even species comparisons and therefore provide an excellent basis for broad phylogenetic comparisons.

Application of social network analysis to welfare in teleost fishes

The network approach has attracted attention also in the applied sciences, and Chapter 11 gives a detailed account of its use in the area of animal welfare. In fish, the network approach has been instrumental in investigating aggressive behaviours in the context of aquaculture. A series of studies used the network approach to quantify intraspecific

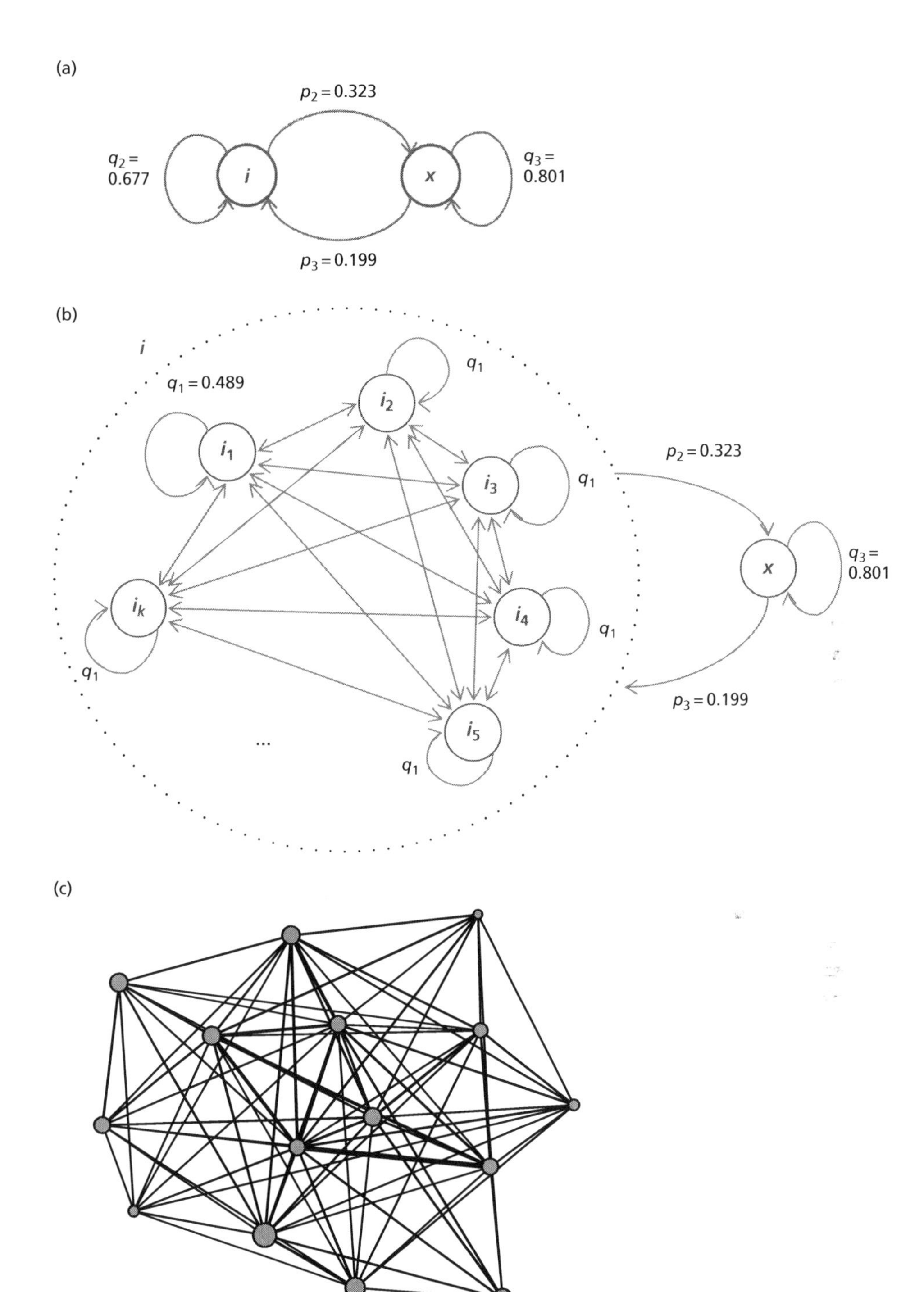

Figure 14.2 Egocentric fission–fusion model of social behaviour in the guppy. (a) A fish can either be social (with a conspecific) denoted by *i*, or alone (no conspecific within four body lengths), denoted by *x*. The variables p_2 and p_3 indicate the respective probabilities for changing states, and $q_2 = 1 - p_2$ and $q_3 = 1 - p_3$ indicate those for retaining current states. (b) A detailed model of state *i* in the presence of *k* potential nearest neighbours (see A. Wilson et al. 2014 for details of the model). An individual is in the substate i_g of *i* if individual *g* is its nearest neighbour. If a focal individual has chosen a nearest neighbour, it stays with it with probability q_1. (c) This model can be used to generate social networks typical of guppies in terms of contact frequencies and individual preferences.

aggressive behaviours in salmon (*Salmo salar*) in different animal welfare contexts such as feed restriction (Cañon Jones et al. 2010), stocking densities (Cañon Jones et al. 2011), and unpredictable feed delivery (Cañon Jones et al. 2012; Table 14.1). The focus of these studies was on the causes and consequences of fin damage in salmon aquaculture. The authors used social network tools to quantify associations (defined by spatial distance) and aggressive behaviour, calculating in- and out-degree (i.e. actions received and initiated by a focal individual; see Chapter 2) for individual fish to reflect the directedness of aggressive behaviours. The latter approach allowed the authors to identify initiators and receivers and then relate this information to fin damage and growth.

Under feed restriction, salmon networks were observed to show a higher density and greater cluster coefficients than control networks did (Cañon Jones et al. 2010). Feed restriction, unpredictable feed delivery, and high densities all resulted in individuals differentiating into receivers and initiators of aggression as reflected in their in- and out-degrees in the aggression network. Initiators showed higher growth rates and less fin damage (Cañon Jones et al. 2010). The work by Cañon Jones et al. made use of directed interactions (where an initiator and a receiver can be clearly defined), in contrast to many studies on guppies or sticklebacks, for which associations are usually assumed to be symmetric (i.e. in the interest of both partners; see Chapter 6).

Network analysis of teleost fishes and elasmobranchs: conclusions and future directions

Use of the social network approach has thus far been limited to a relatively small number of fish species and research topics (see Table 14.1) and is not representative of the breadth of activities going on in fish behavioural research. We hope that this chapter, and this book in general, will contribute to changing this situation by creating a greater awareness of the range of topics for which social networks (and teleost fishes) have relevance. Many important questions and predictions regarding sexual selection, cooperation, collective behaviour, parasite transmission, and social learning could be addressed by carrying out social network studies on fish.

One of the great appeals of working on teleosts is that many behavioural studies can be completed in relatively short time periods and at moderate costs as compared to most other studies on vertebrates. Many species of poeciliids, sticklebacks, damselfish, and cichlids are very amenable for experimental use because they are often relatively easy to maintain and breed in the laboratory and to study in the field. Individuals of these species are often available in very large numbers which allows for replication and manipulation. Also, the entire genome has been sequenced for a number of fish species including the platyfish (*Xiphophorus maculatus*), medaka (*Oryzias latipes*), fugu (*Fugu rubripes*), zebrafish (*Danio rerio*), tetraodon (*Tetraodon nigroviridis*), and threespine stickleback (Gasterosteus aculeatus), and some species are naturally clonal (e.g. mangrove killifish (*Kryptolebias marmoratus*) (Tatarenkov et al. 2009) and Amazon molly (*Poecilia formosa*) (Schlup et al. 1994; Schartl et al. 1995)), providing great opportunities for separating environmental and genetic effects. The zebrafish (*Danio rerio*) has become such an established study system in developmental biology that it is surprising that so little work exists which puts existing knowledge of its biology into context with social network studies (but see Vital and Martins 2011 for an exception).

Recent technological developments make it possible to study the interaction patterns of fish species which cannot be easily directly observed because they live in deeper water and range over greater distances (Krause et al. 2013). We discussed the use of proximity loggers, passive integrated transponder tags and hydroacoustic arrays in 'Techniques for identifying individuals in teleost fishes and elasmobranchs'. These approaches provide an interesting way forward to study the behaviour of large numbers of individual fish and their interactions. Hydroacoustic telemetry has the advantage that, since it provides information about the three-dimensional position of individuals, the proximity of other individuals can be inferred (Cook et al. 2013). Therefore, both the location of an individual and its proximity to conspecifics can be determined. In contrast, proximity loggers do not provide information on where encounters took place.

A potential weakness of both proximity loggers and hydroacoustic arrays is that spatial proximity does not automatically mean that an interaction took place (Krause et al. 2011). Further information about the duration of encounters and the location where the encounters took place is required and additional sensors (which can pick up heart or respiration rates) have to be used, in order to decide whether an encounter resulted in an interaction and to characterize the nature of that interaction (Krause et al. 2013). Implantable microchips for fish are a promising way forward in this direction (Staaks pers. communication).

At the other end of the spatial spectrum, new options have emerged for tracking the behaviour of individual fish in tanks (Herbert-Read et al. 2011; Katz et al. 2011; Herbert-Read et al. 2013). This approach makes it possible to obtain information on social networks in an automated (or at least semi-automated) way. Improved computer vision and tracking software facilitate the accumulation of huge datasets which give accurate information on individual locations in time and can be used for constructing activity profiles, characterize personalities, and infer social interactions to construct networks.

Finally, it has become evident that social networks can have important consequences for evolutionary and ecological processes (Kurvers, Krause, et al. 2014). We can use the network approach to gain an understanding of how and when individuals disperse (Blumstein et al. 2009) and how populations respond to environmental perturbations or invasive species (Beyer et al. 2010; Kurvers, Krause, et al. 2014). This type of research should provide a major growth area in future years, linking studies in animal behaviour with broader topics in ecology and evolution.

Acknowledgements

Financial support was provided to A. D. M. W. by the Alexander von Humboldt Foundation, to J. K. by the Pakt programme for Research and Innovation of the Gottfried Wilhelm Leibniz Gemeinschaft, and to D. P. C by the Leverhulme Trust. We thank Stefan Krause for comments.

CHAPTER 15

Social networks in insect colonies

Dhruba Naug

Introduction

Social insect colonies have been long admired and studied for their remarkable group-level properties such as division of labour, collective foraging, and information sharing, each of which largely emerges from bottom-up processes at the level of the individuals, without any central control or supervision (Camazine et al. 2003). Given that the number of individuals in some social insect colonies is very large, it is indeed impressive that they can act in a coordinated and cohesive fashion to produce work that is both flexible and adaptive to changes in the colony and the outside environment. It is obvious that accomplishing this feat requires a rapid flow of information and materials among the colony members, allowing the different individuals working in a variety of tasks to rapidly assess the state of the colony and the outside environment and then convey it to others. In fact, how the numerous interdependent parts of a group communicate with each other and behave in a coherent fashion is a fundamental question for all biologists as well as students of numerous other disciplines, be it the manner in which neurons interact within a nervous system, how genes interact within a genome, or how a group of robots can accomplish a specific task (Kauffman 1993).

Until quite recently, the large majority of studies have addressed questions regarding colony organization in social insects, with a focus on the characteristics of the individuals that perform the different tasks, and not so much on the details of the interindividual interactions that act as conduits for both materials and information to coordinate the activities of the workers. However, it is being increasingly recognized that the properties of a biological organization cannot be entirely revealed by studying only the constituent units, because the functional integrity of the group is defined by the ability of these units to interact and communicate with each other. This has led to a large increase in the number of studies that explore the social interactions in the colony using a network approach to understand the different processes involved in colony functioning (Naug 2009; Pinter-Wollman et al. 2011; Jeanson 2012). This approach basically consists of first visualizing each individual as a node, and any interaction between two of them as an edge or a vertex between these nodes, and then analysing the properties of the resulting network.

The first hint of what can be seen as a network approach in understanding social insect behaviour can be seen in the study by Oster and Wilson (1978) on the different possible designs of colony social organization. The idea here revolves around how a task consisting of multiple parts (requiring a transfer of material, information or both) can be partitioned among colony members (Figure 15.1), the three possible schemes being (1) series, when all parts of the task must be completed in sequence by a solitary individual, (2) parallel–series, when all parts of the task are still completed in sequence by an individual, but many individuals do the entire sequence in a colony, and (3) series–parallel, when an individual needs to complete only one part of the task, leaving the rest to be completed by someone else.

Animal Social Networks. Edited by Jens Krause, Richard James, Daniel W. Franks and Darren P. Croft.

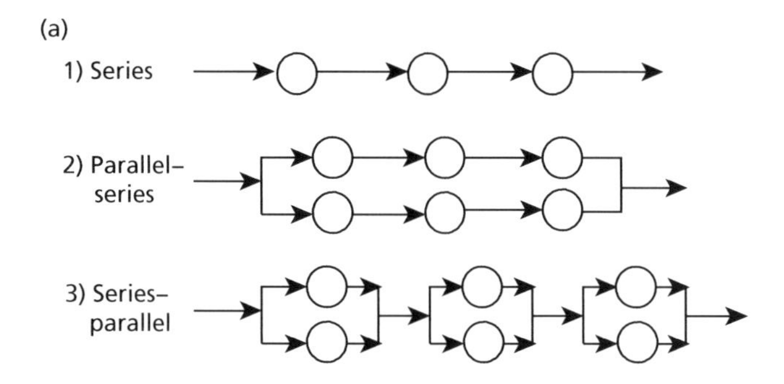

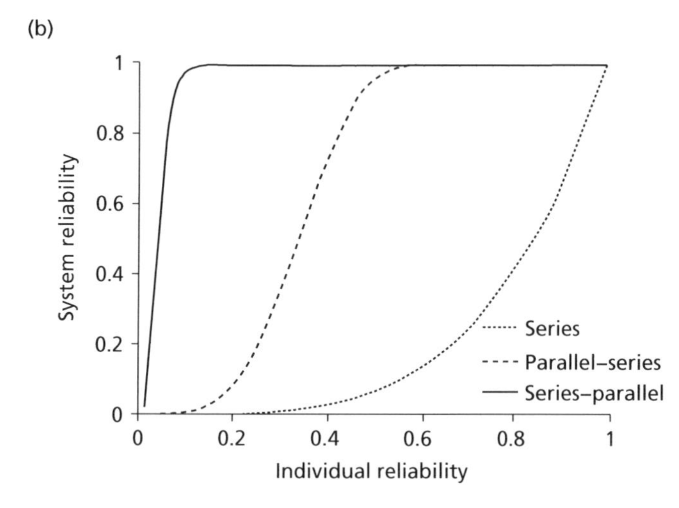

Figure 15.1 (a) Three possible schemes of colony organization and (b) system reliability as a function of individual reliability in these different schemes (after Oster and Wilson 1978; reproduced with permission from Princeton University Press). If individual reliability is defined as the probability p of an individual successfully performing one part of a task successfully, then system reliability (P) or the probability of the entire task sequence being completed in the colony at least once is $P(\text{series}) = (p)^n$,

$$P(\text{series–parallel}) = 1-\prod_{j=1}^{m}(1-]\prod_{i=1}^{n}[p_{ij}) \text{ and}$$

$$P(\text{parallel–series}) = \prod_{i=1}^{n}\left[1-\prod_{j=1}^{m}[(1-p_{ij})\right]]$$

where m is the total number of tasks, and n is the total number of individuals.

With each successive scheme, there is an increase in the reliability of the system for successfully completing the entire task sequence. What allows the redundancy or the potential swapping of task components among individuals is essentially a high level of connectivity among them, or a decentralized network. This kind of an organization was later described as a 'heterarchy' (Wilson and Hölldobler 1988), where the properties of the higher levels in the organizational design affect the lower levels but, unlike a hierarchy, the activity of the lower level units also feeds back to influence the higher levels.

Social interactions and their proximate basis

The connectivity among the individuals in an insect colony is largely based on social interactions that are either antennal or mouth-to-mouth contacts involving the transfer of chemicals, food, or other materials, although even simple physical contact can serve some of the same functions. These interactions can be a powerful mechanism through which individuals can gather information about the global state of the colony, an awareness which otherwise seems well beyond the cognitive capacity of a single individual. For example, the rate of interactions can inform an individual about the colony population density or the size of a task group without actually requiring it to count (Pacala et al. 1996). The time taken to find an appropriate partner and initiate an interaction can inform a worker about the demand level for a certain task (Jeanne 1986; Seeley 1989). Empirical work with honeybees as well as theoretical models have shown that social interactions transferring an inhibitor can also modulate the plasticity of division of labour in the colony (Huang and Robinson 1992; Naug and Gadagkar 1999; Beshers et al. 2001).

As the flow of materials and information among colony members is most pertinent among workers that form either similar or complementary task groups, the connectivity in the colony should conform to a structure that serves this process well, rather than being random or uniform among all individuals across all task groups. For example, a high level of connectivity among foragers is critical for exchanging information about the location of different food sources in order to mount a rapid response to the highly dynamic resource environment. Similarly, information flow between the foragers and the brood-caring nurses is critical for the two groups to provide feedback to each other regarding the environmental supply of resources and the brood demand for the same, allowing the colony to maintain a homeostatic set point based on the ratio between the two. In contrast, there are no such obvious reasons for foragers to maintain a high level of contact with, say, maintenance workers that are involved in keeping the nest clean. One may therefore expect the foragers to be more strongly connected among themselves and with the nurses than with other task groups, leading to an interaction network that is structurally non-random or non-uniform.

If interactions define the social network structure and form the basis of work organization in the insect colony, it is important to inquire into their underlying mechanistic basis, or what makes a specific individual interact with a specific another. The most parsimonious or passive mechanism which might lead two individuals to interact is if they share spatial proximity. If tasks are spatially organized within the colony (Tofts and Franks 1992), it will allow individuals of pertinent task groups to interact. In spite of some criticism regarding its importance as being the main driving force behind the division of labour in the colony, there is good evidence that tasks and the workers performing them are indeed spatially segregated to some extent (Seeley 1982; Naug 2008; Mersch et al. 2013).

Social insect individuals are also known to be responsive to the pheromone or the cuticular hydrocarbon (CHC) signature of their nestmates. Harvester ants have hydrocarbon profiles that are specific to the tasks they perform due to the different physical environments associated with each task (Wagner at al. 2001). Harvester ant individuals of different task groups also have different antennal sensitivities to the odours associated with different tasks (Lopez-Riquelme et al. 2006), thus showing that olfactory cues could be responsible in driving interactions between specific individuals.

In a study specifically directed at understanding the proximate basis of the social network structure in a honeybee colony, workers were found to have age-specific CHC profiles, and the frequency of behavioural interactions among the age groups was correlated to the antennal sensitivity of each group to these age-specific hydrocarbons rather than being simply defined by who were present in its spatial proximity (Scholl and Naug 2011). Young workers, who did not have a strong sensitivity towards the CHC signature of any age group, were indiscriminate in terms of who they interacted with, while middle-aged and old workers had a higher antennal sensitivity towards the CHC signatures of respectively similar-aged workers and preferred interacting with their own kind. In addition, antennal sensitivity for CHCs followed a highly skewed distribution among colony members, with the majority showing a low sensitivity and only a few individuals displaying a high sensitivity. Interestingly, this closely matches the degree distribution (the number of edges per node (see Chapter 2)) of the colony interaction network, in which most individuals interact relatively little and only a few interact a lot, thus suggesting that recognition via chemical cues plays an important role in laying down the structure of the interaction network.

Structure of the colony interaction network

A number of recent studies exploring the interaction networks in different species of social insects (Naug 2009; Pinter-Wollman et al. 2011; Jeanson 2012) interestingly show that all of them share one common structural feature, an asymmetry in connectivity, or a degree distribution with a majority of interactions being performed by a minority of individuals (Figure 15.2). Networks with such right-skewed degree distributions have generated a lot of interest, given the disproportionate influence of a few individuals

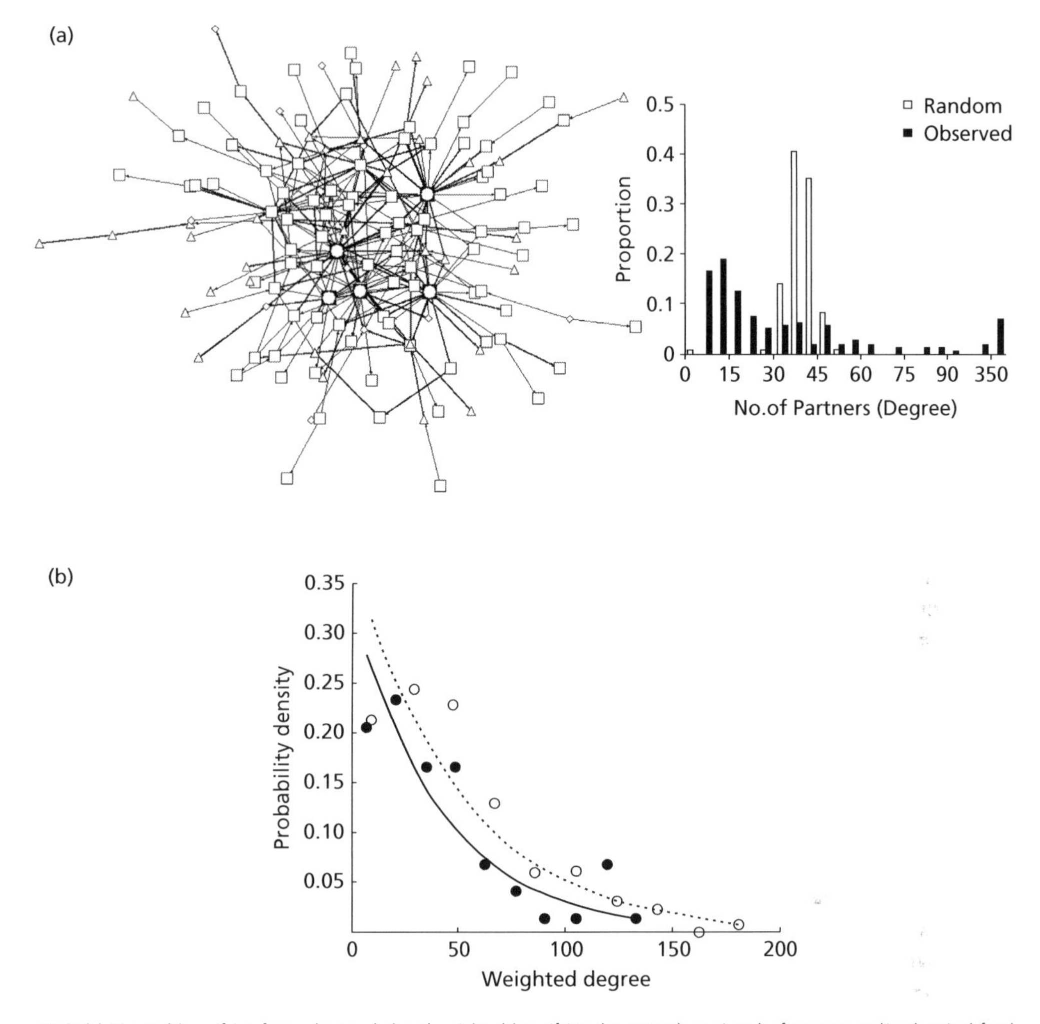

Figure 15.2 (a) Directed (specifying from who to who) and weighted (specifying the strength as given by frequency and/or duration) food transfer interaction network in a honeybee colony, with each node represented by a shape that corresponds to a particular age class: ◘ = foragers, ❑ = oldest within-nest bees, Δ = middle-aged within-nest bees, and ◊ = youngest within-nest bees. The connection between two nodes is represented by an arrow, with the direction and thickness of the arrow corresponding to the direction of food flow and the total duration of contact; the observed degree distribution of the network compared with its random equivalent given as an inset (after Naug 2008; reproduced with permission from Springer). (b) Weighted degree distribution of the interaction network in two harvester ant colonies (● and ○) showing a geometric distribution fit (— and · · · ·) to the data (Pinter-Wollman et al. 2011; reproduced with permission from the Royal Society).

in them, and it is important to understand the possible mechanisms that can give rise to such a connectivity structure. Although individual characteristics such as sensitivity to CHC may be fundamentally responsible for driving the structure of the interaction network, group-level properties such as colony size, by indirectly imposing upon the nature of interactions, can have an important influence on its final shape.

A study of wasp colonies of different sizes (Naug 2009) revealed that, while connectivity is largely complete in smaller colonies (~10 individuals), there is a decrease in connectivity and an increase in the heterogeneity of the degree distribution in larger colonies (~50 individuals). This suggests that an increase in the physical size of the nest, an increase in the number of individuals, or both can constrain the complete mixing of colony members. This can lead to individuals with high spatial fidelity inside the nest having a low connectivity, while those with a high mobility or who occupy areas of high activity being highly connected (Pinter-Wollman et al. 2011; Jeanson 2012). Interestingly however, the decreased connectivity in larger colonies did not result in a significantly higher path length (the shortest chain of individuals through which two individuals are connected (see Chapter 2)) between two individuals. What seems to make this possible are signs of 'small-world' property in these interaction networks, indicated by the emergence of a few 'hub' individuals, those with a disproportionately high connectivity, acting as 'shortcuts' and connecting individuals who would otherwise be separated by a large path length in larger colonies.

The differences in the structure of the interaction network in colonies of different sizes can have important implications for how work is organized in them and can possibly explain well-known patterns such as the trend of increasing behavioural specialization with increase in colony size. In smaller colonies, high and uniform connectivity makes it possible for individuals to have relatively more direct access to information regarding the state of different tasks, potentially allowing them to participate in all of them and be generalists. In contrast, individuals in larger colonies, due to their lower connectivity, are likely to be information limited about the colony state of all tasks, thus leading them towards behavioural specialization on one or a few tasks for which they have the most information. The heterogeneity in the connectivity distribution and the emergence of a few highly connected individuals might also underlie the well-known occurrence of 'elites' in social insect colonies, individuals who perform a given task at extremely high frequencies (Robson and Traniello 1999).

Since the structural design of a network predicts its functional properties, it is instructive to compare the structure of social insect interaction networks with other biological and technological networks. An important study in this direction by Waters and Fewell (2012) points out that interaction networks in insect colonies have structures that are fundamentally different from other social networks because associations in insect colonies are not based on individual success, which forms the basis of most other social networks. Using a novel analytical approach that identifies different types of motifs or sub-graph structures in the network, the study shows that the dominant motif in a majority of ant colonies is one that characterizes a feed-forward loop, rather than a motif that is typical of human social networks. The feed-forward motif, also seen in a number of regulatory networks that modulate information transfer at different biological levels, can maximize the efficiency of information flow, which therefore could be the primary selective driver in generating the structure of social insect interaction networks. Dominance–subordinate interaction networks in insect colonies might provide an alternative framework for comparison with human social networks. One such study found that the dominance–subordinate network of a wasp species where the queen is a docile individual, rather than another species where the queen is the most aggressive individual, is similar in terms of homogeneity to the social network of children in a classroom (Bhadra et al. 2009).

Function of the colony interaction network

Information collection and transfer

A major factor that has contributed to the tremendous success of social insects is their ability to efficiently exploit the environment, which is generally patchy both in space and time. The amazing nature of this ability is apparent to anyone who has seen ants appearing at a newly set-up picnic table in no time, or bees finding a small bloom of crocuses in the middle of emptiness at the first hint of spring. What allows a colony to respond to such ephemeral resources is its ability to gather information from the environment and disseminate it within the colony with impressive rapidity. In a study quite ahead of its time in terms of applying a network approach, Adler and Gordon (1992) examined how a colony of harvester ants accomplishes this. Using an analytical model, they showed that the efficiency of discovery and the speed of information flow are positively correlated to the size of the network, which consisted of ants walking around to collect information and exchanging it with other ants upon encountering them. In this network, the movement patterns of individual workers and the frequency of their encounters determined its structure and efficiency of transmitting information. It is important to note here that, although in principle a higher connectivity in larger groups may allow them to be more efficient in tracking the environment, it also opens up the possibility of an information overload, and individuals might need to regulate their behaviour and rate of social interactions for optimal task performance. While this study was one of the first to explicitly incorporate the idea of networks in the context of social insects, it did not attempt to characterize the detailed structure of the network.

Colony work organization

A worker's decision to perform a specific task can be influenced by the details of its interaction pattern with other workers and what they are doing. Deborah Gordon's research on this question, focussing on the overall interaction pattern (including the rate, the total number, and the interval between interactions) in the colony, has uniquely emphasized the importance of looking at the colony interaction network in its entirety (Gordon 1996, 2010). Considering individual ants as units which are connected to each other via an interaction network and which change their task states depending on their interaction patterns, the view of colony organization as a parallel distributed process (Gordon et al. 1992) provided a quantitative abstraction of earlier concepts of colony organization in terms of parallel work chains and heterarchy. The results of this model show how the performance of different tasks and the number of workers in different task groups in the colony can be interrelated, although not necessarily in a simple, reciprocal manner. One of the nice features of this model is that it considers both the states of the individual units as well as the structure of the interaction network to be dynamic, allowing the work organization to be highly responsive to a perturbation.

Division of labour in the colony can be also considered in terms of a switching network (Page and Mitchell 1991) in which the nodes are individuals that are in one of two states, either ON or OFF, in terms of their likelihood of performing a given task (Figure 15.3). In this model, the nodes are assumed to be densely connected to each other, each node sending an output about its state to all other nodes and receiving inputs from all of them regarding their own states. Each node has a switching (Boolean) function described by its response threshold: the node switches to an OFF state when the number of ON inputs received by it is equal to or exceeds this threshold and switches to an ON state when the number of ON inputs received falls below the threshold. The number of ON and OFF nodes is therefore tightly coupled to the current level for a task in the colony and regulates the behaviour of each individual in accordance with it.

Simulations of this model show that ordered group behaviour can emerge from such a network and that the level of connectivity between the nodes and how information flows between them can have an important effect on the dynamics. If information flows through the network in bursts,

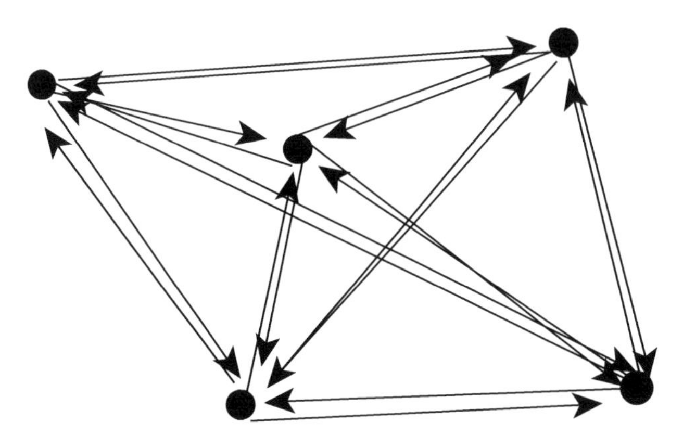

Figure 15.3 A model of division of labour with the colony visualized as a switching network for a case where $n = 5$ individuals (•) and connectivity $K = N - 1$, which means that each individual is connected to every other individual. In this directed graph, each individual represented by a node receives an input edge (•←) from all other $N - 1$ nodes and sends an output edge (•→) to each of them. Each output edge from a node carries an ON (1) or OFF (0) value based on its current state (ON or OFF), and the state of each node is in turn a consequence of the total number of ON and OFF input edges directed to it (Page and Mitchell 1991; reproduced with permission from EDP Sciences).

so that all the individuals receive their inputs simultaneously, all the individuals turn ON or turn OFF at the same time, leading to stable oscillations that are reminiscent of a mass action response seen under many circumstances in social insect colonies. In contrast, if information flow is such that individuals receive their inputs in a sequential or a random manner, the system reaches a steady state equilibrium at which each individual can be either permanently turned OFF or turned ON, which can be interpreted as an extreme specialization for a task, resulting in a strong division of labour and homeostasis.

The scope of these early network models of colony work organization can be expanded by looking at the effects of non-random connectivity in the network, by assuming that individuals of similar genetic backgrounds or task groups are more connected to each other than with others. The effect of such asymmetric connectivity on the regulation of task allocation was pointed out by Fewell (2003) in a formal description of the colony as a network of interacting task groups that communicate with each other regarding the state of the colony and that of the environment (Figure 15.4). In this outline, the different colony tasks form the nodes of a network, with individual workers forming the edges between these nodes and transferring positive and negative feedback signals that control the state of each node. As individuals performing the same task are assumed to form clusters of high connectivity with lower connectivity between different task groups, in this model the so-called weak ties play a critical role in adjusting colony task allocation. Describing the colony network in this fashion, where the tasks rather than the individuals form the nodes, is an uncommon but useful approach, since it is a functional description of the colony, and the central theme of most colony work organization models is to understand the mechanisms that coordinate colony function. Variation in worker connectivity can influence the dynamics of colony work organization also by modulating the response threshold of workers (O'Donnell and Bulova 2007).

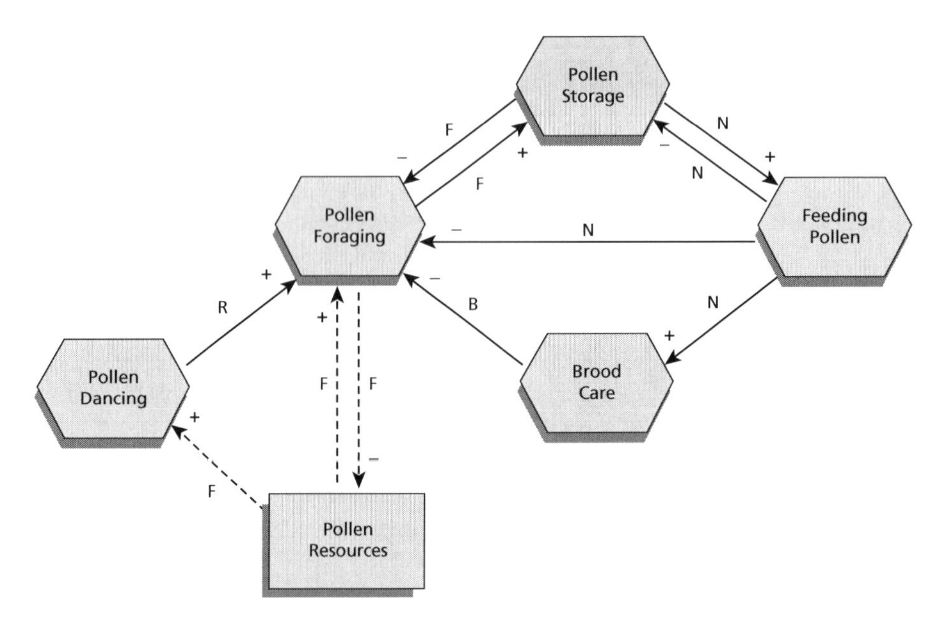

Figure 15.4 Network diagram representing the regulation of pollen foraging in a honeybee colony, with each node corresponding to a task and the edges representing individuals (F = forager; N = nurse; B = brood; R = recruit) transmitting information between them as stimulatory (+) and inhibitory (–) signals. Foragers returning with pollen receive information about pollen storage levels as they store pollen loads in cells, and the amount of stored pollen acts as a negative feedback for pollen foraging. Pollen is removed from cells by nurse bees, who feed it to the developing brood and give the excess to pollen foragers. Receiving pollen from nurses acts as a negative feedback for pollen foraging, and so does the information received from the brood, which produce a hunger pheromone when they are not fed. Pollen dancers provide a positive feedback to pollen foragers based on information on pollen availability and location (Fewell 2003; reproduced with permission from AAAS).

Material transport

While network models comprise a useful set of tools to understand information flow and task allocation within the colony, these models are arguably both more powerful and rigorously testable when one uses them to study the interaction-based transport networks moving various materials within the colony. Some tasks in social insect colonies consist of two or more distinct components involving direct or indirect transfer of material between individuals. For example, a nectar forager in a honeybee colony brings back nectar to the nest and passes it on to receivers, who may pass it along to a next set of receivers, and so on, before it is finally stored in a cell. In leafcutter ants, material transfer could be indirect, through a dump or cache where pieces of leaves are deposited by one set of individuals and retrieved by another. These transport networks could become even more complex when multiple task groups and multiple kinds of transfers are involved (see Ratnieks and Anderson 1999 for a review).

One of the most significant studies regarding how material transfer among individuals can influence the organization of a task is an analysis of nest-building behaviour in a species of paper wasp (Jeanne 1986). The task, requiring the coordination of three sub-tasks—water foraging, pulp foraging, and building, each of which is performed by a different group of workers—involves the transfer of water or pulp from one worker to another (Figure 15.5). Using the theoretical construct developed by Oster and Wilson (1978), the study provides a quantitative model, based on empirical data, which shows how a series–parallel organization of nest construction, possible in colonies with a large number of workers, results in a significantly higher colony work efficiency over the more series or parallel–series like operations in smaller colonies.

The study shows how the size of the transfer network and the redundancy of its nodes can positively influence the flow of materials in the colony. It also makes the important point that, in order for material flow to reach maximum efficiency within such a network, the size of different task groups or nodes has to complement each other well, as any imbalance among them will increase waiting times or queuing delays in proportion to the difference in

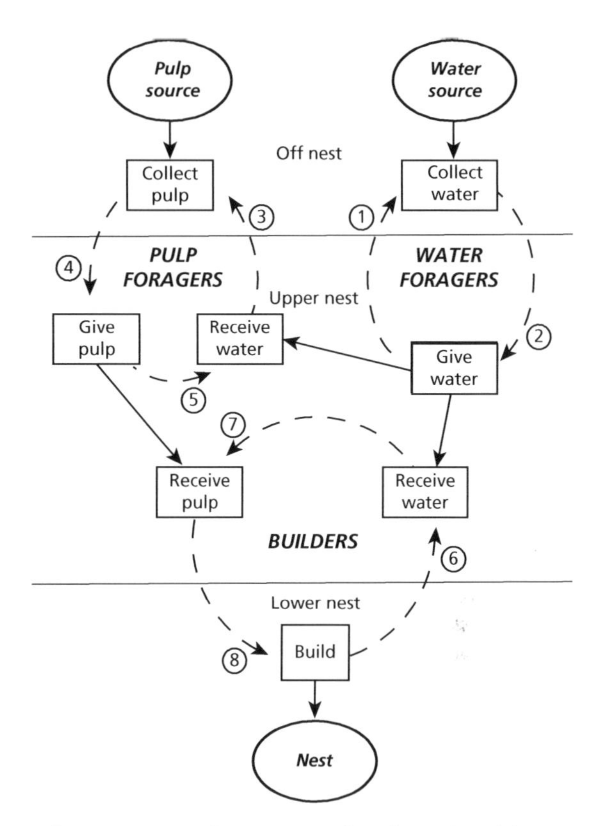

Figure 15.5 Material transport network in a large colony of the wasp *Polybia occidentalis* engaged in nest construction. Materials flow from the top to the bottom of the diagram. Water foragers collect water in the field and return to the nest, where they give it to pulp foragers and builders. After filling their crops with water, pulp foragers collect pulp in the field, using some or all of the water to soften the fibres. At the nest, they transfer the pulp to builders. Builders further wet the pulp mass with water from their own crops prior to adding it to the nest (Jeanne 1986; reproduced with permission from Springer Science and Business Media).

size among these task groups. Therefore, while species with large colonies can gain from organizing a complex transport network in which each node or task group can specialize on a task and increase its performance efficiency, species with smaller colonies are unlikely to make this gain due to the potential bottlenecks that are more likely to arise between the nodes in such groups due to stochastic forces. The adaptive complexity of the network is thus contingent upon colony size (Jeanne 1999).

The network with arguably the highest connectivity in a social insect colony is the one meant for distribution of food. The first detailed structural analysis of a social insect interaction network comes from a study of the food distribution network in honeybees, a comparison of two scenarios of a colony exploiting nectar sources of different qualities). The study shows that the network of food sharing interactions undergoes an adaptive structural change as a function of resource quality. Richer nectar results in a higher overall connectivity, represented by a higher number of degrees per node and a lower clustering coefficient (see Chapter 2 for details), spreading the food and the likely information regarding it more extensively and rapidly through the colony. In contrast, more dilute nectar results in a low and asymmetric connectivity, resulting in it being distributed in a more limited fashion such that only a few individuals receive the bulk of it. The mechanistic basis underlying this seemingly complex difference in network structure is rather simple: a rich sugar solution is retained longer in an individual's crop, thus allowing many small similar-sized transfers, while a dilute solution is not retained as long, thus resulting in large initial transfers followed by very little or no transfers. Interestingly, these results show that simple aspects of individual physiology can drive the social dynamics of the transfer process and in turn can have global effects on the structural and functional properties of the interaction network.

The effect of individual physiology on the structure of the interaction network is well illustrated by how the hunger levels of individuals can significantly impact the structure of the food distribution network. Higher individual hunger has been shown to cause changes in the nature of food distribution interactions in both ants and honeybees such that the resulting transfer network has a more uniform structure, which in turn leads to a more rapid and universal distribution of food within the colony (Cassill et al. 1999; Feigenbaum and Naug 2010; Sendova-Franks et al. 2010). This global change in network structure is brought about by a change in individual behaviour, such as a higher overall movement rate of individuals in a hungry colony, resulting in higher levels of mixing and consequently a higher number of interactions. While the possibility of such rapid structural changes in the interaction network makes functional sense and shows its efficiency in dealing with a crisis such as starvation, it also brings to light a somewhat less recognized and an adverse functional consequence of an efficient transfer network.

Any design feature of the interaction network which increases ergonomic efficiency is also likely to raise the vulnerability of the colony to an infectious disease if a pathogen can move between individuals through the same network (see also Chapter 10). Given that large social insect colonies are routinely infected with a range of infectious diseases, it seems quite likely that the social interaction network is quite efficiently exploited by pathogens and parasites. The exposure and susceptibility risk to disease should therefore play an important selective force on the design of the colony social network (Schmid-Hempel 1998; Naug and Camazine 2002), and social insect colonies face the challenge of how to be ergonomically efficient as well as be resistant to an infectious disease at the same time. Some recent work has, however, suggested possible mechanisms by which, under certain conditions, an increasing complexity in social interaction patterns can mitigate the risks of pathogen pressure in complex social organizations (Hock and Fefferman 2012).

The speed with which a disease spreads in a colony is directly related to the density of the interaction network, and an individual's position in the network determines that individual's risk of infection (Otterstatter and Thomson 2007). As social insect interaction networks are largely characterized by a skewed degree distribution, the minority of individuals responsible for the majority of interactions could therefore assume the role of what are known as superspreaders, individuals who are disproportionately responsible for spreading an infectious disease. However, a more detailed network analysis might be in order before making such a conclusion. For an individual to act as a superspreader, the high number of interactions must translate to a high out-degree (outward interactions from a node) rather than a high in-degree (inward interactions to a node; see Chapter 2). In fact, if an individual has a high in-degree and a low out-degree, that individual can absorb pathogens without passing them on to others, thereby acting as a sink that insulates the rest of the network to some extent from a spreading infection (Naug 2008). Individuals, which accumulate any food that is passed on to them in ant colonies and so are sometimes called 'living silos' (Sendova-Franks et al. 2010), could serve such a role, and it would be interesting to test if the presence of such 'silos' could reduce the spread of a food-borne infectious disease.

As incoming food moves from the outside to the inside of the colony in a centripetal fashion. Older workers, generally located at the outermost parts of the colony, occupy the initial and the most highly connected parts of the food distribution network, while younger workers and the queen are located at the interior regions of the nest and are more sparsely connected at the terminal end of the network. This structural property makes a pathogen entering the colony through food more likely to contact and infect the older workers, thus providing some measure of insulation to the more valuable younger workers and the queen, a property that has been termed organizational immunity (Naug 2008). The queen, who is also surrounded by a relatively constant subset of workers termed her retinue, which confines her interactions to a certain extent, may also benefit from this additional layer of insulation from the wider part of the network.

Any segregation in the network, however, increasingly starts breaking down after a period of starvation, as individuals move and mix around in order to facilitate the distribution of food. Therefore, periods of starvation followed by a large food influx seem to be vulnerable points when a pathogen can exploit the interaction network to rapidly invade and spread inside the colony. Interestingly, however, the organizational immunity remains relatively stable even during this general increase in connectivity (Feigenbaum and Naug 2010).

As the structure of the colony social network is a global property that results from the individual behavioural properties of the colony members, any host–parasite interaction that alters the behaviour of the individual can also alter the structure of the colony social network, in a manner that is either adaptive to the host or the pathogen (Naug and Gibbs 2009; Campbell et al. 2010). Research regarding host pathophysiology and the behaviour of infected individuals can therefore be revealing in terms of how the colony interaction network could be the central arena for host–pathogen co-evolution.

Conclusions and future directions

Although questions regarding important group-level processes such as division of labour, dominance hierarchies, and the transfer of pheromone, food, information, and disease could benefit considerably from using network approaches, such studies are still relatively rare. While the early studies using this approach have shown the importance of understanding the detailed structure of the interaction network in the colony, the technical and statistical challenges involved have hindered the rush into this new area of research to some extent. While one may think that the ability to track individuals easily in social insect colonies makes it quite easy to study their interaction networks, the reality is that the large number of individuals in the colony and the equally large number of interactions, most of which are also brief in nature, makes it challenging to both record them with precision and analyse them.

The sheer volume of data one can amass is both a blessing and a curse in this case. The availability of relatively inexpensive video cameras that can record in high definition has made it easier to capture the interactions but the larger and more difficult challenge lies in classifying interactions and assigning them to specific individuals. The ability of automated software to track uniquely tagged individuals is somewhat limited, and more traditional methods involving direct observations or the transcription of video recordings take exorbitant amounts of time. In order to circumvent this problem, researchers have been investing more and more in developing algorithms and software that can record, categorize, and analyse such interactions (Kimura et al. 2011; Moreau et al. 2011; Mersch et al. 2013), but such methods are still from being perfect, and a substantial amount of human effort is still required to record and analyse such data. The technical advance by Mersch et al. (2013) in automating behavioural recordings and generating interaction networks from them is a particularly promising step forward in this direction.

These technical limitations have also resulted in almost all social insect interaction networks being constructed by pooling all the interaction data into single, static networks, even though there is an increasing recognition that dynamic networks, that is, how network structure changes in time, can reveal patterns that are not otherwise apparent. In a notable example of this, Blonder and Dornhaus (2011) looked at information flow between individuals using time-ordered networks. Assuming the upper limit of information flow in a group to be when all individuals move and interact randomly, they found that actual information flow in ant networks is significantly different from this predicted maximum in two contrasting fashions: it is slower on a long-time scale but faster on a short-time scale. The restricted flow of information at global scales arises from an individual's spatial fidelity and the relatively long delays between interactions. The authors speculate that a fast local flow and a slow global flow might generate a best-of-both-worlds scenario, where the performance efficiency of a given task is increased only within the small spatial scale where the task is located while enabling the colony to restrict the flow of potential harmful agents, such as pathogens, across the entire colony.

A potentially promising new research direction that could see the use of social network theory is the investigation of interindividual variability (Jeanne 1988), which has recently gained a lot of popularity as behavioural syndromes (Sih et al. 2009). While the study of collective behaviour in social insects has typically minimized the role of individual behavioural variance for obvious reasons of simplicity, it is nonetheless recognized that biologically meaningful variation exist at the individual level. The existence of a few 'elites' that perform a task in an atypically high frequency and that of a few individuals that are responsible for the majority of interactions raise interesting questions such as whether one's position in the social network has any role in driving this heterogeneity. Response threshold models propose a mechanism by which behavioural specialization can emerge through enhancement of the intrinsic variability in individual behaviour via positive feedback mechanisms that are related to one's social context (Plowright and Plowright 1988; Theraulaz et al. 1998). While there is empirical evidence that such social enhancement can occur over the lifetime of an individual (Pankiw et al. 2001), one could also ask complementary questions regarding how certain behavioural types might end up occupying specific positions in the social network (see also Chapter 6).

A second novel area of possibilities concerns the relationship between the organizational structure of a colony and its metabolism. There has been a lot of recent interest in metabolic scaling theory and how it impacts life history trajectory in social insects (Hou et al. 2010; Shik 2010; Waters et al. 2010; Cao and Dornhaus 2013). While these studies have mainly addressed how colony size can affect individual- and colony-level behavioural and physiological traits, it might prove promising to look at the scaling properties of the social and transport networks across a range of colony sizes and ecological contexts to determine how network structure impacts colony metabolism. An interesting question to ask is whether the scaling property of the colony interaction network is adaptively optimized to colony function or if it acts as a constraint.

The strength of social insect colonies as model systems has largely been the ability to do experimental manipulations. This same feature can hold in good stead for developing testable hypotheses regarding the structure and function of social networks in the context of information and material flow, disease dynamics, etc. Behavioural studies determining the structural properties of a network can be complemented with studies using radioactive or other tracers that can outline the functional outcome of these networks in terms of their flow characteristics (Buffin et al. 2009; Feigenbaum and Naug 2010). However, one must also caution against getting carried away by the novelty of this area and ending up either just describing the social network structure of a multitude of different species or just restating some well-known facts in this new language. The focus should be on testing novel and interesting biological hypotheses rather than just applying a complex statistical or analytical technique borrowed from other disciplines. Our greatest challenge, and it is by no means an easy one, lies in identifying important biological questions that can benefit from the use of a network analysis and reveal details that otherwise remain hidden to more traditional approaches.

CHAPTER 16

Perspectives on social network analyses of bird populations

Colin J. Garroway, Reinder Radersma, and Camilla A. Hinde

Introduction to social network analysis in birds

Studies of birds have played a particularly important role in the development of evolutionary and ecological theories and their subsequent empirical verification. For instance, Darwin's ornithological natural history notes from his travels on the *Beagle* coloured his evolutionary thinking and contributed to his development of the theory of natural selection (N. Barlow 1963). Niko Tinbergen, Konrad Lorenz, Julian Huxley, and Robert Hinde, studying bird behaviour, were among the first to recognize that behaviour is subject to selection, and their bottom-up, mechanistic, and comparative approaches to its study led to the development of ethology (Birkhead et al. 2014). Also working on avian models, Nick Davies, John Krebs, Amotz Zahavi, and others became interested in the evolutionary basis and adaptive nature of behaviour and helped formalize the discipline of behavioural ecology (Birkhead et al. 2014). Finally, the broad evolutionary and ecological approaches of Robert MacArthur, David Lack, and Ernst Mayr were developed, or inspired, while studying avian populations and helped focus ecology on rigorous quantitative hypothesis testing and questions about how individual behaviours can influence population level processes (Birkhead et al. 2014). An important thread connecting many parts of this pioneering work was the recognition of a central role of social interactions. Whether they be competitive, cooperative, exploitative, or related to mating, social interactions can structure populations and influence the process of evolution. Consequently, understanding the ecological and evolutionary processes that cause variation in social behaviour, and the emergence of social structure, is a major research theme in evolutionary ecology and its various subdisciplines (W. Hamilton 1964a,b; Maynard Smith and Szathmáry 1995; Krause and Ruxton 2002; Whitehead 2008a; Bourke 2011).

The important contributions of the above-mentioned pioneers and many others significantly advanced our knowledge of the form, function, and evolution of many aspects of avian social behaviours. For instance, we now have a firm understanding of the adaptive significance of the extensive variation in mating systems in birds, and we are able to quantify many of the ecological life-history trade-offs underlying individual behaviours. However, relative to this vast literature, there has been little research emphasis on understanding the social structure of birds in terms of the pattern, nature, and quality of social relationships within populations. Many bird species are social, and the extent and duration of social behaviours vary tremendously both across species and over the course of life cycles. Understanding both individual and across-species variation in the tendency to be social is key to understanding how and why different degrees of sociality evolve and the extent to which individual phenotypes relate to their bearers' participation in social behaviour. It is also important to understand how social structures emerge, because these structures determine the extent to which social behaviour might mediate selection on both social and other phenotypic characters.

Animal Social Networks. Edited by Jens Krause, Richard James, Daniel W. Franks and Darren P. Croft.

The lack of work on bird social structure that builds from the level of individual associations has not been due to a lack of interest or effort. Quantitatively rigorous examinations of vertebrate social structures at this level require detailed data on interactions among identifiable individuals (Hinde 1976; Croft et al. 2008; Whitehead 2008a,b), and such interactions have been very difficult to collect for birds. Largely due to the generally high vagility of many avian species, it has been logistically difficult to monitor interactions among a sufficient number of marked individuals over timescales appropriate for delineating social structure. Then, having marked and identified individuals, defining individuals as associated or not in a meaningful manner is also often not straightforward. Many methods for following individual birds require extremely intensive field work (e.g. 9288 hours spent watching colour-ringed manakins (D. McDonald 2007)) to follow what is sometimes, statistically speaking (Whitehead 2008b), few individuals. Thus, questions of flock social structures have most often been addressed with presentations of measures of mean group size and demography.

Hinde (1976) proposed a conceptual framework for the study of primate social systems that has been adopted for use while studying animal social systems in general (Whitehead 2008a). His approach is bottom-up and considers dyadic interactions to be the fundamental unit of social analysis. Relationships are defined by the content (what a pair are doing), quality (how they do it), and patterning of successive interactions. A social structure can then be inferred from the nature, quality, and patterning of social relationships. At each level we can abstract social properties not apparent at the level below; relationships emerge from the pattern of interactions, and social structure from the pattern of relationships (Figure 16.1). Almost four decades since its publication, Hinde's framework continues to be influential, having led to rich and novel insights into the ecology and evolution of complex and sometimes cryptic animal societies (Goodall 1986; Cheney 1987; Dunbar 1988; Whitehead 1997; Whitehead 2008a).

Of the studies that have assessed pair-wise associations among individually identifiable birds, and thus implicitly or explicitly followed Hinde (1976), social structure has been best studied among temperate non-migratory passerines (e.g. Nilsson and Smith 1985; Ekman 1989; Elena et al. 1999; Brotons 2000; Kraaijeveld and Dickinson 2001; Drent 2003; Griesser et al. 2009; Liker et al. 2009; Michler et al. 2011). The typical life history for this group consists of (1) a breeding season, during which pairs, sometimes together with helpers, defend territories and rear offspring; followed by (2) a brief period of continued offspring dependence post-fledging and a breakdown of breeding territoriality; and finally (3) winter flock formation. The social structures described for these species are broadly similar, at least superficially, to many other fission–fusion social systems in that fluid associations occur among pairs either within largely closed social units or across unbounded socially fluid populations. A third category of social structure, multi-tiered hierarchical associations among social groups at multiple levels (e.g. African elephants (Wittemyer et al. 2005)), has to our knowledge not been identified for birds. Within closed-unit fission–fusion systems, social flocks typically comprise of fewer than ten individuals who occupy exclusive ranges and are most often characterized by linear dominance hierarchies among group members (Ekman 1989). In unbounded fission–fusion systems, individuals form loose flocks that mingle and thus exchange individuals with other flocks in non-exclusive home ranges. Within these groups, dominance is typically site related (e.g. Krebs 1982; Brawn and Samson 1983; Yasukawa and Bick 1983; S. Smith 1984).

There has also been some focus on understanding the structure and consequences of sociality during breeding (e.g. Davies and Lundberg 1984; Davies and Hartley 1996; Double et al. 2005; D. McDonald 2009; Grabowska-Zhang et al. 2012; Hatchwell et al. 2013). For instance Grabowska-Zhang et al. (2012) used a 41-year dataset to show that great tits that had previously bred on adjacent territories and were thus familiar with their neighbours had greater breeding success. This suggests that social relationships on breeding territories can also be important.

Much of this previous work does not explicitly consider temporal aspects of social relationships, a key component of Hinde's framework. One of the first studies to do so in any taxa was a study

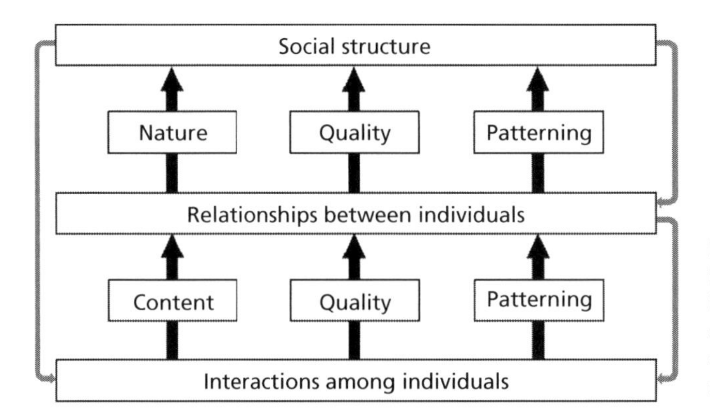

Figure 16.1 A summary of Hinde's (1976) framework for the analysis of animal social structures, after Figure 1.3 in Whitehead 2008a. Black arrows indicate characteristics of the lower level that are used to define the higher-level organization. Grey arrows indicate potential feedback mechanisms.

of the social structure of sanderlings (*Calidris alba*) in California (Myers 1983); incidentally, this work is also perhaps the first to formally compare association patterns to a null expectation of random associations, a practice that is now standard (Croft, Madden, et al. 2011). The research findings in this regard, however, were summed up well in the title which tells us that 'sanderlings have no friends'. Thus, while it is clear that mated pairs can remain associated across years, we still know very little about the persistence of other types of social relationships over time for most avian species. Incorporating time directly into network analyses is not straightforward, but progress is being made to identify appropriate techniques (Blonder et al. 2012; Hobson et al. 2013).

The general dearth of research into bird social structures, together with recent technological and analytical advances that allow for high-resolution data collection, presents researchers with an exciting opportunity to address consequential and broad questions about social evolution. In the rest of this brief overview of avian social network analysis, we highlight network building techniques and discuss recent work. In addition, we outline new research directions which illustrate that avian social network analyses are well placed to expand our understanding of social evolution and ecology.

Building avian social networks

Recent technological and analytical advances have alleviated some of the difficulties associated with monitoring sufficient numbers of individuals for detailed analyses of social structure and now allow for large-scale quantitative social analyses of wild bird populations. However, more low-tech approaches are still valuable for many questions and can have advantages over other approaches in some instances. In this section, we discuss the primary approaches used to build avian social networks and briefly highlight advantages and disadvantages of each (Table 16.1).

Ringing and observation in avian social networks

Constructing social networks based upon colour ringing is straightforward; unique colour combinations of rings on birds allow individuals to be visually identified by an observer (e.g. D. McDonald 2007; Farine and Milburn 2013) (Figure 16.2a). A major advantage of this method is that interactions can be directly observed rather than inferred from associations, as is often the case with the automated methods described below. However, there is still scope for subjectivity, and particular types of interactions may often be unobservable. Observing direct interactions can provide an important context for social network construction and analyses. For example, directionality can be introduced into the network structure, and agonistic interactions can be differentiated from non-aggressive interactions for more rich descriptions of social networks.

One drawback of this approach is that it is difficult to observe many highly vagile species in the wild. It is also difficult to identify a great number of individuals at one time. Finally, the quantity and

Table 16.1 Comparison of the pros and cons of different methods for building avian social networks.

		Colour ring observations	Capture and recapture	Passive integrated transponder tags	VHF telemetry	Satellite and GPS telemetry	Encounternet
Data characteristics	Distinguish directionality and/or type of interaction	Yes	No	No	No	No	No
	Gambit-of-the-group assumption	No	Yes	Yes	Yes	Yes	Yes
	Sample size (no. individuals)	Small	Medium	Large	Small	Small	Small/medium
	Temporal resolution	Low	Low	Medium	Medium/high	Medium/high	High
Costs	Labour intensity	High	High	Low	High/low[1]	Low	Low
	Costs of marks/tags	Low	Low	Low	Medium	High	Medium
	Costs of observation method	Low	Low	Medium	Low/high[1]	Medium/high	Medium
Longevity	Lifetime of marks/tags	Long	Long	Long	Short	Short/medium	Short
Biases	Risk of observer bias	Yes	Low	No	No	No	No
	Observation method affects distribution of individuals	Possibly	Possibly	Yes	No	No	No

[1] Triangulation with handheld antennae is labour intensive, but antennae are relatively cheap. In contrast, using antennae towers is less labour intensive but much more expensive.

quality of the data strongly depends on the skills of the observers, and the ambiguous nature of interactions might result in variations between observers. Overall, colour ringing and subsequent visual identification of associations is a useful strategy to address questions that require detailed information on a subset of individuals that occur together (e.g. constructing social networks to describe mating systems; D. McDonald 2007). It is generally less suited to address questions about social networks at the population level due to limits of observations and sample size.

Observations based on captures of birds with standard numbered rings (Figure 16.2) can also be used to construct social networks. Here, individuals trapped within the same timeframe are considered

Figure 16.2 (a) A wild New Caledonian crow fitted with a harness-mounted miniature proximity logger ('Encounternet') as well as colour rings and a standard numbered metal ring (photo by Dr James St Clair; reproduced with permission from Rutz et al. 2012, *Current Biology*). (b) A wild great tit with an integrated passive integrated transponder tag in a plastic split ring on its left leg, and a standard numbered metal ring on its right leg (photo by Nicole Milligan).

to be a part of the same flock. Relative to some of the automated methods, it is difficult to obtain large sample sizes with this approach; however, often more individuals can be sampled than via colour ringing and observation. Therefore, this method can begin to address population level questions about social structure (Oh and Badyaev 2010).

This approach begins by making a number of assumptions common to many types of studies of social structure in order to construct the network. For example, because we do not directly observe interactions among individuals, we assume individuals that are associated in 'flocks' (i.e. are caught together) interact to the same degree with every other individual in the flock. This assumption is termed the 'gambit-of-the-group' (Whitehead and Dufault 1999) and can be useful when sample sizes are appropriately large and weighted network measures are used for analyses (Franks et al. 2010).

There is also a degree of arbitrariness associated with assumptions made here about the time window within which individuals are grouped that needs to be considered. We know a priori that the choice of a fixed time window within which to define individuals as grouped will not be ideal. For example, environmental conditions vary across days and seasons, and individuals mature; both of these factors are likely to lead to variation in social tendencies. Nevertheless, for many species, biologically informed, fixed time windows may serve purposes well.

Passive integrated transponder tags in the study of avian social networks

Passive integrated transponder tag (sometimes 'passive inductive transponder tag'; PIT tag, Gibbons and Andrews 2004) monitoring systems are increasingly being used for the automated study of animal social structure (e.g. bats (Patriquin et al. 2010), flying squirrels (Garroway et al. 2012), and birds (Aplin et al. 2012)). PIT tags are microchips that are activated when they pass through the electromagnetic field of a logger antenna, whereupon the unique chip ID is sent from the tag to the logger and recorded, together with the date and time. Typically, PIT tags are encased in 'bio-compatible' glass and so can be implanted subcutaneously (Nicolaus et al. 2008; Schroeder et al. 2011). Implantation is straightforward, and detailed instructions for implantation can be found elsewhere (Gibbons and Andrews 2004; Nicolaus et al. 2008). Glass-encased PIT tags have also been glued to colour rings with no ill effects (as reported for house sparrows (Schroeder et al. 2011)), although potential detrimental effects likely vary among species.

PIT tags have also recently been incorporated into plastic split rings (e.g. IB Technology, UK, and Dorset ID, the Netherlands) (Figure 16.2b) that are suitable for some bird species. A benefit of using rings with encased PIT tags is that only one worker is required to mark an individual, whereas implantation is usually a two-person job. In addition, ringing is generally favoured from an ethical point of view, because it is regarded as being less invasive than implantation. The use of PIT tags incorporated into standard plastic split rings can usually be covered by an amendment to standard bird ringing licences.

A major advantage of PIT tag logging systems is that PIT tags are not battery powered and so, barring defective tags, they are expected to last the lifetime of the bird. The logging systems do, however, need power. PIT tags are also relatively cheap; thus, many birds can be marked. Loggers can range from very inexpensive systems (Bridge and Bonter 2011) to quite expensive, custom-built systems (e.g. selective feeders, and problem-solving devices).

One potential drawback that should be considered is that, when loggers are affixed to feeders, they act as attractants; so, care is needed during study design to disentangle the effects of non-social association at a limiting resource from those of social association. Another potential drawback of feeders at fixed locations is that it can be difficult to separate social from spatial effects. Possible solutions for both issues involve scheduling logging sessions to minimize attraction, and randomly moving systems throughout an area to reduce their predictability and thus decouple social from spatial effects. However, a counter-argument is that the use of fixed points, under a controlled sampling regime, allows a degree of control over the type of observations collected, which might be of particular use when carrying out experimental analysis. We should also note, from personal experience, that feeders also act as attractants for other animals which may cause expensive damage to systems (e.g. feral grey squirrels at our study site in the UK).

A number of companies will build bespoke PIT tag data-logging systems, and low-cost instructions have been published for those that wish to build loggers themselves (Bonter and Bridge 2011; Bridge and Bonter 2011). Thus, marked birds can be monitored via PIT tags in a variety of ways. Typically, social associations are inferred from joint visits to feeding stations (although they might also be extracted from records of individuals passing through fixed points on the landscape). This gives a temporal data stream of individuals visiting a feeder. However, it is not straightforward to extract social groups and infer associations from this type of data.

Making the 'gambit-of-the-group' assumption here again, in this case the assumption that individuals that are temporally associated interact socially, it may sometimes be biologically meaningful to simply set a time window within which one defines groups or associations. An advantage of fixed time windows is that they are easy to calculate. However, they rely on an arbitrary cut-off for defining associations, and this cut-off is assumed to be fixed in space and time, which again is unlikely. Psorakis et al. (2012) suggest that a more rigorous method for extracting social groups from temporal data is to search for and extract regions of dense bird activity from the data stream of feeder visits. They term these regions 'gathering events' and these can be used to define social groups. From these groups, association indices can be calculated, networks constructed, and the duration of gathering events inferred from the data. This method avoids choosing a time window, which tends to have a large effect on the network measures extracted from the data (Psorakis et al. 2014).

Telemetry tracking in the study of avian social networks

Telemetry tools for tracking individuals are well developed and in wide use across taxa. There are now four basic types of telemetry tracking systems: systems based on VHF radio-telemetry, systems that use satellite transmitters, systems that use GPS receivers, and the recently developed Encounternet system, which is based upon digital radio-telemetry. All of these technologies can give information on the spatial locations of individuals.

In principle, these location data could be used to build social networks based upon measures of individual proximity in space. In practice, however, this is easier for some systems than others. The spatial location of individuals affixed with VHF radio-telemetry tags are calculated via triangulation, which can be labour intensive and requires more than one observer (more observers leads to greater accuracy). It will rarely be possible for a sufficient number of marked individuals to be tracked concurrently and for sufficient spatial resolution of positions to be calculated to construct social networks with this method.

Alternatively, a grid of antennae towers can be used to monitor individuals with VHF radio tags continuously. This approach has not to our knowledge been used for birds, and examples in other taxa are scarce (e.g. Crofoot et al. 2008; Lambert et al. 2009), probably due to the high cost of towers and limited spatial coverage. However, it is possible to follow focal individuals via VHF telemetry to their flocks, identify those colour-ringed individuals that are present, and then construct social networks from these data (e.g. Templeton et al. 2012). Satellite tracking uses a high-powered transmitter to transmit data to satellites that calculate the spatial location of individuals and send that information to researchers over the Internet (e.g. Scherr et al. 2010; Trierweiler et al. 2013). GPS tracking involves affixing GPS receivers to individuals, which either need to be retrieved to recover data (e.g. Weimerskirch et al. 2002; Guilford et al. 2008) or send data via radio transmission on prescheduled times to a receiver (e.g. Van Gils et al. 2007). For gathering social data on birds, both of these techniques suffer from constraints associated with the cost and size of these instruments as well as occasionally inaccurate location determination.

The current costs of these technologies are likely to be prohibitively high for most researchers wishing to monitor the social behaviour of a considerable number of marked individuals, although costs are falling and accuracy improving, driven by developments in technology areas such as the mobile phone industry. In addition, the power requirements of these techniques require the use of large batteries that make the size and mass of the units too big for many species of birds. The current smallest commercially available GPS tracker for birds weighs 12.5 g, so it is suitable for use with birds that weigh >250 g (following the '5% rule of thumb'; but see Barron et al. 2010). These problems seem likely to dissipate in the future as technological advances decrease both the size and cost of equipment, making these technologies potentially more useful for social analyses.

Encounternet is a recently developed digital radio-telemetry technology that holds great promise for providing detailed data that can be used to construct social networks (Mennill et al. 2012; Rutz et al. 2012; Krause et al. 2013). Encounternet tags (Figure 16.2a) transmit pulses that contain a unique identifier together with date and time stamps. Importantly, they also log pulses from other tags within range, and the received signal strength of pulses, which can be used to estimate the distance between individuals. Ultimately, the tags communicate these data to base stations from which researchers can remotely download and reprogram tags. Thus, detailed information about associations can be recorded on tags carried by the individuals (Rutz et al. 2012).

Similar to the challenge of defining individuals as associated or not in time with PIT tag data, it is necessary to choose distance thresholds to define social associations in space using Encounternet (Rutz et al. 2012). Appropriate distance thresholds will be species- and context-dependent, and indeed various distance thresholds are likely to be appropriate for different social processes within the same species. Mennill et al. (2012) have posted a useful video demonstrating the use of Encounternet for tracking individuals as a supplement to their recent paper introducing the technology (<http://youtu.be/47XIZdxGOpU>).

Choosing the best method for studying avian social networks

For most species, it is now possible to use automated monitoring systems for the collection of high-resolution data about social associations among individuals. Particularly, we feel that Encounternet and PIT tag rings show great promise for adding insight into the social networks of birds in particular, by increasing the scale and resolution of data from which social networks are typically sampled. Which methodology will be best to use will of course depend on the scientific questions of interest.

PIT tags should last the lifetime of the bird and are cheap. However, in many cases, loggers require an attractant. Encounternet tags provide continuous data about social associations. However, Encounternet tags are more costly than some other marking methods, so often fewer birds can be marked. Encounternet tags are also larger than other kinds of tags (although still suitable for many small, passerine-sized birds) and are battery powered, so they have a limited lifespan. Finally, although the sample sizes obtained by direct observation will be smaller than those obtained by the automated systems, there will certainly be many instances when direct observation of interactions among individuals will be preferable to using the automated systems, as direct observation can provide contextual information that would be unavailable using the automated remote methods.

Exploring avian social networks

Using the methods described above, social network approaches are increasingly used to investigate avian sociality. However, the literature using a social network approach is still sparse. In this section, we present findings from a sample of this literature to illustrate the methodologies described in 'Building avian social networks' and present the scope of some questions being addressed by current research. The studies presented first investigate how the individual is affected by social structure, before taking a broader view to look at studies describing population level networks.

Two pioneering studies of avian social networks have suggested that social lability can potentially influence individual fitness (D. McDonald 2007; Oh and Badyaev 2010). In the first study, D. McDonald (2007) used a long-term dataset of colour-ringed long-tailed manakins (*Chiroxiphia linearis*) to show that early social lability, as measured by 'information centrality', predicted future reproductive success in this species. Long-tailed manakins have an unusual lek-mating system, in which pairs of males cooperate at leks to perform dance and song courtship displays. Before forming this alliance, males spend up to 8 years interacting with many other males, thus forming dynamic social networks. During this time, males strive to rise in social status from a non-dancer, to a dancer, and finally to a beta or alpha male (the cooperative pair).

D. McDonald found that, for individual males, current social position was not related to social rise; however, the male's social position 5 years earlier was a strong predictor of social rise. It seems likely that young males seek to maximize their chance of rising in status to alpha male, and consequently achieving maximum copulatory success, by establishing relationships in leks where, relative to other males, they have the best chance of rising.

Similarly, Oh and Badyaev (2010) found that socially labile house finches (*Carpodacus mexicanus*), as determined by 'betweenness centrality', may also select their social environments to increase fitness. Male house finches display colourful breast plumage, ranging from pale yellow to deep red/purple, which is important for mate choice by females. During the non-breeding season, individuals interact in flocks within which mate selection, based upon plumage ornamentation, is thought to occur (Oh and Badyaev 2006). When Oh and Badyaev (2010) constructed a social network of associations during this non-breeding period, they found that less well-ornamented but socially labile males were able to increase their pairing success by choosing social groups according to the attractiveness of their ornamentation relative to that of others (see also Bateson and Healy 2005). Highly ornamented males increased their pairing success by having low lability, and less well-ornamented males did best to move between groups until they found a group within which their ornamentation compared favourably to that of the other male members. Together, these two studies provide evidence that animals can actively manipulate their social environments and thus the associated social selection pressures.

It is expected that individuals use social information to find food, and there have been several documented examples of this use (e.g. Lachlan et al. 1998). Aplin et al. (2012) recently showed that social information can be a benefit of social connectivity. In their study of PIT-tagged, mixed-species tit flocks (family Paridae), they showed that not only do individuals use social information to exploit ephemeral food resources but that social network position predicted access to information. Specifically, network-based diffusion analysis (Franz and Nunn 2009) showed that social network position predicted the order of arrival at artificial food patches, suggesting that social tendencies influence access to information. Further, the order in which information is acquired can to some extent be predicted from the measured network.

In addition to identifying ways in which individuals stand to benefit from their position within their social network, network analysis has also given more general insight into how social structure interacts with population processes. Here, we describe four examples; in each case social network structure has provided insight into individual relationships. First, Templeton et al. (2012) used radio-telemetry to follow focal juvenile male song sparrows and subsequently identify their associates via colour rings, in order to build their social network. The authors found that the density of the network varied throughout the season and that seasonality was related to the individual's number of associates of different age and sex classes. Juvenile males formed stable relationships, often with other juvenile males, and these alliances had long-term consequences, because males were more likely to hold future territories near each other and acquire similar song types.

Second, D. McDonald (2009) used the long-tailed manakin dataset described earlier to investigate social, spatial, and genetic structures and determine commonalities between them. Although he found that genetic relatedness decreased with increasing social distance, there was no evidence for kin selection in his study, because all social distances were less than expected by chance. Such a comparison of social structure with genetic relatedness gives unique insight into population processes, as well as individual association strategies.

Third, Aplin et al. (2013) have recently shown that for great tits, individual variation in personality traits underlies some aspects of social structure. Finally, social networks have also been used to investigate between species interactions. Farine and Milburn (2013) studied a mixed-species social network based upon associations within thornbill (*Acanthiza* spp.)-dominated flocks. The authors built their network via visual identification of colour-ringed individuals. They found a rich mixed-species structure similar to that found within species and identified a shift in foraging niche by thornbills depending upon flock composition. Beyond population level analysis, there is also scope for social network analyses of mixed-species flocks that would make important contributions to our understanding of the evolution of species interactions (Farine et al. 2012).

This section was not at all an exhaustive survey of avian social network research. While there are some emerging themes, such as social niche selection (D. McDonald 2007; Oh and Badyaev 2010), it is clear that social network analysis as a tool has broad scope to address many interesting questions. Additional topics within which network analysis has been used as a tool include disease transmission (MacGregor et al. 2011), the fitness consequences of within family social interactions (Royle et al. 2012), mating systems (Ryder et al. 2008; Ryder et al. 2011), mate choice (Henry et al. 2013), communication (Miller et al. 2008), and collective navigation (Bode, Franks, Wood, et al. 2012), among many others.

Social network analysis of avian societies

A small but growing body of literature has examined avian social networks, and recent technological advances will increasingly facilitate detailed studies. Thus, future avenues of avian social network research are exciting. Here, we discuss a few particular lines of research that we believe will be particularly fruitful for addressing questions about social evolution in birds in the near future.

Eco-evolutionary processes affected by the structure of avian societies

In general, across taxa, we have very little understanding of the importance of social interactions and social structure for ecological and evolutionary processes. In part, this is a result of difficulties associated with characterizing social behaviour at suitable scales in terms of sample sizes of individuals, and numbers of generations (Clutton-Brock and Sheldon 2010). Relative to many other taxa, birds are particularly amenable to detailed multigenerational study of large, sometimes whole, populations (e.g. Kluijver 1951; Lack 1964). Long-term studies have typically focussed upon relatively easy-to-monitor species but have come from across diverse avian taxa including passerines (e.g. Kluijver 1951; Lack 1964; Grant 1999), seabirds (Dunnet et al. 1979), waders (Harris 1970), waterfowl (F. Cooke and Rockwell 1988), and raptors (Newton 1985). The number of individuals monitored annually in these studies ranges between hundreds to thousands of individuals, and, very often, detailed life-history data pertaining to reproductive effort, timing, and success, together with environmental data associated with breeding locations, individual survival, morphology, and other measures can be collected. These types of data also allow for the construction of detailed pedigrees.

Such long-term studies are certainly not unique to birds (e.g. Goodall 1968; Connor and Smolker 1985; Festa-Bianchet 1989). However, the life histories of many birds species make them particularly tractable for such detailed data collection on a very large number of individuals and the subsequent monitoring of social structure. With detailed multigenerational data on life histories and large sample sizes, studies of avian population wide social networks can be constructed. These studies could have important implications for our understanding of the life-history costs and benefits of sociality.

Social phenotypes and gene × environment interactions in avian societies

An interesting question, although one that is difficult to address, is the degree to which genetics and environment interact to produce an individual's social phenotype. Teasing apart these processes is likely not possible without experimental manipulations. This is because parents, offspring, and siblings all share similar environments; such conditions lead to genotype × environment correlations. The life history of many bird species also makes them particularly amenable to experiments in the wild. Cross-fostering experiments, where eggs or chicks are taken from their genetic parents and reared by unrelated foster-parents according to an appropriate experimental design, are straightforward to conduct and widely used in wild avian populations. Cross-fostering experiments among parents and environments could be used to account for the effects of environmental and genetic sources of similarity between relatives (Kruuk and Hadfield 2007).

Another reason to perform cross-fostering experiments is that the expression of an individual bird's social phenotype is likely to depend on the social phenotypes of the birds it is interacting with. When offspring tend to stay close to their natal area, they are more likely to interact with relatives; thus, the genetic and shared environment effects on the social phenotype may be overestimated. This will be the case when the probability that an individual initiates an interaction correlates with the probability it responds to an interaction (e.g. individuals initiating aggression towards conspecifics are also more likely to respond with aggression towards aggressive encounters initiated by conspecifics). With experimental approaches such as cross-fostering, we can begin to address questions that are currently wide open regarding the heritability and strength of selection on social network position.

The social characteristics of individuals, and emergent network structure in avian societies

Recent advances in monitoring technology, and the development of PIT tag loggers in particular, enable direct experimentation on social networks. The automated detection of individuals with PIT tags can be used as the basis for applying experimental treatments at the individual level, in order to test how individual traits influence the structure of social networks. For instance, individuals might be selected to receive audio playback, or visual stimulus (such as predator presentation), to determine how individual variation in the perception of the environment feeds through to the emergent properties of the population. Alternatively, and more speculatively, linking detection to food delivery offers the possibility of targeting certain individuals with specific dietary supplements, hormones, or psychotropic drugs, and thus asking how changes in individual nutritional or physiological state influences social structure. Finally, loggers can be designed so that they require birds to solve a complicated task (e.g. Morand-Ferron and Quinn 2011) in order to gain some reward, thus enabling researchers to monitor innovation and learning on social networks.

The role of social context in natural and sexual selection in avian societies

Studies of social processes will benefit from being able to put the individual into its broader social context (see Oh and Badyaev 2010; G. McDonald et al. 2013). It is generally unrealistic to assume that a focal individual is affected similarly by all other individuals in a population. It will therefore often be desirable to account for the importance of associations between specific other individuals within the context of the entire population.

For instance, with a network approach, studies of sexual selection no longer need to assume that each individual has a choice of all other individuals in the population but rather can assume, more realistically, that each individual has a choice of just those individuals that are within its social network (see Benton and Evans 1998; Kasumovic et al. 2008). An important paper by G. McDonald et. al (2013) highlights this, stressing the importance of considering the social network at a number of different levels. For example, the sexual network limits the population to the individuals a focal bird copulates with (these ideas are expanded upon in detail in Chapter 4) and thus is expected to be important for sperm competition, maternal investment, and sexually transmitted disease transfer. Similarly, the social network limits the population to the individuals a focal bird has contact with, revealing a realistic subset of the population under consideration for, for example, mate choice.

Underlying social structure and collective behaviour in avian societies

Another area in which bird social structure has been quantified is the context of collective behaviour. Collective behaviour is the study of how coordinated group-level patterns emerge from interactions among individuals. The fields of animal social network analysis and collective behaviour are developing somewhat separately but are perhaps more closely aligned than is often considered (Krause and Ruxton 2010). Social networks and the data used to generate individual-based models of collective behaviour are often derived from similar sources and so, in some cases, social networks can be seen as static summary representations of collective social group behaviour (Krause and Ruxton 2010). Perhaps one of the most striking images of collective behaviour is that of European starling (*Sturnus vulgaris*) murmurations. Three-dimensional analyses of murmurations suggest that the impressive coordinated movement emerges from individuals monitoring the nearest seven or so birds (Ballerini et al. 2008).

Jolles et al. (2013) recently showed that heterogeneous social structure related to dominance hierarchies and mated pairs in mixed-species corvid flocks shaped collective flight behaviour. Nagy et al. (2013) recently incorporated hierarchical social networks into analyses of collective motion in domestic pigeons (*Columba livia*) to show that dominance and leadership hierarchies were independent of each other in this species. Taken together, these two studies suggest that there is information to be gained about collective processes by accounting for social relationships among individuals with social networks, rather than by assuming that individuals are interchangeable and all follow the same simple rule in the same manner.

Individual interactions and social structure in avian societies

Most studies of social networks in birds are based on gambit-of-the-group observations (i.e. it is assumed that individuals that are part of the same foraging flock directly interact with all other individuals); relatively few studies have used social network analyses to investigate interactions per se. An exception is the work on communication networks (Naguib et al. 2004; McGregor 2005; Matessi et al. 2008; also see Chapter 9). In communication networks, directionality is known, and the meanings of interactions are often known as well. The identities of the receivers are, however, not always clear, as the travel distances of communications often vary with environmental conditions (Matessi et al. 2008).

One advantage of studying interactions is that they can more easily be translated into particular types of behaviour, such as competition for mates. Competition for mates is probably best studied by means of so-called sexual networks (G. McDonald et al. 2013). Other competition-based networks can be constructed as well; however, competitive interactions are generally harder to measure than cooperative ones (e.g. avoidance).

For example, avian social networks can be constructed by measuring the exploration or exploitation of resources by different individuals so that the exploitation or exploration of the same resources is counted as an interaction. Visits to the same nest cavity can be monitored via PIT tags, and competition-based bipartite networks built based upon visits to potential nest sites. Alternatively, aggression between individuals can be measured. Indeed, there is a rich literature exploring dominance hierarchies, and a recent emphasis placed upon exploring dominance hierarchies within a network context (D. McDonald and Shizuka 2012; Shizuka and McDonald 2012; also see Chapter 7).

The conservation and management of social units in avian societies

Delineating management units is difficult, due to well-known problems associated with choosing the appropriate definition of a population. These problems are typically related to genetic structure, space use, and population size. Once a population is defined, management actions are often directed towards individuals as the constituents of the population. However, active association among individuals implies a benefit of being social. It seems likely that, for some social species, the appropriate target for conservation within populations should be social groups rather than individuals, if social relationships are important for survival and reproduction.

For instance, many passerine species flock during winter. Benefits of flocking are thought to be associated with predator avoidance and dilution as well as increased foraging efficiency (Sridhar et al.

2009). Targeting management at solely the individual level may ignore important features associated with maintaining flock cohesion, potentially inducing something akin to Allee effects (Allee 1938): a negative relationship between social group size and reproduction. Network analyses provide efficient methods for delineating complex social structure (Newman 2006) and so could be particularly important for understanding conservation needs relative to social structure.

Genetic determinants of variation in social phenotypes in avian societies

Although sociality can, like other phenotypic traits, affect fitness (Silk 2007) and therefore be under selection, very few studies have explored the relation between genetics and individual variation in the tendency to be social. Some avian studies have explored the heritability of socio-behavioural traits such as dispersal (e.g. McCleery et al. 2004; Charmantier et al. 2011; Doligez et al. 2012) and personality (e.g. Dingemanse et al. 2003) but, to our knowledge, only a few studies of mammals have explored the heritability of the tendency for individuals to occupy certain social positions within broader social structures (Fowler et al. 2009; Lea et al. 2010; Brent, Heilbronner, et al. 2013; Christakis and Fowler 2013). The next step is to link individual social tendencies and genetics to fitness but, again to our knowledge, this has only been done for humans (Christakis and Fowler 2013). Potential reasons for this are that (1) social traits can be more difficult to measure than morphological traits; (2) sociality is expected to be more prone to additional sources of variation than morphological traits; and (3) social traits are thought to be more plastic than morphological traits and are therefore more likely to respond flexibly rather than through evolutionary changes. Studies of the heritability of social positions have found wide variation ($h^2 = 0.10 - 0.84$) (Lea et al. 2010; Brent et al. 2013). The availability of long-term studies on birds with pedigrees (i.e. for quantitative genetics models and to assess fitness (Clutton-Brock and Sheldon 2010)), the rapid technical advances and reduction of costs of molecular genotyping of non-model species (i.e. for marker-based genetic models (Ellegren 2008; Garvin et al. 2010)) and the advances in techniques for collecting social network data (as discussed extensively in this book) open up exciting new possibilities for understanding the genetic basis of sociality and for linking the ecological significance of sociality to its evolutionary dynamics.

The interaction of genetic and social structures in avian societies

In landscape and population genetics, the importance of social processes such as dispersal, kin structure, and social segregation are widely accepted and integrated into analyses (Sokal and Oden 1978; Fortin and Dale 2005; Fowler et al. 2009). Studies often tend to either correct for those effects or model them as spatial autocorrelation, which serves as a receptacle for many social and non-social biotic processes such as genetic drift and shared population histories (Diniz-Filho et al. 2012). In animal breeding, quantitative genetic models incorporating indirect genetic effects have recently been developed to take social structure into account. Until now, indirect genetic effects were based on either dyadic responses (Sartori and Mantovani 2013) or group or cage effects (i.e. multiple unweighted but fully saturated networks (Bijma et al. 2007)). There is scope for introducing social networks in quantitative genetics models, similar to the way indirect genetic effects are modelled by introducing weighted social networks as variance–covariance matrices in models resembling techniques for incorporating spatial structure (e.g. Stopher et al. 2012).

Another interesting approach for studying the interdependence of social and genetic structures is the use of eigenvector-based methods for canonical ordination of graphs. These methods can be used to model the effect of social structures on multivariate datasets (such as genetic markers), similar to the way those methods are currently used to model spatial structures (Legendre and Legendre 2012). Both social and spatial structures can be modelled simultaneously to separate spatial from social effects and thus are particularly useful for studies on wild populations (Radersma et al. 2014).

Social network analysis in birds: conclusions and future directions

The importance of social processes in evolution has long been recognized, and the study of avian social networks has the potential both to refine existing theory and to contribute new insights. New technology now allows for detailed social network construction based upon many more individuals than had previously been feasible. The types of opportunities for research discussed here are not uniquely avian; however, many will be particularly amenable to avian systems.

Acknowledgements

We thank Lucy Aplin, Ross Crates, Antica Culina, Damien Farine, Josh Firth, Ada Grabowska-Zhang, Lindall Kidd, Nicole Milligan, Ioannis Psorakis, Ben Sheldon, Brecht Verhelst, and Bernhard Voelkl for many conversations and ideas that contributed to this chapter. We thank Ben Sheldon, David McDonald, Jens Krause, and an anonymous reviewer for insightful comments on early versions. The authors were supported by an ERC Advanced grant to Ben Sheldon (AdG 250,164).

CHAPTER 17

Networks of terrestrial ungulates: linking form and function

Daniel I. Rubenstein

Introduction to terrestrial ungulate social networks

Individual actions shape the structure and functioning of animal societies. Yet these actions are themselves shaped by the environments in which individuals live and, thus, by the actions of others. Vigilance by one influences the vigilance of neighbours, just as the behaviour of one when acquiring mates and other ecological resources is influenced by that of others. The collective outcome of these actions in turn shapes the dynamics of populations via the spread of genes, memes, and disease, as well as the transfer of information among individuals about abilities, values, personalities, and needs. Just as game theory provided insights into understanding the evolutionary dynamics of individual decision-making (Maynard Smith 1982), social network analysis is beginning to provide insights into how decisions and social actions structure societies and govern their social dynamics.

The play and the players in terrestrial ungulate social systems

Nowhere is the link between environment and behaviour more apparent than in the terrestrial African hoofed mammals—the ungulates and their close relatives. Jarman's (1974) classic study illustrates how ecological factors interact with physiological constraints to shape the core elements of ungulates' social systems. Predicated on the notion that ungulates of different body sizes require different amounts and quality of vegetation (Schmidt-Neilson 1972) and face different types and degrees of predation, Jarman argued that different species perceived and utilized similar landscapes differently. Since differences in sex, reproductive state, and age also shaped these needs, Rubenstein and Wrangham (1986a,b) constructed a general framework that argued that females were primarily selected to solve ecological problems of acquiring resources and avoiding predators, and that males were selected to maximize mating opportunities. Depending on the abundance and distribution of key resources that are affected by habitat features and essential physiological needs, a variety of core social structures could emerge.

According to this framework, the smallest bodied ungulates, such as dik-diks (*Madoqua* spp.), duikers (*Cephalophus* spp.), suni (*Neotragus* spp.), and klipspringers (*Oreotragus oreotragus*), will require high-quality forage, but in small amounts. Given that such food tends to be widely scattered within relatively moist habitats, females will be forced to spread out, thus preventing males from associating with more than one. Under these conditions, monogamy should result. Facing a broad spectrum of predators, surreptitious movements by such small-bodied species should also be favoured, and such movements should also reinforce strong pair-bonding.

For large-bodied species, both dietary and predator constraints will change, and various forms of polygyny and polyandry should emerge. Since large-bodied species can consume low quality vegetation, competition should be reduced, fostering group living. Once individuals live in groups,

Animal Social Networks. Edited by Jens Krause, Richard James, Daniel W. Franks and Darren P. Croft.

a host of anti-predator benefits are likely to arise. As a result, females in most mid-sized species, such as impala, Grants' (*Gazella granti*) and Thompson's gazelles (*Gazella rufifrons*), wildebeest (*Connochaetes taurinus*), hartebeest (*Alcelaphus buselaphus*), and waterbuck (*Kobus ellipsiprymnus*), are expected to live in groups that should be attracted to the best patches of vegetation.

Since these patches vary in size, the groups occupying any patch at one time should also vary in size. Because these patches also vary in quality, males should compete vigorously for the best ones, which can host the most females. The males' defence of these patches should lead to what Emlen and Oring (1977) defined as 'resource defence polygyny'.

As animals reach even larger sizes, the landscape is likely to be perceived as if it were a large, relatively homogenous sward capable of sustaining very large groups. Males no longer should be able to defend such large areas or the very large groups such areas support. In some species, such as the Cape buffalo (*Syncerus caffer*), males will be forced to wander within these large groups, identifying females ready to mate; and, if of high-enough rank, such males should consort and mate with these females before moving on to search for others. Alternatively, males could position themselves in areas called leks, where female movements converge. This happens in topi (*Damaliscus lunatus*), as male topi defend small non-resource-based territories, displaying to attract the attention of females (Gosling 1986).

Jarman's comparative analysis successfully linked social and mating behaviour of males and females has been applied to other African ungulates with great effect. The societies of elephants (*Loxodonta africana*) and giraffes (*Giraffa camelopardalis*), two species qualifying as very large-bodied herbivores, are variants on the 'dominance defence polygyny' theme (Emlen and Oring 1977). Similarly, Grevy's zebra (*Equus grevyi*), as a large-bodied species, exhibits 'resource defence polygyny' (Klingel 1977; Rubenstein 1986). And while plains zebras (*Equus quagga*), another type of large-bodied herbivore, exhibit a unimale, multi-female core social system (harems), they do so because, as they are hindgut fermenters, competition for forage is dramatically reduced. Because maximizing the time spent grazing takes precedence over acquiring high-quality food, females attach themselves to males that can reduce sexual harassment, scan for predators, and provide females extra time (Rubenstein 1986, 1994; Linklater et. al 1999).

As useful as this framework has been in accounting for the evolutionary and ecological differences among the species, the patterns that emerge are attributed to actions that benefit broad classes of individuals. Details about individual interactions and relationships remain hidden. Yet, it is individual interactions that shape the class-based strategies that will determine and affect the frequency, form, and function of the competitive and cooperative relationships that develop within each of these societies.

Interestingly, despite the fundamental differences in the core social systems of African meso- and mega-herbivores, all but the monogamous species show fluidity and flexibility in the associations that form between individuals or classes of individuals. At some level, the disparate societies of terrestrial African ungulates exhibit fission–fusion dynamics. As a result, social network analysis can be used to determine why the structures within each of these societies take on the shapes that they do, as well as to decipher the functional consequences. Although interactions and associations have been measured for other ungulates such as buffalo (*Syncerus caffer*) and deer (*Elaphurus davidianus*), most have been dyadic and used for examining the nature, causes, and consequences of group structure (Melletti et al. 2010) In this section, only those ungulate species for which association indexes have been analysed with clustering algorithms or used to construct networks are discussed.

The script, the data, and methodologies for ungulate social systems

Constructing networks and assessing how relationships determine a species' social structure requires the accumulation of a large amount of data from a large number of individuals. Understanding the types of data, and ways in which data are gathered and are aggregated, is essential at the outset of any study, since different types of data produce different types of networks and thus limit the inferences they provide. Constructing networks requires collecting repeated actions and associations from known individuals. Identifying individuals is a difficult and often tedious process. Not surprisingly, only for a handful of species whose core social units and their ecological determinants does sufficient interaction data exist for performing social network analysis.

Originally done by visual inspection of stripes, spots, and other unique morphological features, computer software has made the task of recognizing individuals both easier and more efficient without sacrificing accuracy (Hiby and Lovell 1990; Crall et al. 2013; Mayank et al. 2011; Bolger et al. 2012). In fact, improvements in recognition algorithms will soon make it possible to identify individuals in real time and in places where humans have found it difficult to reach. Unmanned aerial vehicles, otherwise known as drones, that can be programmed to fly predetermined routes, strategically positioned 'camera traps', and even photographs from armies of citizens will expand the pool of species that can be studied and thus expand the ability to link remotely sensed ecological and association data.

While GPS tracking data and GIS analyses have helped identify how and why individuals move about landscapes, associations derived from tagged individuals should be used with caution when drawing inferences about associations. Rarely are all members in groups, or all groups in a population, tagged with GPS units. And unless they are, the associations that emerge represent a biased subset of those that exist. Networks missing invisible but potentially important individuals will yield misleading metrics.

Perhaps the most powerful feature of social network analysis is the way it moves beyond characterizing societies by dyadic relationships. Instead, networks help visualize and quantitatively characterize direct and indirect connections among every individual in a society. Assessing centrality is a powerful way of measuring an individual's importance, but there are many ways of calculating centrality (Croft et al. 2008). Whereas degree centrality measures the number of direct connections an individual displays, page rank centrality and eigenvector centrality measure the connectivity by accounting for the contacts of a focal individual's direct contacts. As such, these metrics measure an individual's global reach, as does both closeness centrality and betweenness centrality. Whereas closeness centrality measures the shortest path between an individual and all other individuals in a society, betweenness centrality measures the number of shortest paths between all pairs of individuals that flow through a focal individual.

Each measure reveals a different facet of a society's structure but, when taken together, they can be used to better understand how societies function. This will be especially true when nodes are labelled by phenotype and their network metrics are compared. Then, potential societal roles can be identified if correlations between centrality metrics and phenotypes emerge.

'Personality' is one phenotypic feature that is likely to play an important role in structuring societies. Animals can often be arranged along a continuum of high to low risk taking or high to low levels of leadership or followership. Such traits are often heritable, and mixtures of phenotypes are maintained within societies by frequency-dependent selection, much like 'producers' and 'scroungers', a common set of behavioural alternatives. But individuals are just as likely to display personalities that develop as a response to their social milieu.

Such 'social personalities' are likely to be condition dependent and more malleable than those typified by boldness or leadership. Nonetheless, their importance should not be underestimated; their appearance or disappearance from societies can profoundly affect the functioning of those societies. Experiments in which individuals of known social personalities are added or removed are likely to identify which types of individuals play the greatest role in governing the functioning of particular societies.

Social networks can be constructed from various types of data. Depending on the way interactions are characterized, the networks that emerge will vary. For example, characterizing associates based on repeated interactions produces weighted networks, whereas associations described by single contacts generate unweighted networks. Associations based on membership in a group produces 'gambit-of-the-group' networks (Whitehead and Dufault 1999), in which connections are assigned to everyone in the group, whereas associations derived from dyadic interactions yield more nuanced networks. Sometimes interactions are directed—emanating from, or going to, particular individuals—but sometimes they are not, and the networks they produce provide different lenses for assessing how relationships structure societies. Last, data used to characterize associations can be aggregated or binned over time, or the associations can be treated as a time series of interactions.

These different data processing schemes produce static or dynamic networks, respectively. No type of data is superior to another, and often the type collected is constrained by a species' biology and ecology. But each data type produces social networks that offer different perspectives on the way sociality emerges from social interactions. While weighted graphs illustrate the varying strength of relationships that can exist in a social system, unweighted graphs and their metrics reveal the diversity of neighbours with whom individuals interact. Just as ecological diversity indexes blur the distinction between richness, the number of items being counted, and evenness (a measure of the variation in the relative abundance of each item) weighting networks fails to separate the number of different individuals with whom a focal individual associates, and the strength of those relationships.

Thus, for a complete picture of social dynamics, many social network studies on ungulates present both types of networks. Weighted networks are usually constructed from matrices of association indexes that are sometimes subjected to randomization procedures to create 'preferred networks'. The existence of non-random patterns, however, need not reveal preference; see Croft et al. (2008) for a detailed discussion of hypothesis testing in social networks.

Associations are usually undirected and produce proximity networks. When they are based on pair-wise associations, especially those of nearest neighbour, they foreshadow prosocial intentions; affiliative behaviour, for example, usually occurs between close neighbours. Directed graphs, however, reveal actual interactions, and networks of affiliative behaviour typically are built from grooming 'in' and 'out' interactions, whereas those depicting agonistic networks are typically build from 'in' and 'out' aggressive actions. Correlations among such networks can be very informative, and expected negative relationships between networks of 'in-degree' and 'out-degree' aggression do not always materialize because not all social relationships are transitive (Crofoot et al. 2011).

Most social network studies on ungulates and other animals accumulate association or interaction data over time and aggregate it before computing association indexes. The graphs that result thus represent average patterns of social structure and, by doing so, necessarily hide the temporal vicissitudes that characterize waning or waxing strengths of social bonds. New approaches are emerging that keep each observation separate and use algorithms for finding dynamic structures that retain temporal flows and generate parallel, but more powerful dynamic metrics for comparing societies (Rubenstein et al. in review). Ungulate societies have been subject to both approaches.

Equids as model organisms

Because most equids, especially zebras, are boldly marked, they represent an ideal group to connect the broad-scale patterns of sociality described above with the fine-grained details of how individual relationships build societies when actors, actions, and environments change. Moreover, equids exhibit a variety of core social structures and since the various species show broad similarities in body plan and physiology, social differences are more likely to be shaped by social and ecological interactions rather than evolutionary constraints. And even when some of the core structures consist of strong bonds among members, each equid population exhibits fission–fusion social dynamics at some level (Rubenstein and Hack 2004). In addition, a variety of populations within species have been studied for long periods of time in different locales and under different demographic regimes (Rubenstein et al. 2007; Cameron 2009; Kearnes 2009). From these comparative studies, network analyses have revealed four important features: (1) similar social systems can differ in subtle but important ways that represent different adaptations to different ecologies; (2) when hierarchical societies form, they are characterized by strong bonds at lower levels and more fluid bonds at higher levels, and this structure has important consequences for shaping strategic routes to adulthood in males; (3) identifying communities, always a difficult task, can be aided by using dynamic rather than static social network analysis; and (4) structure does shapes function.

Similar social structures but different social networks among equids

African (*Equus africanus*) and Asiatic (*Equus hemionus*) wild asses and Grevy's zebras (*Equus grevyi*) are the two most arid-adapted equids, and both species show fission–fusion social dynamics at the individual level. This is not surprising because, once food and water become widely segregated, females in different reproductive states weigh the needs of acquiring forage and water differently (Rubenstein 1994). Moreover, as Sundersean, Fischhoff, and Rubenstein (2007) have shown, females nursing young foals facing a postpartum oestrus need to associate with particular males to reduce harassment that could harm them and their infants. Thus, females in different states range differently, and as a result, bonds with particular males and females that are strong in their closest evolutionary relative and congener, the plains zebras, are weakened.

However, when social network analysis is applied to Grevy's zebras and asses, major differences in the patterns of associations appear (Sundaresan, Fischhoff, Dushoff, et al. 2007). For both unweighted and preferred (weighted) networks, segregation or modularity in Grevy's zebras is greater than in the wild asses as measured by the number of connected components, cliquishness (strength of cluster coefficients), and average path length. The analysis shows that Grevy's zebras are divided into more components, exhibit higher cluster coefficients typically approaching 1, and display shorter path lengths than wild asses. Overall, these results suggest that when fissions occur in wild asses, they tend to be based on individual actions whereas, in Grevy's zebras, fissioning occurs when small groups of moderately strong associates break off as a group from larger groups. Given that the ecological risks of predation and not finding water are slightly higher on Grevy's zebra landscapes, selection appears to favour larger and more cohesive groups in Grevy's zebras as opposed to wild asses in India.

Even among populations within the same species, social network analysis has identified subtle but important social differences: two populations of wild asses, one in the more arid Negev desert of Israel and the other in the more human-moderated dry lands of the Little Ran of Kutch in India, have been shown by social network analysis to differ in structural ways that, again, appear to represent adaptive responses to ecological conditions. In the drier Negev, where food patches and watering points are harder to access, both unweighted and weighted networks of onagers show more connections than the khur of India, even when the half-weight threshold of association is set high (Rubenstein et. al 2007). For populations of similar size, not only do those of onagers exhibit more connected components, they also are characterized by higher cluster coefficients and thus higher cliquishness. Moreover, the onagers showed stronger association scores among individuals of similar reproductive states than did the khur. As the population of onagers grew, both degree and cliquishness declined as new immigrant males and maturing females joined the ranks of breeders. Clearly, social network analysis helped identify key differences in structure from which inferences could be drawn about social dynamics.

Fission–fusion functionality in hierarchical equid societies

Plains zebras and horses live in uni-male, multi-female groups in which membership remains constant for years, if not lifetimes. Yet, whereas horse harems tend to avoid each other unless drawn to a common resource, plains zebra harems actively seek each other out and form herds that persist for hours, if not days. Of the three factors that have been hypothesized to favour the formation of zebra herds—lowering predation risk, foraging enhancements, and harassment reduction—only harassment reduction emerged as a key force in a full multivariate model (Rubenstein and Hack 2004): males that formed coalitions were better able to keep bachelor males away from their females, thus enhancing their foraging efficiency. In areas with high numbers of bachelors, herds were larger than in areas where bachelor males were scarce.

In this study, stochastic dynamic programing models were used to identify the best ontogenetic strategy by which bachelor males could become breeders and overcome stallion dominance. When predation pressure was moderately high and developing superior fighting ability to dominate peers and impress females was highly valued, the best route selected favoured bachelors that associated with herds of harems rather than just other groups of bachelors. In this way, bachelor males maximized their chances of dominating both other bachelors and stallions while at the same time showing off their talents to observant females.

Social network analysis shows that bachelor males behave as predicted. Repeated censuses of over 600 zebras twice a day for 30 days yields the graph of male associations illustrated in Figure 17.1. Clearly, three modules emerge, each consisting of a core group of harem stallions strongly connected to a subset of bachelor males. While many of the stallions associate with males from other modules, a few of the bachelors do so as well. In agreement with the predictions of the stochastic dynamic programing model, bachelor males B64 and B72 exhibited some of the highest levels of bachelor male degree centrality and were the only ones during this period attempting to establish harems by stealing young primaparous females emigrating from their natal groups. Such potential fitness outcomes underscore the benefits to individuals of developing wide-ranging and diverse linkages in social networks that carry over from one ontogenetic state to another and are likely to enhance fitness (Sih et al. 2009).

Identifying communities among equid societies

Everyone knows a community when one is part of one, but defining or identifying a community is difficult. Humans belong to towns, religious groups, social orders, professional associations, and other organizations. Often these communities are defined by a set of common values, traditions, or ad hoc characteristics. Even when members interact with only a small subset of the other members, they consider themselves part of the community. Such communities evolve gradually over time and often have an inertia that maintains them even as membership changes.

Imputing communities as separate from groups in fission–fusion societies is harder in animal societies than in human societies. It is impossible to ask animals to identify to which communities they belong. Animal communities are characterized by groups of individuals that are more connected among themselves than with other individuals in the social network. Computing the strength of these connections is not trivial, and many techniques have been developed (Croft et al. 2007; J. B. W. Wolf et al. 2007). One way of improving the process is to define communities dynamically as a collection of individuals who interact consistently and frequently. By assuming that individuals interact more often with members of their own communities, occasionally move to and interact with members of other communities, and rarely make whole scale switches between communities, and by assigning costs to switching and visiting other communities, Tantipathananandh and Berger-Wolf (2011) developed algorithms that produced a community structuring that minimized total costs of the recorded visits and moves.

As shown above, the static or aggregate network of Grevy's zebras consisted of three connected components, one of which was large and included most member of the population. Once the dynamic community identification algorithm was applied and the community identities were superimposed on the static graphs, the large cluster was shown to consist of two communities whose members came together and split apart on a regular basis. One community consisted of lactating females that tended to stay near water, and the other was composed of non-lactating females that tended to wander more widely in search of better pasture and did not need to drink as frequently.

The dynamic community identification procedure also generated dynamic metrics that added a time dimension to the traditional static metrics. Whereas the average shortest path in a connected component in a static network is the mean number of edges connecting two nodes, the average shortest temporal path in a dynamic community

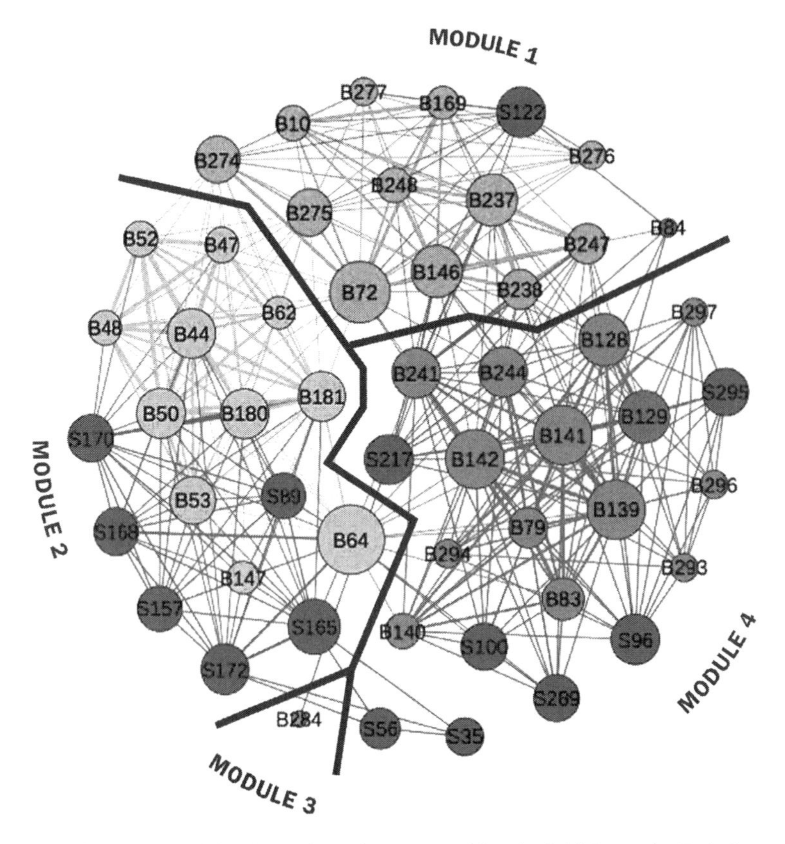

Figure 17.1 Static network of harem and bachelor plains zebra males constructed from herd sighting on the Mpala Conservancy during the summer of 2013. Breeding stallions the dark circles demarcated with an 'S'. Bachelor males are denoted with a 'B', and their grey shade varies depending on the network module assigned by the clustering algorithm. Three modules exist and are delineated with lines. While most bachelor males show strong connections mostly to other bachelor males and stallions within their assigned modules, some show connections to breeding stallions and bachelor males in other modules as well.

is the mean number of time steps to connect two nodes. As Table 17.1 shows, the small differences among the static metrics that initially identified differences in the deep structures of Grevy's zebra and wild ass societies are magnified when temporal dynamic metrics are employed. The dynamic analysis revealed that onager society had many small communities, with individuals switching affiliations more often than those in Grevy's zebra society. Overall, Grevy's zebra communities were cohesive yet flexible, whereas onager communities were constantly changing; this observation suggests that onagers may be incurring lower costs of switching affiliations than Grevy's zebras. Presumably, the lower cost structure emerges because the predictability of finding water and

Table 17.1 Standard static and dynamic social network metrics for Grevy's zebras and onagers. Both populations were similar in size, but they differed in many structural dimensions. Small social differences emerge from the static network metrics, but these are magnified by the dynamic analogues.

	Grevy's zebras	Onagers
Number of individuals	28	29
Density	0.3	NA
Dynamic density	0.52	0.24
Average shortest path	1.8	1.7
Average temporal path	4.8	7.5
Diameter	4	3
Cluster coefficient	0.9	0.7
Dynamic cluster coefficient	0.1	0.03

avoiding predators are both higher for onagers than for zebras.

Structure shapes function in equid societies

Once the structures of societies are identified, then 'what if' simulations can be run using those structures to assess how different societies function in terms of facilitating or blocking the spread of genes, memes, or disease. Using the dynamic community networks of Grevy's zebras and onagers, and a simple linear cascade model of spread in which a single contact is sufficient to infect a node on the network, Habiba (2013) and colleagues (Habiba, Rubenstein and Berger-Wolf, unpublished data) showed that, when transmission thresholds were high, it was more difficult for disease to spread in onager societies than in Grevy's zebra societies, because onager communities are small, often consisting of solitary individuals; one set of simulations is shown in Figure 17.2. Yet, when it comes to blocking the spread of disease, the removal of just a few individuals from the onager community had a greater impact than removing a similar number from the Grevy's zebra community. The removal of each onager, even if randomly chosen, had more impact on destroying the connectivity in the onager network than the removal of a Grevy's zebra had on the zebra network.

But what is even more striking is that the 'social personality' of those individuals removed mattered. Removing individuals with high node-degree or cluster-coefficient scores blocked the spread of disease more effectively than removing individuals at random did. For onagers, by the time the top three such individuals were removed, the spread of the disease was effectively blocked. In Grevy's zebra communities, the removal of a much larger fraction of the highest ranked individuals was required in order to produce the same impact. Yet, if selection were to favour the retention of knowledge rather than thwarting disease spread, then the more modular, more cohesive, larger, and more coordinated communities of Grevy's zebras would be favoured. Thus, no one network structure is universally best at solving all environmental problems; only by first identifying the structure of communities and the most important individuals in those communities

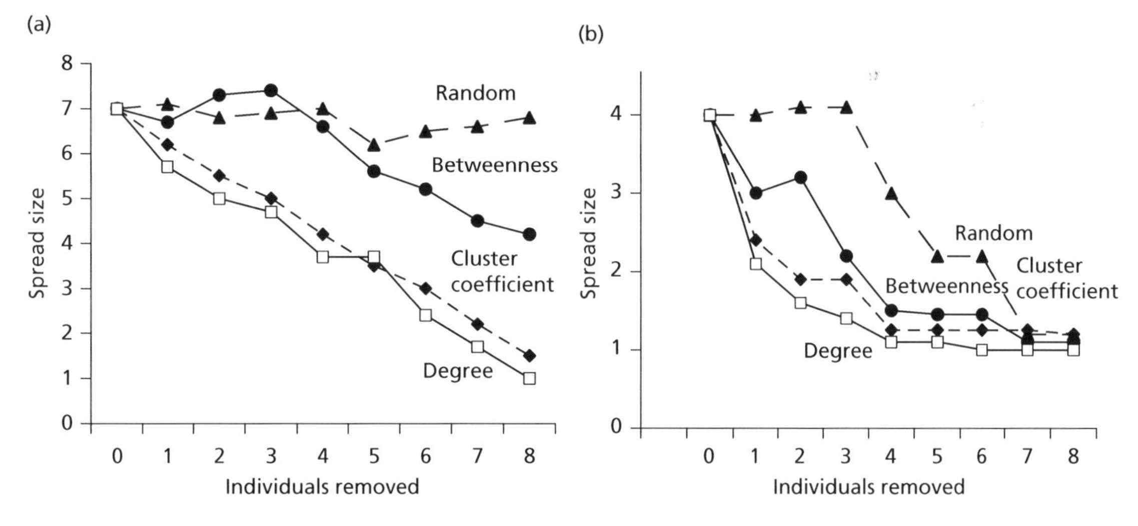

Figure 17.2 Spreading and blocking processes in (a) Grevy's zebras and (b) onagers. The *y*-axis depicts the extent of spread for an intact network as well as after individuals are removed sequentially from the network, either at random or by scores on various network metrics (social personality) from high to low. The extent of spread as well as the ability to block spread on the networks of Grevy's zebras and onagers differs. When ease of transmission of disease or ideas is moderately easy, spreading is slightly faster in Grevy's zebras then in onagers. As individuals are removed one by one, however, it is much easier to block spread and in essence disrupt network functioning among onagers than among Grevy's zebras. Whereas it takes the removal of only three onagers to eliminate spread, it takes the removal of nine—approximately a third of the population—to do so for Grevy's zebras. The most effective equids for blocking spread are those with high cluster coefficients and node degrees.

based on their social personalities is it possible to gain insights into the adaptive functionality of those structures and conjecture how selection was likely to have led to their evolution.

Generalizing to other ungulates

Wild and free-ranging ungulate species

Horses

As the closest relatives of plains zebras, horses live in harems that exhibit relatively strong bonds among members. Yet detailed analysis of nearest neighbour associations, as well as directed dyadic agonistic and grooming interactions, reveals that the network structures among groups vary and that these variations have important fitness consequences. For example, Cameron et al. (2009) demonstrated that increased social cohesion among free-ranging New Zealand female horses increased female reproductive success by collectively reducing sexual harassment by stallions.

A similar result emerged for a population of feral horses that live on a barrier island off the coast of North Carolina and whose population is managed via the use of the immunocontraceptive porcine zona pellucida. One of the unintended consequences of this type of fertility control is that females change groups more frequently, with a majority showing associations with three or more males during the breeding season (Nuñez, Adleman, and Rubenstein 2010; Madosky et. al 2010). Typically, females changing groups are harassed both by the males whom they are leaving and by the males whose groups they are joining (Rubenstein and Nuñez 2009). Only those females with high degree centrality were able to lower the rate at which they were harassed (Kearnes 2009).

Differences in strength of associations also can impact survival. On the same island, prior to the advent of fertility control, a large cull took place to dramatically lower the size of the population. During the removals, about half the foals that remained on the island did so as orphans and without any other members of their natal group. All the foals with at least one surviving parent survived until the age of independence. However, survival among the orphans was affected by the degree of their social connectivity before the social upheaval occurred. Only those with high unweighted degree survived, suggesting that the experience of building a broad social network early in life potentiates doing so again when parents are not around to help modulate social interactions in beneficial ways (Nuñez, Adelman, and Rubenstein 2014).

In other free-ranging horse populations, social connectivity has also been shown to play a significant role in structuring societies. In a population of over 300 Esperia ponies inhabiting the Italian Abruzzi Mountains, individuals intervene and break up social interactions occurring among other ponies. Females engaged in this behaviour significantly more often than males and virtually all interventions disrupted affiliative rather than agonistic interactions. While dominant females initiated most of the interventions, they directed them towards those with whom they were not strongly bonded and mostly targeted females interacting with other females with whom the intervening female was strongly bonded. Given that such interventions are likely to incur small risks, such behaviour appears to be an investment in preventing the dilution of exclusive bonds. Third-party interventions keep social partners from exchanging affiliative interactions with others (Schneider and Krueger 2012). As the previous studies on New Zealand and North Carolina horses show, maintenance of strong bonds can have future fitness benefits.

Elephants

Elephants, both African and Asian, live in hierarchical structured societies. Smaller group are embedded in larger groups, some of which fission at certain times of year, only to reform later as groups coalesce (Moss and Poole 1983). In African savannah elephants, seasonal rains enable them to come together at the best parts of the landscape, thus fostering the formation of large herds; this is also when the most dominant bulls come into musth and use their dominance to displace lower males to mate with as many females as possible. During drier periods, however, African savannah elephants retreat to water-rich refuges, which tend to be scattered and small and thus cause groups to splinter.

Using association indexes and clustering techniques to quantify the degree of hierarchical structuring in a central Kenya population and the level of cohesion displayed by groups at each hierarchical level, Wittemeyer et. al (2005) showed that African elephant societies exhibit four tiers, the first being the mother–calf unit. Aggregations of many mother–calf pairs produce 'core' or 'family' groups that are relatively stable. These groups are highly kin connected, and their stability suggests that they are relatively immune to changes in ecological conditions. Non-random associations among family groups produce 'bond groups' (the kin groups of Moss and Poole 1983), which are less cohesive than family groups and are the groups that split during the dry season when, presumably, competition outweighs the benefits of allomothering and collective defence against predators. Bond groups often merge to produce a fourth tier that is the least cohesive of the groups.

What shapes the dynamics of these groups is not clear. However, low correlations with changing ecological conditions and observations of mutualistic interactions between groups at this tier suggest that the dynamics of fourth-tier groups are driven more by social as opposed to ecological factors. The fact that these groups show high spatial overlap lends support to this interpretation.

Similarly, Archie et al. (2005) examined the hierarchical structure of African elephants in a southern Kenya population to determine the extent to which kinship and genetic relatedness structured the hierarchy of levels. Using both nuclear and mitochondrial DNA, they showed that family groups consisted of highly related individuals and that, when these groups split, associations among mothers and daughters and maternal sisters were preserved. At higher levels, though, genetic relatedness was much lower; this observation suggested that the factors binding individuals at lower levels differ from those binding individuals at higher levels. At higher levels, dominance as well as reciprocal and mutualistic interactions comes into play.

Asian elephants also exhibit hierarchical structuring, but their fission–fusion dynamics appear to differ from those of African savannah elephants. As de Silva et al. (2011) showed, Asian elephants in Sri Lanka formed the largest groups during the dry season, which is the opposite of what occurs in African elephants. The authors suggest that mutualistic support is needed in intergroup competition for the defence of dwindling food resources. Asian elephants in general exhibited weaker associations than African elephants, with some individuals showing strong associations with a few individuals and others showing weak associations with many individuals.

What is surprising about the hierarchical structuring in Asian elephants is that social instability is highest at both the lowest (dyads) and highest levels (clusters) of organization. Social units (affiliated sets of individuals) and 'ego' units (the network of direct links between a focal elephant and her associates) showed the most cohesion. Perhaps this novel hierarchical pattern of cohesion results from the fact that the habitats used by Asian elephants are more closed than those used by African elephants. Individual Asian elephants maintain contact via long-distance vocal and chemical communication; this observation suggests that these elephants are still connected even when both elephants and human observers cannot see connected groups. If this is true, then it serves as a cautionary tale: connectivity based solely on close associations may underestimate the interactions and associations that particular species consider as biologically relevant. The generality of these findings will emerge as association data accrues from populations of African forest elephants.

Giraffes

As another large-bodied, wide-ranging ungulate, it is not surprising that giraffes exhibit dominance defence polygyny as the foundation of their society. But until recently, it has been assumed that social associations among adults were weak and formed somewhat randomly (Pratt and Anderson 1982). A recent study by VanderWaal, Wang, et al. (2014) on the reticulated giraffes that inhabit the semi-arid areas of central and northern Kenya, however, shows that associations are highly structured for at least adult females and subadult males and in fact reveal hierarchical structuring with some novel features not seen in the hierarchical communities of plains zebras and elephants.

Community structure algorithms identified three levels of clustering. Social cliques were identified as the core units; within cliques social association indexes were extremely high. Social cliques were then embedded within subcommunities where indexes of association were moderately high, and the subcommunities themselves were embedded within communities. This type of structure mirrors that exhibited by African elephants. Although males search for and consort with females, males and females essentially occupy different social networks, since over 90% of the social cliques consisted only of members of one sex. Male social cliques consist solely of bachelor males, a pattern that is very different from the community network of plains zebras.

As with African elephants, giraffe associations appear to be driven somewhat by spatial structure, especially for females. Over 90% of females assigned to the same spatial neighbourhoods were members of the same social cliques. But spatial proximity does not shape all of giraffe sociality. On the one hand, spatial separation between female subcommunities occurred despite the absence of any physical barriers to movement; yet, at a finer scale, the home ranges of different female social cliques overlapped. For males, the relationship between network structure and shared space use is even weaker. Thus, for the species as a whole, social structure is not solely the result of space use. At least at the level of social cliques, strong kinship ties exist; but it is not yet known whether genetic relatedness operates at higher levels as it does in African elephants (Archie 2005).

Hyraxes

Although much smaller than elephants, their close relatives, hyraxes (*Procavia capensis*), live in societies also built around females. As plural breeders, many females help care for each other's young (Borocas et al. 2011). Networks have been created for the Ein Gede, Israel, population that depict the changing nature of relationships from year to year. Group living has been shown to affect survival by either reducing the risk of predation, increasing foraging efficiency, or both in many species (Alexander 1974) and has been shown to vary with social system (Clutton-Brock et al. 2001) in cooperative breeders.

In hyraxes, individual prospects of survival are shaped more by the distribution of associations across the entire network than by the number of associations maintained by individuals. In the Ein Gede population, individual longevity increased with a decrease in the variance of individual centrality. It appears that individuals do better if they inhabit networks with more equal associations among its members. Perhaps egalitarian associations, rather than the number or strength of individual associations, reduce mortality risk, by eliminating stress or enabling more efficient collective actions in networks.

Domestic and captive animals

Animals living with humans live in unnatural settings, where space is often constrained, daily time budgets are altered, and normal patterns of associations are disrupted. As we have seen, free-ranging African elephants live in complex societies with many societal tiers, each with different network properties. Elephants in zoos live under very different conditions. Rarely are associates in zoos kin; yet, a variety of networks can be created to characterize the nature of sociality that develops in zoos (Coleing 2009). Perhaps the most informative networks are those that reveal cliques and nature of prosocial connections. Since management periodically requires removals and introductions, visualizing current structures in relation to previous ones can serve as a guide on how to up- or downregulate specific types of associations.

For livestock, the use of GPS collars or data loggers that communicate to regularly positioned wireless sensors is also enabling ranchers to not only monitor movement but also gather important data on social relationships. Networks can reveal the formation of cliques, identify leaders, or even detect when and which females are forming consorts with males (Handcock et al. 2009). Virtually continuous monitoring of proximity networks where the phenotypes of nodes are known, in conjunction with knowledge of movements and habitat use, will enable ranches to optimize sex- and reproductive state-specific paddock rotations to maximize rangeland health and livestock body condition.

Terrestrial ungulate social networks: conclusions and future directions

It is clear that social network analyses have played important roles in identifying important details about social structure both in terms of their causes and consequences. Even when ungulate species exhibited similar mating and core social systems, network structures were shown to diverge. This suggests that local conditions shape the interactions and relationships of individuals and that they do so differently for males and females, as well as for members of the same sex that are in different reproductive states.

Even when core social structures differ among species—some being more fluid in some species than others—hierarchical structures often appear. Identifying these hierarchical structures revealed the importance of clustering algorithms. They not only identified distinct levels of social organization in elephants and giraffes, but also helped quantify the degree of cohesion within each level. Sometimes this appeared driven by changes in ecological conditions, but at other times social factors were the primary determining factors.

Virtually every network analysis on ungulate populations faced the problem of identifying communities. Many methods exist for identifying substructures, subcommunities, or modules on aggregated networks, but the ability to account for temporal change—a key component of what makes communities so hard to identify—can be facilitated by using dynamic network analyses. And they are likely to pay big dividends. Animal behaviourists naturally collect data over time, either by sight or remotely with sensors. Tools that exploited the richness of these temporal sequences yielded new insights into the causes and functioning of zebra and wild ass societies and are likely to do so for elephants, giraffes, and other ungulate species as well.

Adding nodes of individually recognizable individuals or marked livestock to networks of wild species would further expand the diversity of associations and edges. Thought experiments played out using simulations in which individuals of known phenotype were added or removed, or in which disease agents were injected into mixed wildlife–livestock networks, would allow pre-emptive assessment of virtually unlimited and replicated co-management schemes that ordinarily are employed without comparative analyses. Evaluating the accuracy of such predictions will require gathering fine-grained associations over space and time and will require large datasets, high-throughput data analysis, and next-generation natural world sensing. For such data-driven interventions to become a reality, multiple data gathering techniques involving many eyes, camera traps, GPS, and proximity sensor tracking tags and drones must be combined.

With the development of new individual recognition algorithms, more and more ungulate species whose core social systems are well characterized should soon be open to having their social networks probed. At the moment, all species with stripes, spots, protruding veins, or patchy coloration can be studied. With different recognition approaches, even the physically most bland species will qualify for study.

When such 'big data sets' emerge, they are likely to provide insights into how best to manage populations under threat of extinction or at risk from rapidly spreading disease. Simulations on actual networks of equids (Habiba et al., unpublished data) or on spatial–social networks of populations of Pere David deer (*Elaphurus davidianus*) (Getz et al. 2006) illustrate the role that understanding the structure and workings of networks can play in guiding public health interventions. Constructing networks of cattle farms was instrumental in halting the spread of hoof and mouth disease in the UK. Such networks helped identify which British farms had the highest betweenness scores and led to designing and implementing efficacious and effective quarantine interventions to stop the disease's spread and limit the long-term economic damage associated with the disease.

The same approach of using networks to identify key areas or key individuals based on their 'social personalities' within populations should help strengthen conservation interventions as well. As the example with giraffes illustrated, there can be strong connections between spatial and social networks. When there are, then 'hotspots' and 'hotshots', at either the individual, community, or population level can be identified and they can then be used to evaluate different proposed schemes and invest in protecting those areas or biological entities for which the marginal gain and network reach will be greatest.

So far investigations into the structure and functioning of networks have focussed on partitioning individuals into traditional phenotypic categories. Adding other, more social traits such as personality to the list of phenotypes to be accounted for, whether of the fixed or more socially induced 'social' variety, should be encouraged. In this way, important connections between network structure and function will emerge. Individuals and the roles they play in terms of leadership or coordination should be among the first to be identified. And when these structures and the roles individuals play within them are coupled with fine-grained environmental data, the link between environment and sociality will be enhanced.

Acknowledgements

As the ideas for this chapter came together, conversations with Siva Sundaresan, Ilya Fischhoff, Chayant Tantipathananandh, Mayank Lahiri, Habiba, and Tanya Y. Berger-Wolf on many aspects of networks have helped crystalize my thinking. Comments from two anonymous reviewers and Jens Krause have improved the manuscript. The people and government of Kenya as well as the Mpala Conservancy, the Ol Pejeta Conservancy, and the Lewa Conservancy have allowed me to work on zebras, and financial support came from the NSF (IBN-9874523; CNS-025214, and IOB-9874523) as well as Princeton University's Environmental Institute, especially the 'Grand Challenges Program'.

CHAPTER 18

Linking lizards: social networks in reptiles

Stephanie S. Godfrey

Introduction to social networks in reptiles

Reptiles have a reputation for being on the more 'unsociable' end of the scale of sociality, particularly when viewed against more conspicuous forms of sociality in other taxonomic groups, such as birds, mammals, and fish. Therefore, it is not surprising that social networks have had a relatively sparse application to this taxonomic group. However, this does not mean that social networks are not applicable to reptiles.

Reptiles display a diverse array of forms of social organization, including central-place territoriality (Fenner and Bull 2011), pair living (Leu, Bashford, et al. 2010), nuclear family groups (O'Connor and Shine 2003), and larger social groups composed of genetically related members (Duffield and Bull 2002). Reptiles also possess sensory capabilities necessary to maintain complex forms of social structure. Some lizards show the ability to recognize and discriminate kin and familiar conspecifics through the use of olfactory cues (Main and Bull 1996; Bull et al. 2000; O'Connor and Shine 2006). Thus, reptiles can display complex forms of social organization that are similar to those in other taxonomic groups (such as primates (Chapter 12) or birds (Chapter 16)).

However, the utility of the network approach means that it may also be applied to interactions not normally defined as 'social', such as 'solitary-territorial' systems where interactions mainly occur in the context of competition for food, space, and mates (Kaufmann 1983). Scent marking is used by some species as a passive territory marking cue (Carazo et al. 2007), while posturing and visual signals are frequently used as active social signals in maintaining territory boundaries (M. Watt and Joss 2003), and colouring of males can be used to indicate social status or fitness (C. Thompson and Moore 1991). Reptiles are also capable of social learning (A. Wilkinson et al. 2010; K. Davis and Burghardt 2011), can use social information (Kiester 1979; Ord and Stamps 2008), and display behavioural types (Cote and Clobert 2007). Thus, reptiles display highly developed and complex forms of social structure and behaviour, and there exists great scope for the use of networks to understand reptile behavioural ecology. Yet, only a handful of studies have applied a network approach towards understanding this behavioural complexity.

The application of social networks to understanding reptile behavioural ecology

Reptiles have served as model organisms for a variety of ecological and evolutionary questions; yet, there has been limited application of social network analysis to the study of reptilian behavioural ecology. To date, ten social network studies have been conducted on four different species of reptile (one rhynchocephalian and three squamates (Scincidae)), spanning a variety of ecological and evolutionary questions (see Table 18.1). The majority (six) of these studies have used social networks to understand how social structure influences the transmission of parasites. Other research questions include

Animal Social Networks. Edited by Jens Krause, Richard James, Daniel W. Franks and Darren P. Croft.

Table 18.1 Summary of social network studies in reptiles and the ecological and evolutionary questions posed.

Study	Species	Method of tracking	Network type	Conceptual question
Bull et al. 2012	Sleepy lizard (*Tiliqua rugosa*)	Micro GPS loggers continuously tracked movement of lizards	Social contact inferred from synchronous GPS locations	How does behaviour influence parasite transmission?
Fenner et al. 2011	Pygmy bluetongue lizard (*Tiliqua adelaidensis*)	Direct observation of refuge use at monthly intervals	Index of burrow proximity, cumulative across study period	How does behaviour influence parasite transmission?
Godfrey et al. 2009	Gidgee skink (*Egernia stokesii*)	Direct observation of refuge use at irregular intervals (5–40 days)	Asynchronous refuge sharing within infectious period of parasite	How does behaviour influence parasite transmission?
Godfrey et al. 2010	Tuatara (*Sphenodon punctatus*)	Direct observation of home-range use, daily throughout study periods	Index of overlap in home-range area	How does behaviour influence parasite transmission?
Godfrey et al. 2012	Sleepy lizard (*Tiliqua rugosa*)	Micro GPS loggers continuously tracked movement of lizards	Social contact inferred from synchronous GPS locations	Where do different personality types fit in a social network?
Godfrey et al. 2013	Sleepy lizard (*Tiliqua rugosa*)	Micro GPS loggers continuously tracked movement of lizards	Social contact inferred from synchronous GPS locations	How stable are social networks over varying ecological conditions?
Leu, Bashford, et al. 2010	Sleepy lizard (*Tiliqua rugosa*)	Micro GPS loggers continuously tracked movement of lizards	Social contact inferred from synchronous GPS locations	Is social organization different from random expectations?
Leu, Kappeler, et al. 2010	Sleepy lizard (*Tiliqua rugosa*)	Micro GPS loggers continuously tracked movement of lizards	Asynchronous refuge sharing within infectious period of parasite	How does behaviour influence parasite transmission?
Leu et al. 2011	Sleepy lizard (*Tiliqua rugosa*)	Micro GPS loggers continuously tracked movement of lizards	Social contact and synchronous refuge use inferred from GPS locations	Does refuge use influence social behaviour?
Wohlfeil et al. 2013	Sleepy lizard (*Tiliqua rugosa*)	Micro GPS loggers continuously tracked movement of lizards	Asynchronous refuge sharing within infectious period of parasite	How does behaviour influence parasite transmission?

describing social organization and the factors shaping it, and exploring how personality influences the position of individuals in a social network. The breadth of these research questions reflects the utility of the network approach to improve our understanding of the behavioural ecology of reptiles.

Detecting and describing social organization in reptiles

To understand the factors that shape social organization in reptiles, it is important to first distinguish between social structure and random behavioural associations. This is particularly true for reptiles, which do not normally display overt forms of sociality or gregariousness. Thus, it can be difficult to detect more cryptic forms of social organization among reptiles (Shine et al. 2005). Social networks provide a framework for quantifying behavioural associations and testing hypotheses about the factors structuring these associations.

Null models of association can be compared with observed association patterns to test whether the observed social network is different to what we would expect if individuals associated at random (R. James et al. 2009). For example, Leu, Bashford, et al. (2010) generated a null model network which assumed sleepy lizards (*Tiliqua rugosa*) moved around their home range (and consequently encountered each other) at random, and compared this with the observed social network. There were significantly fewer connections in the observed network than the null network, indicating that individuals contacted fewer lizards than expected by chance (Leu, Bashford, et al. 2010), and this 'avoidance' was consistent across years, even when ecological conditions varied substantially (Godfrey et al. 2013). The observed network revealed that sleepy lizards were predominantly pair living, with pair bonds persisting beyond mating (Leu, Bashford, et al. 2010). Thus, a network approach can allow insights into more subtle (or cryptic) forms of social structure.

Similarly, the advantage of a network approach for studying reptile behaviour is that it can be applied to less gregarious forms of social organization, such as territorial systems. Territoriality is the predominant form of social structure in reptiles, and as a consequence, reptiles have become a model system for understanding the establishment, maintenance, and structure of territories (Stamps and Krishnan 1995, 1997, 1998). Although social networks provide an ideal framework for quantifying the structure of territorial systems, they have only been used to describe the structure of territories in tuatara (*Sphenodon punctatus*) (Godfrey et al. 2010), and the central-place territorial system of pygmy bluetongue lizards (*Tiliqua adelaidensis*) (Fenner et al. 2011). Although these studies were specifically aimed towards describing pathways for parasite transmission (Table 18.1), they illustrate the utility of this approach for describing the structure of territorial systems and exploring the factors that influence the structure of these systems.

What shapes social organization in reptiles?

Understanding the factors that influence the evolution of different forms of social organization is a central theme of behavioural ecology common to most taxonomic groups (Alexander 1974). Due to the relative infrequency of more gregarious forms of social organization, and the behavioural (general absence of parental care, except for the crocodilians (J. Reynolds et al. 2002)) and physiological (ectothermy) differences between reptiles and other taxonomic groups, reptiles provide a unique viewpoint on the evolution of social structure. While some factors that shape social organization in reptiles are synonymous with social structure in other taxa (such as kin-based sociality), others are more specific to reptiles (thermal and ecological needs).

Thermal and ecological requirements for reptiles

As ectotherms, reptiles are often dependent on their habitat (and particularly, the availability of refuges) for their thermoregulatory requirements, with many requiring some form of refuge (e.g. burrows, hollow logs, rock crevices) to thermoregulate (Heatwole and Taylor 1987) and provide protection from predation for diurnal species (J. Martin and Lopez 1999). Thus, in some species, particularly where suitable refuges are limited, individuals can appear less territorial and even form aggregations, which are mainly driven by their thermal needs (Graves and Duvall 1987; Shah et al. 2003). While this is hypothesized as a precursor to the evolution of sociality in reptiles (While et al. 2009a), distinguishing between associations driven by physiological needs versus social associations is important in understanding the behavioural complexity of reptiles.

Social networks provide a framework for testing these hypotheses. For example, Leu et al. (2011) identified the social contacts and refuge sites of sleepy lizards to develop three different networks; one based on observations of social association during their diurnal activity period, one based on observations of synchronous overnight refuge sharing, and one based on hypothetical random refuge sharing. Comparisons among the different networks revealed that lizards shared refuges as frequently as expected by chance, and although the refuge-use network was correlated with the social network, lizards associated more frequently while active than when inactive under refuges (Leu et al. 2011). For the sleepy lizard, social behaviour is not a by-product of refuge use, and lizards generally avoided conspecifics while actively seeking out social associations with their mate (Leu, Bashford, et al. 2010, 2011). Thus, networks provide a useful framework for testing hypotheses about the importance of habitat availability in influencing social behaviour.

Extending this to comparisons among populations would allow broader insights into the role of habitat structure in shaping social organization. A number of studies on lizards from the scincid genus *Egernia* have suggested that variation in habitat structure can influence aggregation patterns (M. Gardner et al. 2007; While et al. 2009b; Michael et al. 2010). Michael et al. (2010) found that aggregations of tree skinks (*Egernia striolata*) were more likely at sites where large rock masses were present, and with more crevice microhabitats. In contrast, M. Gardner et al. (2007) found consistent patterns of aggregation among populations of gidgee skinks (*Egernia stokesii*) despite differences in habitat structure and refuge availability. Habitat structure is also hypothesized to influence the structure of territorial systems, where habitat visibility can affect territory display behaviour (Calsbeek and Marnocha 2006; M. A. Johnson et al. 2009). A network approach would allow insights beyond traditional group-based metrics, such as group size, to examine how different aspects of social structure (including both direct and indirect associations) vary across ecological gradients.

Genetic structure and kin-based sociality in reptiles

A major theory behind the evolution of sociality is that the benefits of group living are enhanced by sharing those benefits with kin (W. Hamilton 1964a). Thus, the genetic relationships among individuals are of importance to understanding the evolution of sociality, both broadly (Ross 2001) and for reptiles (While et al. 2009a). Numerous studies have described the genetic structure of social groups of reptiles and normally find that groups consist of highly related members (M. Gardner et al. 2001; Fuller et al. 2005; Chapple and Keogh 2006; A. Davis et al. 2011; McAlpin et al. 2011; Duckett et al. 2012). However, these 'group-level' analyses may obscure finer-scale patterns of relatedness within groups.

A network approach would allow finer-scale insights into the genetic structuring of reptilian social organization, including analyses at the dyadic level, and in relation to the 'social distance' between individuals. In addition, this approach could easily be applied across a range of forms of social structure. Similarly, social networks may allow insights into more cryptic forms of social structure by exploring kin structuring in less gregarious societies. Although a network approach has been exploited in other taxa to explore these questions (J. B. W. Wolf and Trillmich 2008; Wiszniewski et al. 2010; Rollins et al. 2012), it has yet to be applied to reptiles.

Proximate mechanisms of social organization in reptiles

While understanding the ultimate causes of social organization is of interest from an evolutionary perspective, identifying the proximate mechanisms that determine behaviour and social structure are also necessary to understand the factors that shape social organization. In reptiles, olfactory and visual signals are important mechanisms in influencing social interactions in a range of forms of social organization, from territorial systems (Carazo et al. 2008) to gregarious social systems (Bull et al. 2000). Similarly, hormone levels play an important role in influencing behaviour and social interactions in reptiles (DeNardo and Licht 1993; DeNardo and Sinervo 1994). More generally, differences among individuals, both in individual quality (e.g. mass or reproductive capacity) and behaviour (behavioural types and personality) can also impact directly on social interactions. Social networks allow the quantification of these characters in respect to the position of individuals in networks and therefore provide a powerful framework for understanding how these factors shape social organization within populations.

Visual and olfactory signalling and communication pathways in reptiles

Communication is a necessary component for the evolution of complex forms of social organization. Among reptiles, the most apparent form of communication is visual displays, usually involving colour and movement as signals to conspecifics, and normally in the context of territory defence and maintenance (Ord et al. 2002; Simon 2011). Many reptiles possess a well-developed olfactory system, which enables complex forms of olfactory communication (Mason and Parker 2010) and allows individuals to discriminate among relatives, familiar individuals, and non-familiar individuals (Main and Bull 1996; Bull et al. 2000; O'Connor and Shine 2006; Pernetta et al. 2009; Ibanez et al. 2012). Olfactory signals can include scent marking (Carazo et al. 2011; Hews et al. 2011; Font et al. 2012) and scat piling (Bull et al. 1999; Fenner and Bull 2010), both of which are often used as territory markers. Some species are also social information users, in that information gained from either visual or olfactory cues can influence individual decisions and behaviour (Kiester 1979; Aragon et al. 2006; Cote and Clobert 2007). In both visual and olfactory communication systems, communication is often used as a means of regulating social associations, either in the maintenance of exclusive territories or in the cohesion of social groups.

As for other taxonomic groups, networks can provide a means for gaining deeper insights into the role of communication in the function of social structures for reptiles (J. Foote et al. 2010; Matessi et al. 2010; also see Chapter 9). This is of particular relevance to the study of territorial systems, where communication plays a central role in mediating territory boundaries and territorial interactions (Mason and Parker 2010; Font et al. 2012). By developing communication networks, either through observations of visual signalling between conspecifics, or the placement of scent marks or scat piles, networks could be developed that represent the pathways of communication. Comparisons of these networks with the social status or reproductive success of individuals may provide insights into the role of signalling in influencing social structure. Similarly, it may allow insights into how intraspecific variation in signal strength affects the structure of communication networks, and the flow on effect of this onto the social networks. Networks provide a potentially valuable contribution to this area that is yet to be exploited.

Individual differences and network position in reptile societies

One of the advantages of the network approach is that it allows us to explore how different factors influence social organization on an individual level, that is, how individuals vary in their position in networks in respect to different factors. This allows for a wide array of questions to be asked of the proximate factors that influence social interactions and organization. This may range from simple questions, such as how individual sex, size, or condition relates to social position or embeddedness in the network, to more sophisticated questions about how individual differences (such as hormone levels or behavioural types) influence social network structure.

In reptiles, the effect of hormones on social dominance and the outcomes of aggressive encounters (T. Baird and Hews 2007; Husak et al. 2007) are the central questions underpinning studies of territoriality in reptiles. Comparisons of networks derived from social or spatial relationships among territory holders, with signal size or hormone levels, would enable testing of hypotheses about the factors shaping territory structure. Similarly, experimental manipulations of hormone levels, such as those carried out by Grear et al. (2009) in wild mice social networks, would allow insights into the impact of hormone levels in shaping social structure in reptiles. Social networks in this context offer a new perspective on understanding the factors shaping territory structure, particularly in terms of considering the broader 'social neighbourhood'.

Behavioural types and personality have long been of interest to the study of reptile behavioural ecology (Lopez et al. 2005; Cote and Clobert 2007; A. Carter et al. 2010); however, to date, only one study has applied a social network approach to understanding how personality is associated with social structure in reptiles (Godfrey et al. 2012). Godfrey et al. (2012) found that less aggressive male sleepy lizards (lovers) were more strongly connected to females and, in particular, spent more time with their female partners than more aggressive males (fighters) did. Studies comparing behavioural types and social network position will provide deeper insights into the social consequences of personality and allow insights into how personality influences social organization in reptiles.

Consequences of social networks: parasite transmission among reptiles

Evaluating the costs of parasites to the evolution of group living involves understanding how host behaviour influences the transmission of parasites (Moller et al. 1993; Altizer et al. 2003). More recently, the question of how behaviour influences parasite transmission has also become a focus for understanding the factors generating heterogeneities in the transmission of diseases through wildlife populations (Craft and Caillaud 2011; Tompkins et al. 2011). Social networks provide an ideal framework for answering these questions, as they can represent how social interactions and space use generate pathways for the transmission of parasites within host populations. Although this question is common to all taxonomic groups, it has dominated social network studies in reptiles, with this research bias mainly driven by the research interests of the group from which this work originates (Prof. Mike Bull at Flinders University).

Nonetheless, reptiles have clearly become a good model for exploring these questions. They provide an alternative view on the role of host behaviour in influencing parasite transmission, since compensatory behaviours such as self-grooming and allogrooming are infrequent. Being mainly terrestrial, the movement of reptiles in space and time is relevant to the transmission of a range of parasites. Reptiles also display a variety of forms of social structure, which allows insights into how host behaviour influences parasite transmission across different social structures.

The main question underpinning these studies is whether network structure is associated with parasite infection patterns. Depending on the transmission method of the parasite being considered, networks have been constructed from overlap in refuge use (nidiculous parasites (Godfrey et al. 2009; Leu, Kappeler, et al. 2010)), overlap in home range area or spatial proximity (parasites with infectious stages in the off-host environment (Godfrey et al. 2010; Fenner et al. 2011)), or social contacts (contagious infections, or those with short infectious periods (Bull et al. 2012)). These studies have considered different forms of social structure, from solitary (Godfrey et al. 2010) and central-place territoriality (Fenner et al. 2011), to pair living (Leu, Kappeler, et al. 2010; Bull et al. 2012) and group living (Godfrey et al. 2009). Comparisons between network degree or strength and infection patterns have revealed mainly positive associations between the connectivity of individuals in the network and their parasite burden (Godfrey et al. 2009; Godfrey et al. 2010; Leu, Kappeler, et al. 2010). An exception to this was observed by Bull et al. (2012), who instead found that adjacent individuals with stronger connections (higher edge weight) on the social network were more likely to share the same strain of *Salmonella*.

Comparisons of infection patterns with different network-based parameters have allowed the testing of alternative hypotheses about the role of host behaviour in parasite transmission. Godfrey et al. (2010) developed separate node-based measures to describe the connectedness of individuals to males (male strength) and females (female strength) in a network of tuatara. Male strength in tuatara was positively correlated with tick loads, while female strength was unrelated to tick loads, suggesting that males may play a predominant role in the transmission of these parasites (Godfrey et al. 2010). Similarly, Fenner et al. (2011) found that nematode infection patterns of pygmy bluetongue lizards were more closely related to disperser strength (sum of connections to dispersing individuals) than resident strength (sum of connections to resident individuals), suggesting that dispersers may be more important in the transmission of nematode worms. The ability of a social network approach to quantify the epidemiological consequences of host behaviour has allowed deeper insights into the role of host behaviour in influencing parasite transmission (this is discussed in more detail in Chapter 10). While the studies noted above involve fairly simple empirical comparisons, they have laid a foundation for improving our understanding of how host behaviour generates heterogeneities in the transmission of parasites in wildlife populations.

Advantages of using networks in understanding reptile systems

As outlined in the preceding sections of this chapter, social networks provide a means of exploring a range of conceptual questions about the behavioural ecology of reptiles. However the main advantage of a network approach for understanding reptile behavioural ecology lies in its ability to quantify social structure across a variety of forms of social organization. Because the predominant form of social structure among reptiles is territoriality, networks provide a new method for quantifying these less gregarious forms of social structure. In addition, many studies of reptile behaviour infer social organization from patterns of space use and overlap (Osterwalder et al. 2004; Hagen and Bull 2011; C. Frost and Bergmann 2012; Qi et al. 2012). A social network approach allows this to be quantified, and analytical methods allow insights into whether the observed social network is different from random, and therefore represents true social structure. Studies of social structure in reptiles are usually limited to describing social structure among defined social units (groups or dyads); however, social networks can quantify structure at different levels of interest (individual, dyadic, and community/network), and allow both direct (social interactions) and indirect (social environment) associations to be considered.

Although the study of reptile behavioural ecology has a lot to gain from the application of networks, reptiles themselves can offer some advantages to the use of a network approach. While many species may be unsuitable for adopting a network approach (due to their size and cryptic nature), where these issues are not a problem, reptiles can make good study organisms for the study of social networks. Many reptile species (except perhaps for some notable exceptions, i.e. crocodiles) are normally easy to handle, so the capture and marking or tracking of individuals is straightforward. Reptiles are also normally tolerant to handling procedures, unlike some birds and mammals that quickly get stressed during the capture experience (Putman 1995).

Reptiles can have large population sizes (compared to vertebrates) which can be useful for studying population biology and enable good sample sizes for social network analysis. Being largely terrestrial and usually having high spatial fidelity means that reptiles are easy to track in space. While the lack of gregariousness is problematic in some contexts, in others it makes it easier for data analysis due to a reduced 'gambit-of-the group' effect (R. James et al. 2009) encountered in most other systems. Similarly, the lack of gregariousness means that many networks in reptiles are likely to be spatially structured and thus may also enable insights into the spatial ecology of reptilian species. Thus, reptiles can make a good model organism for the application of a network approach.

Challenges of applying a network approach to reptiles

Social networks offer new opportunities to gain insights into the behavioural ecology of reptiles. However, there are several challenges involved in collecting data appropriate for developing representative networks. Although many of these challenges are similar to those encountered in other taxonomic groups, there are several that are particularly prevalent among reptiles.

Many reptile species either have cryptic lifestyles, spending much of their time in refuges or hiding under vegetation, rocks, or logs, or are easily disturbed by human presence. This makes it difficult to directly observe the behaviour of individuals, particularly during their social interactions. Relying on direct observation may also create an observation bias, particularly in territorial systems, where more dominant males are often more visible. Overlaid upon this, human presence can impact reptile behaviour (Kerr, Bull, and Mackay 2004); thus, simple observation techniques are often impractical or may provide limited information on behavioural associations.

Reptiles do not usually display overt gregariousness, so quantifying social structure relies on capturing infrequent associations among individuals. Thus, 'snap shot' observations of associations that may be meaningful in other systems (e.g. Croft et al. 2004) may not capture sufficient information on the social structure of reptiles, particularly for more solitary species. Although infrequent social interactions can be important even in group-living species, the relative infrequency of social contact in reptilians systems may require more detailed and continuous observations.

The study of juvenile reptiles can be particularly challenging, since juveniles are often highly cryptic and therefore difficult to directly observe. In addition, they may be too small to mark by conventional means (besides toe clipping) and often have a fast growth rate, such that they outgrow any attached devices relatively quickly. This means that the study of juveniles in free-living populations can be limited, which is of particular importance to the study of kin structuring in social networks.

Finally, except for testudines, crocodilians, and some larger members of the squamates, the majority of reptiles are small in body size (<1 kg). Thus, monitoring behaviour, particularly at the level of detail required for capturing sufficient information for developing social networks, is challenging. Although technology is improving, the cost and size of tracking units remain a limitation for most study systems. Similarly, the body shape of some reptiles (e.g. snakes) requires more invasive methods of tracking, such as surgical implants, to minimize the impact of tracking on the behaviour of the animal. As technology improves, the cost-effectiveness and size of tracking technology will hopefully become more accessible to a broader range of study systems.

Marking, identifying, and monitoring reptiles

Marking individuals is an essential part of any ecological study, and there are a range of methods available for marking reptiles. The method used will depend on the types of behavioural associations of interest, including those that depend on capturing individuals, visual observation of individuals, and remote monitoring of behaviour. It will also depend on the permanency of the mark required for the study. There are many good references about reptile marking (Blomberg and Shine 2006; McDiarmid et al. 2012) so, for the purposes of this chapter, I will outline those of most use to social network studies.

Identification of reptiles upon capture

For social network studies where repeated capture locations can be used to develop a network (e.g. Godfrey et al. 2009; Fenner et al. 2011), marking individuals so they can be recognized upon subsequent capture is a useful technique. Toe clipping (where a combination of toes are 'clipped' according to a numerical code) is a traditional method of marking reptiles and, in cases where it does not predispose individuals to higher mortality (e.g. for species that are terrestrial, and do not dig their own burrows), it can be a useful means of permanent identification (G. Perry et al. 2011). Other methods of marking are becoming available, including microchipping and the use of sub-dermal fluorescent dyes (S. Petit et al. 2012) (Figure 18.1a). Many reptile species possess their own unique markings (scale configurations) which can be used to recognize individuals (Figure 18.1b). Although in the past this has been done manually, advances in computer programming are beginning to enable the automation of this process (Sacchi et al. 2010); taking a photo of an individual upon capture allows subsequent identification of the individual.

Visual identification of reptiles

For social network studies that require observation of individuals to record the nature and length of social interactions, means of marking that are recognizable from a distance are required (e.g. Godfrey et al. 2010). Some researchers paint their subjects, either with different colour combinations or with letters or numbers (Figure 18.2a). While these are effective in the short term for observation, they are not permanent and, depending on the nature of

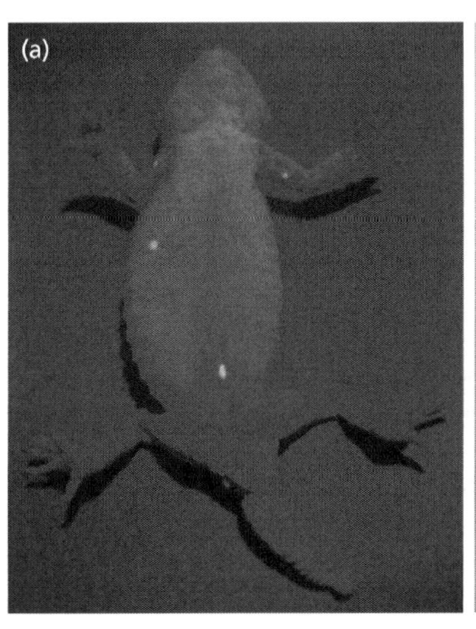

Figure 18.1 Examples of marking and identification techniques for recapturing individuals. (a) Sub-dermal fluorescent dyes on a knob-tailed gecko (*Nephrurus levis*) (photo courtesy of Annabel Smith). (b) Unique scale patterns and configuration in a pygmy bluetongue lizard (*Tiliqua adelaidensis*).

Figure 18.2 Examples of marking reptiles for visual observation. (a) Paint markings on a gidgee skink (*Egernia stokesii*) (photo courtesy of Aaron Fenner). (b) Coloured beads inserted through the nuchal crest in a tuatara (*Sphenodon punctatus*).

the species (particularly of those species that shed their skin), markings may need to be replenished on a semi-regular basis. An advantage is that this method is applicable to a range of species (that is, it is not subject to the size, shape, or anatomical features of the species), so long as it does not increase their chances of predation or alter their behaviour. A more permanent means of visually marking individuals is by inserting beads through the nuchal crest (a fleshy crest on the top of the head, present in some reptiles) (Figure 18.2b). Different colour combinations can be used to mark individuals and enable recognition of individuals from a distance using binoculars.

Remote monitoring of associations among reptiles

Methods that enable the remote collection of almost continuous streams of information on the behaviour of individuals are ideal for social network studies, and there are a number of methods that could be applied to the study of reptile social networks. Micro GPS loggers have been used for monitoring the behaviour of the sleepy lizard (*Tiliqua rugosa*) (Figure 18.3a), and this technology has allowed detailed insights into different aspects of their behaviour (e.g. Leu, Bashford, et al. 2010, 2011; Bull et al. 2012; Godfrey et al. 2012). These loggers collect GPS locations of lizards and measure the activity of the lizard via an on-board 'waddle-o-meter' (Kerr, Bull, and Cottrell 2004), providing additional insights into activity levels. By synchronizing the time that locations are acquired, social associations can be inferred from individuals being at the 'same place at the same time', and the collection of GPS locations at regular intervals allows insights into patterns of home range, habitat, and refuge use. GPS loggers have been fitted to a range of reptiles with different methods, including a backpack 'vest' fitted onto bluetongue lizards (Price-Rees and Shine 2011) and glue-on GPS units on lace monitors (Flesch et al. 2009). While GPS loggers are becoming smaller, sleepy lizards are still considered large in terms of lizard size (~750 g) so thus far this technology remains out of reach for most other reptile systems.

Another method of gaining information on social contacts, activity patterns, and range use is video surveillance. This method is only suitable for reptiles that are relatively sedentary and have small home ranges (usually around a central point such as a refuge). While video recording of behaviour is not a new technology and has provided insights into reptile behaviour for many years, advances in technology have allowed larger 'networks' of cameras to be placed in wild populations of lizards. For example, Fenner et al. (2012) used a network of cameras to monitor activity and burrow use of Slater's skink (*Liopholis slateri*), mounting cameras outside the entrances of burrows. Larger scale networks of

Figure 18.3 Examples of remote monitoring of reptile behaviour. (a) A GPS activity logger attached to a sleepy lizard (*Tiliqua rugosa*). (b) A network of video cameras used to observe a wild population of pygmy bluetongue lizards (*T. adelaidensis*).

cameras have also been used on pygmy bluetongue lizard populations (Ebrahimi and Bull 2012; Ebrahimi et al. 2012) and have allowed insights into the movement and behaviour of individuals within populations.

A limitation of this method is that, unless individuals can be visually tracked through video recordings, a visual method of marking is required to identify individuals from video footage. Reviewing the footage is also a time-consuming process. However, a large amount of information on the behaviour of individuals may be gained via this method.

For social network studies where social interactions are the focus, contact loggers are normally used (Hamede et al. 2009). However contact loggers are usually too large for most reptiles (current models start at 50 g). A new range of micro-contact loggers that can log the proximity of other tagged individuals are emerging; these loggers provide the same function as the larger contact loggers, but at a fraction of the size. Although these loggers have been applied to birds (Rutz et al. 2012), they are yet to be applied to reptiles. A limitation of these loggers is that they do not record the spatial location of the contacts, which may or may not be important depending on the research question of interest.

For social network studies where patterns of refuge or habitat use are of interest, the use of automated passive integrated transponder (also known as PIT) logging stations are among the most promising of technologies. These stations allow the time and identity of a microchipped animal that comes into proximity (usually 10–20 cm) of the station to be recorded. Thus, antennas can be placed at points of interest (e.g. entrances to burrows) and log the identity of individuals that use them. Commercial versions of these are available (usually designed for livestock and aquaculture, although wildlife applications are becoming more widespread), but they can also be made 'in house', which offers a cost-effective alternative to commercial options, and increased customization in terms of antenna dimensions. Although this technology has yet to be applied to reptiles, it offers great potential for remote monitoring the way individuals use specific, spatially fixed, resources.

Finally, radio telemetry is the most traditional method of tracking for most wildlife, including reptiles, and can provide valuable information on habitat use, home range area, and refuge use of individuals. This method of tracking is among the most versatile, with improving technology enabling smaller individuals to be tagged, and surgical implants enabling species such as snakes to be tracked. Radio telemetry is, however, quite time consuming and will not necessarily provide information on social contacts. Automated radio telemetry systems have been trialled (Kays et al. 2011), allowing regular fixes to be collected from a series of towers, although this technology remains limited and is logistically difficult to set up. However, even for the most basic reptile social network study, radio telemetry is a good option for most species.

Social networks in reptiles: conclusions and future directions

While the application of social networks to understanding reptile behavioural ecology promises to allow new insights into the ecology and evolution of social structure in this unique taxonomic group, there are plenty of challenges that remain to the application of this approach. While technology continues to improve, in the meantime there is great value in finding good model systems to work with. The sleepy lizard (*Tiliqua rugosa*) is clearly a good model system that continues to provide detailed insights into the ecology and behaviour of these animals, with the collection of unprecedented amounts of detailed data. Similar systems that are well suited to both the technology available and the ecological and evolutionary questions of interest will provide equally revealing insights into the structure and function of social networks in reptiles.

The application of social networks to a broader range of reptiles will advance insights into the behavioural ecology of this unique taxonomic group, and many representatives are good candidates for such studies. Many of the crocodilians are large enough to track and display interesting behaviours, such as maternal care (J. Reynolds et al. 2002). Similarly, turtles and tortoises are often large in size and, while the aquatic nature of some members of this group provides an additional challenge to their study, for terrestrial species there is certainly scope to apply social networks, particularly in light of recent studies of social learning among members of this group (A. Wilkinson et al. 2010; K. Davis and Burghardt 2011). Similarly, the squamates, which are a diverse group, could certainly benefit from more attention to their behavioural ecology.

Increasing awareness and accessibility of a network approach may help to stimulate this field of research among herpetologists. Social networks are often met with trepidation; the data-hungry nature and the seemingly complex analytical side of social networks is enough to scare off many reptile behavioural ecologists. However, as these methods become more accessible, and analysis methods become more established and routine, this will encourage more people to take a risk and delve into the world of social networks.

Acknowledgements

I thank Julien Cote for constructive feedback on a previous version of this chapter, and the Australian Research Council for supporting this research.

SECTION 4

Animal Social Networks: Conclusions

CHAPTER 19

Animal social networks: general conclusions

Jens Krause, Richard James, Daniel W. Franks, and Darren P. Croft

The application of network thinking to the study of animal social systems is not new. As early as the 1950s, primatologists were drawing sociograms of behavioural interactions among primates (see Chapter 12). However, it is only in the last decade that the modern approach of the quantitative analysis of social networks has been regularly applied to the study of social behaviour in non-human animals. The social network approach is now a well-established and a popular means by which to study aspects of the behaviour, ecology, and evolution of organisms. In contrast to our first book on the network approach (Croft et al. 2008), which was very much methods driven, we have focussed here on the biology and on what has been learned since the social network approach was introduced to the discipline of animal behaviour.

The network approach has added significantly to our understanding of major biological issues (see Section 2) and has found applications in almost all major vertebrate groups; in addition, it is beginning to be more widely used in studies of social insects (see Section 3). While we have learned a great deal through the application of social network theory to animal societies, we are still to unlock the full potential of this approach. Many of the chapters in this book contain excellent ideas for future research and how social network analysis can be used to address long-standing problems and questions in animal behaviour. In this chapter, we draw general conclusions and highlight key areas for future research.

The unifying feature of all previous studies that have measured social network structure in animals is that the resulting network structure is non-random (Chapters 12–18). The recognition that social interactions are highly structured has had significant implications for our understanding of processes such as sexual selection (Chapter 4), disease transmission (Chapters 10, 15, and 17), information diffusion (Chapters 5, 9, 12, 13, and 15), collective behaviour (Chapters 8 and 14), and the evolution of cooperation (Chapter 3). While early models of such processes often assumed that populations were well mixed and lacked social structure, the inclusion of social structure has been shown to have a major influence. For example, local assortment in social networks of cooperators can maintain cooperation without the need for complex cognitive mechanisms (Chapter 3); social preferences (e.g. based on kinship, phenotype, or familiarity) can dramatically alter the collective movement of social groups (Chapter 8); and heterogeneity in contact patterns can significantly change the transmission dynamics of infectious diseases (Chapters 10, 15, and 17). Finally, by considering social network structure, we can significantly improve our ability to measure both sexual selection (Chapter 4) and social influence on the diffusion of novel behaviour in animal populations (Chapter 5). One of the challenges for future work is to test the predictions and assumptions of such models empirically, and there are many excellent ideas for future research outlined in the chapters in Sections 2 and 3.

Social interactions and thus social networks are dynamic. These dynamics result from changes in the behaviour of individuals that may be driven, for example, by changes in the ecological environment (E. Foster et al. 2012) or population composition

Animal Social Networks. Edited by Jens Krause, Richard James, Daniel W. Franks and Darren P. Croft.

(Darden et al. 2009). Social network data thus have a very strong temporal structure element and it is important to consider this when studying population processes (e.g. information diffusion and disease transmission).

Understanding the drivers and mechanisms that underpin these dynamics is an exciting area for future research. Individuals may manipulate their social environment to gain adaptive benefits. For example, in a wild population of house finches (*Carpodacus mexicanus*), Oh and Badyaev (2010) demonstrated that males of low attractiveness can increase their relative attractiveness by moving between different groups (see Chapter 4). As has been highlighted by a number of recent reviewers (Blonder et al. 2012; Pinter-Wollman et al. 2014) and in several places in the preceding chapters, studying the temporal dynamics of social networks is going to play a major role for future animal social network studies.

For some populations, social structure (and social dynamics due to the addition and removal of individuals) is under the management of humans. The importance of social housing and group stability for domestic animals has long been recognized and can significantly impact both production and welfare traits including feeding rate, growth rate, and mortality rate (L. Keeling and Gonyou 2001). What we know very little about is how social relationships within a group contribute to productivity and welfare and how to manipulate the social environment to improve welfare (see Chapters 11 and 13). Given the ubiquitous social housing for many domestic species (L. Keeling and Gonyou 2001) and the significance of social interactions for economic and welfare traits (Bijma et al. 2007; Bergsma et al. 2008; Ellen et al. 2008), this represents an important priority for future research.

Many studies have measured the structure of social networks but we know comparatively little regarding the mechanisms that underpin this structure. For example, few studies have examined the relationship between the ecological environment and social network structure (E. Foster et al. 2012). Individuals may differ in social traits that drive variation in network position among individuals (Chapters 6 and 13). While understanding the patterns, mechanisms, and functions of consistent interindividual variation in behaviour (often referred to as personality) has been of considerable interest over the last decade, we know very little about the degree to which social network traits show repeatability at the level of the individual (see Chapter 6). Similarly, we know little about the partner-choice mechanisms that underpin social ties. Determining these mechanisms is key for understanding the link between social network structure and evolutionary processes. As outlined in Chapter 3, for example, determining the mechanisms that drive the assortment of cooperators in social networks is essential for understanding how cooperation has evolved among non-kin.

One area where network thinking has been present for some time is in animal communication. Communication networks are typically defined as several individuals within a signalling and receiving range of one another. As outlined in Chapter 9, there are important parallels between animal social networks and communication networks. Indeed a number of studies have defined social interactions based on patterns of communication and, in some species, there is a very close link between communication and social complexity (Chapter 13). There is clearly great potential to integrate the work on animal communication with animal social networks, both in terms of defining edges in networks and in studying the dynamics of information flow on social networks.

The changing nature of animal social network data

It cannot go unnoticed by a reader of this book that there has been a distinct shift in the nature and volume of animal social network data that are now being gathered and analysed. Ten years ago, a typical network dataset for a wild population of animals was sparse; associations (more likely than interactions) were point sampled, usually via a protocol involving manual recording, and often recorded as a by-product of some other behavioural observations. There would usually be but a handful of samples. Perforce, the preoccupation of the analyst was the reliability of the data as a representation of any meaningful social connection. Justified caution led to conservative statistical schemes for the extraction of a biological signal (Croft et al. 2008; Whitehead 2008a). For many of the questions outlined in the preceding chapters, however, we need to have data on a much finer scale, and the advent of reality mining (defined as the collection and analysis of machine-sensed data regarding animal social behaviour with the goal of modelling behavioural patterns (Krause et al. 2013)) offers an exciting development for the automated collection of animal social network data.

Individual identification, which lies at the heart of most network studies, can often be achieved with new biometric approaches such as face and pattern recognition, technologies which allow automated data collection on individuals and their interactions in different settings (Kühl and Burghardt 2013). For example, in studies on marine animals, semi-automated recognition is already routinely used (Speed et al. 2007; Holmberg et al. 2009). Individual recognition of phenotypes used to be limited to a few species with obvious stripe or spot patterns. However, recent technological developments in computerized systems have extended the capability of recognition systems to a wider range of species. For example, Herbert-Read et al. (2013) reported that individual mosquitofish (*Gambusia holbrooki*) could be recognized by their movement characteristic when swimming in a tank. When in a larger group, however, these characteristics disappeared because the fish began to make compromises with the locomotion patterns of other individuals.

In cases where individual recognition is not possible (or not straightforward) from phenotypic appearances, a wide range of tracking algorithms and electronic tags are potential alternatives. Biologging methods exist which consist of attaching GPS- and RFID-based devices or hydroacoustic tags (for underwater applications) to animals and allow tracking over long distances in some cases (Krause et al. 2013). Usually position data with a time stamp can be obtained from these devices, and based on these data, the spatial proximity of animals can be detected. If combined with other sensors regarding physiological data or with small cameras (Rutz et al. 2007), more information can be obtained that provides details on interactions between individuals (Krause et al. 2013).

The advantage of automated data acquisition systems is that huge datasets on the spatial and social dynamics of animals can be collected using standardized methods based on objective criteria. The fully or semi-automated data collection methods outlined above enable us to gather vastly more dense data, so that it makes sense to consider the exact timing and order of behavioural events forming 'temporal networks' (Blonder et al. 2012) in which either the structural topology of the social structure, the flow of a process on that structure, or both are changing on the timescales being monitored. Reality mining also opens up more realistic chances of replication and manipulation, even in the field.

While this approach might help deal with the traditional bottleneck of data acquisition, it does not offer a panacea. It will still often be the case that proxies for the real behaviour are being recorded, and there will still be imperfect sampling (of animals, for example). It will still be necessary to decide the key time (and spatial and social) scales that are appropriate to address a particular biological question, and we will need to consider new issues such as the calibration of equipment used to log proximity (Boyland et al. 2013; Krause et al. 2013). Reality mining generates new challenges typical of dealing with 'big' data, the analysis of which often requires interdisciplinary collaborations with computer scientists, applied mathematicians, and physicists. Several chapters give detailed examples of how interactions can be recorded remotely to obtain social networks for different taxonomic groups (Chapter 14 on fish, Chapter 16 on birds, Chapter 17 on terrestrial ungulates, and Chapter 18 on reptiles) and discuss the methodological challenges involved.

The changing nature of animal social network analysis

A constant theme of the animal social networks literature of the past ten years or so has been the struggle to find analysis methods that do justice to the effort required to collect relational data, and the ambitions we have for a networks approach to social phenomena. The key problem, of course, is that the non-independence of network data means that we fall foul of the assumptions of many conventional statistical methods. It is nonetheless a recurring disappointment that we often appear to be able to do little more than test a null hypothesis, or look for correlations against a single variable, and that even these sorts of analyses still require considerable care (Whitehead 2008a; Croft, Madden, et al. 2011).

The impatience with this situation is reflected in the recent review by Pinter-Wollman et al. (2014) and the flurry of comments it excited. The review calls upon the animal social networks community to make more use of sophisticated statistical tools developed for analysis of human social networks. They advocate, sensibly, a move from methods of design-based inference, in which we control for network structure—usually via the use of constrained randomization of the data—to those of model-based inference, in which we model network structure more directly (Carrington et al. 2005; Snijders 2011). Many of the methods of model-based inference are based on the family of 'exponential random graph models' (Robins et al. 2007). Certainly such approaches deserve our attention, but these models should surely only be used if our data and question satisfy the assumptions they make (such as equal confidence in each edge or lack of an edge), and if the restrictions they impose (such as a need for binary edges) makes the most of the data collected. As Rendell and Gero (2014) state in one of the invited commentaries on Pinter-Wollman et al. (2014), 'just because we can *run* an analysis does not necessarily mean that we can *interpret* it correctly'.

Rendell and Gero go on to express the view that 'the most exciting advances are likely to be made in collaborations between experts who work directly on these analytical methods and experts who have a deep understanding of their study system and the limits of the data they are collecting. Naïvité in either of these areas is likely to lead to problems'. We concur. The happy news is that there appears to be a small but growing group of statistical modellers who are willing to engage with the animal social networks community and to help us develop methods and models that make the most of the data we have. All of this offers an exciting and challenging set of problems for the analyst. As long as we are guided primarily by the biology of the question at hand, and as long as we continue to invest in analysis that is as thoughtful and careful as the design of our data collection, the future of animal social network analysis is surely a bright one.

References

Abell, J., Kirzinger, M. W. B., Gordon, Y., Kirk, J., Kokes, R., Lynas, K., et al. (2013). A social network analysis of social cohesion in a constructed pride: implications for ex situ reintroduction of the African lion (*Panthera leo*). *PLoS ONE*, 8, e82541.

Abramson, G., Trejo Soto, C. A., and Oña, L. (2011). The role of asymmetric interactions on the effect of habitat destruction in mutualistic networks. *PLoS ONE*, **6**, e21028.

Adler, F. R., and Gordon, D. M. (1992). Information collection and spread by networks of patrolling ants. *American Naturalist*, **140**, 373–400.

Akçay, C., Tom, M. E., Campbell, S. E., and Beecher, M. D. (2013). Song type matching is an honest early threat signal in a hierarchical animal communication system. *Proceedings of the Royal Society B: Biological Sciences*, **280**, 20122517.

Aktipis, C. A. (2004). Know when to walk away: contingent movement and the evolution of cooperation. *Journal of Theoretical Biology*, **231**, 249–260.

Aktipis, C. A. (2008). *When to walk away and when to stay: cooperation evolves when agents can leave unproductive partners and groups*. PhD dissertation, University of Pennsylvania, Philadelphia.

Aktipis, C. A. (2009). Is cooperation viable in mobile organisms? Simple Walk Away rule favors the evolution of cooperation in groups. *Evolution and Human Behavior*, **32**, 263–276.

Albert, R., Jeong, H., and Barabasi, A. (2000). Error and attack tolerance of complex networks. *Nature*, **46**, 378–382.

Alexander, R. D. (1974). The evolution of social behavior. *Annual Review of Ecology and Systematics*, **5**, 325–383.

Allee, W. C. (1938). *The social life of animals*. Norton, New York.

Allee, W. C., and Dickinson, J. C. (1954). Dominance and subordination in the smooth dogfish *Mustelus canis* (Mitchill). *Physiological Zoology*, **27**, 356–364.

Allen, J., Weinrich, M., Hoppitt, W., and Rendell, L. (2013). Network-based diffusion analysis reveals cultural transmission of lobtail feeding in humpback whales. *Science*, **340**, 485–488.

Almeida-Neto, M., P., Guimarães, P. R.,Loyola, R. D., and Ulrich, W. (2008). A consistent metric for nestedness analysis in ecological systems: reconciling concept and measurement. *Oikos*, **117**, 1227–1239.

Almeida-Neto, M., and Ulrich, W. (2011). A straightforward computational approach for measuring nestedness using quantitative matrices. *Environmental Modelling and Software*, **26**, 173–178.

Altizer, S., Nunn, C. L., Thrall, P. H., Gittleman, J. L., Antonovics, J., Cunningham, A. A., et al. (2003). Social organization and parasite risk in mammals: integrating theory and empirical studies. *Annual Review of Ecology, Evolution and Systematics*, **34**, 517–547.

Anderson, R. M., and May, R. M. (1991). *Infectious diseases of humans: dynamics and control*. Oxford University Press, New York.

Andersson, M. (1994). *Sexual selection*. Princeton University Press, Princeton, NJ.

Ansmann, I. C., Parra, G. J., Chilvers, B. L., and Lanyon, J. M. (2012). Dolphins restructure social system after reduction of commercial fisheries. *Animal Behaviour*, **84**, 575–581.

Antal, T., Ohtsuki, H., Wakeley, J., Taylor, P. D., and Nowak, M. A. (2009). Evolution of cooperation by phenotypic similarity. *Proceedings of the National Academy of Sciences*, **106**, 8597–8600.

Antunes, R. (2009). *Variation in sperm whale (*Physeter macrocephalus*) coda vocalization and social structure in the North Atlantic Ocean*. PhD dissertation, University of St Andrews, St Andrews.

Aoki, I. (1982). A simulation study on the schooling mechanism in fish. *Bulletin of the Japanese Society of Scientific Fisheries*, **48**, 1081–1088.

Apicella, C. L., Marlowe, F. W., Fowler, J. H., and Christakis, N. A. (2012). Social networks and cooperation in hunter-gatherers. *Nature*, **481**, 497–501.

Aplin, L. M., Farine, D. R., Morand-Ferron, J., Cole, E. F., Cockburn, A., and Sheldon, B. C. (2013). Individual personalities predict social behaviour in wild networks of great tits (*Parus major*). *Ecology Letters*, **16**, 1365–1372.

Aplin, L. M., Farine, D. R., Morand-Ferron, J., and Sheldon, B. C. (2012). Social networks predict patch discovery in a wild population of songbirds. *Proceedings of the Royal Society B: Biological Sciences*, **279**, 4199–4205.

Aquiloni, L., and Gherardi, F. (2008). Crayfish females eavesdrop on fighting males and use smell and sight to recognize the identity of the winner. *Current Biology*, **79**, 265–269.

Aragon, P., Massot, M., Gasparini, J., and Clobert, J. (2006). Socially acquired information from chemical cues in the common lizard, *Lacerta vivipara*. *Animal Behaviour*, **72**, 965–974.

Araújo, M. S., Guimarães, P. R., Svanbäck, R., Pinheiro, A., Guimarães, P., dos Reis, S. F., et al. (2008). Network analysis reveals contrasting effects of intraspecific competition on individual vs. population diets. *Ecology*, **89**, 1981–1993.

Arave, C. W., and Albright, J. L. (1976). Social rank and physiological traits of dairy cows as influenced by changing group membership. *Journal of Dairy Science*, **59**, 974–981.

Archie, E. A. (2005). *The relationship between kinship and social behavior in wild African elephants* (Loxodonta africana*)*. PhD dissertation, Duke University, Durham, NC.

Arimura, G., Shiojiri, K., and Karban, R. (2010). Acquired immunity to herbivory and allelopathy caused by airborne plant emissions. *Phytochemistry*, **71**, 1642–1649.

Arnbom, T. (1987). Individual identification of sperm whales. *Report of the International Whaling Commission*, **37**, 201–204.

Arnbom, T., and Whitehead, H. (1989). Observations on the composition and behavior of groups of female sperm whales near the Galapagos Islands. *Canadian Journal of Zoology*, **67**, 1–7.

Arnold, K. E., Owens, I. P., and Goldizen, A. W. (2005). Division of labour within cooperatively breeding groups. *Behaviour*, **142**, 11–12.

Arnold, S., and Wade, M. (1984). On the measurement of natural and sexual selection: theory. *Evolution*, **38**, 709–719.

Arnqvist, G. (1992). Spatial variation in selective regimes: sexual selection in the water strider, *Gerris odontogaster*. *Evolution*, **46**, 914–929.

Arzoumanian, Z., Holmberg, J., and Norman, B. (2005). An astronomical pattern-matching algorithm for computer-aided identification of whale sharks *Rhincodon typus*. *Journal of Applied Ecology*, **42**, 999–1011.

Atton, N., Hoppitt, W., Webster, M. M., Galef, B. G., and Laland, K. N. (2012). Information flow through threespine stickleback networks without social transmission. *Proceedings of the Royal Society B: Biological Sciences*, **279**, 4272–4278.

Aureli, F., Schaffner, C. M., Boesch, C., Bearder, S. K., Call, J., Chapman, C.A., et al. (2008). Fission–fusion dynamics: new research frameworks. *Current Anthropology*, **49**, 627–654.

Axelrod, R., and Hamilton, W. D. (1981). The evolution of cooperation. *Science*, **211**, 1390–1396.

Bacher, K., Allen, S., Lindholm, A. K., Bejder, L., and Krützen, M. (2010). Genes or culture: are mitochondrial genes associated with tool use in bottlenose dolphins (*Tursiops* sp.)? *Behavior Genetics*, **40**, 706–714.

Baigger, A., Perony, N., Reuter, M., Leinert, V., Melber, M., Grünberger, S., et al. (2013). Bechstein's bats maintain individual social links despite a complete reorganisation of their colony structure. *Naturwissenschaften*, **100**, 895–898.

Bailey, N. T. J. (1957). *The mathematical theory of epidemics*. Griffin, London.

Baird, R. W., and Whitehead, H. (2000). Social organization of mammal-eating killer whales: group stability and dispersal patterns. *Canadian Journal of Zoology*, **78**, 2096–2105.

Baird, T. A., and Hews, D. K. (2007). Hormone levels in territorial and non-territorial male collared lizards. *Physiology and Behavior*, **92**, 755–763.

Bajec, I. L., and Heppner, F. H. (2009). Organized flight in birds. *Animal Behaviour*, **78**, 777–789.

Ballerini, M., Calbibbo, N., Candeleir, R., Cavagna, A., Cisbani, E., Giardina, I., et al. (2008). Interaction ruling animal collective behavior depends on topological rather than metric distance: evidence from a field study. *Proceedings of the National Academy of Sciences*, **105**, 1232–1237.

Bang, A., Deshpande, S., Sumana, A., and Gadagkar, R. (2010). Choosing an appropriate index to construct dominance hierarchies in animal societies: a comparison of three indices. *Animal Behaviour*, **79**, 631–636.

Bansal, S., Grenfell, B. T., and Meyers, L. A. (2007). When individual behaviour matters: homogeneous and network models in epidemiology. *Journal of the Royal Society Interface*, **4**, 879–891.

Barclay, P., and Willer, R. (2007). Partner choice creates competitive altruism in humans. *Proceedings of the Royal Society B: Biological Sciences*, **274**, 749–753.

Barclay, R. M. R. (1982). Interindividual use of echolocation calls: eavesdropping by bats. *Behavioral Ecology and Sociobiology*, **10**, 271–275.

Barlow, G. W., and Ballin, P. J. (1976). Predicting and assessing dominance from size and coloration in the polychromatic Midas cichlid. *Animal Behaviour*, **24**, 793–813.

Barlow, N. E. (1963). Darwin's ornithological notes. *Bulletin of the British Museum (Natural History) Historical Series*, **2**, 201–278.

Barocas, A., Ilany, A., Koren, L., Kam, M., and Geffen, E. (2011). Variance in centrality within rock hyrax social networks predicts adult longevity. *PLoS ONE*, **6**, e22375.

Barrett, L., Henzi, P. S., and Lusseau, D. (2012). Taking sociality seriously: the structure of multi-dimensional social networks as a source of information for individuals.

Philosophical Transactions of the Royal Society B: Biological Sciences, **367**, 2108–2118.

Barron, D. G., Brawn, J. D., and Weatherhead, P. J. (2010). Meta-analysis of transmitter effects on avian behaviour and ecology. *Methods in Ecology and Evolution*, **1**, 180–187.

Barta, Z., McNamara, J. M., Huszár, D. B., and Taborsky, M. (2011). Cooperation among non-relatives evolves by state-dependent generalized reciprocity. *Proceedings of the Royal Society B: Biological Sciences*, **278**, 843–848.

Bartlett, M. Y., and DeSteno, D. (2006). Gratitude and prosocial behavior: helping when it costs you. *Psychological Science*, **17**, 319–325.

Bascompte, J., and Jordano, P. (2007). Plant–animal mutualistic networks: the architecture of biodiversity. *Annual Review of Ecology, Evolution, and Systematics*. **38**, 567–593.

Bascompte, J., Jordano, P., Melián, C. J., and Olesen, J. M. (2003). The nested assembly of plant–animal mutualistic networks. *Proceedings of the National Academy of Sciences*, **100**, 9383–9387.

Bateson, M., and Healy, S. D. (2005). Comparative evaluation and its implications for mate choice. *Trends in Ecology and Evolution*, **20**, 659–664.

Bayly, K. L., Evans, C. S., and Taylor, A. (2006). Measuring social structure: a comparison of eight dominance indices. *Behavioural Processes*, **73**, 1–12.

Beeton, N., and McCalllum, H. (2011). Models predict that culling is not a feasible strategy to prevent extinction of Tasmanian devils from facial tumour disease. *Journal of Applied Ecology*, **48**, 1315–1323.

Beilharz, R. G., and Mylrea, P. J. (1963). Social position and behavior of dairy heifers in yards. *Animal Behaviour*, **11**, 522.

Beisner, B. A., Jackson, M. E., Cameron, A., and McCowan, B. (2011a). Effects of natal male alliances on aggression and power dynamics in rhesus macaques. *American Journal of Primatology*, **73**, 790–801.

Beisner, B. A., Jackson, M. E., Cameron, A. N., and McCowan, B. (2011b). Detecting instability in animal social networks: genetic fragmentation is associated with social instability in rhesus macaques. *PLoS ONE*, **6**, e16365.

Bell, A. M., and Sih, A. (2007). Exposure to predation generates personality in threespined sticklebacks (*Gasterosteus aculeatus*). *Ecology Letters*, **10**, 828–834.

Bell, A. M., and Stamps, J. A. (2004). Development of behavioural differences between individuals and populations of sticklebacks, *Gasterosteus aculeatus*. *Animal Behaviour*, **68**, 1339–1348.

Belmonte, J. M., Thomas, G. L., Brunnet, L. G., de Almeida, R. M. C., and Chaté, H. (2008). Self-propelled particle model for cell-sorting phenomena. *Physical Review Letters*, **100**, 248702.

Bengtsson, L., Lu, X., Thorson, A., Garfield, R., and von Schreeb, J. (2011). Improved response to disasters and outbreaks by tracking population movements with mobile phone network data: a post-earthquake geospatial study in Haiti. *PLoS Medicine*, **8**, e1001083.

Benton, T. G., and Evans, M. R. (1998). Measuring mate choice using correlation: the effect of female sampling behaviour. *Behavioral Ecology and Sociobiology*, **44**, 91–98.

Berdahl, A., Torney, C. J., Ioannou, C. C., Faria, J. J., and Couzin, I. D. (2013). Emergent sensing of complex environments by mobile animal groups. *Science*, **339**, 574–576.

Bergmüller, R., Schürch, R., and Hamilton, I. M. (2010). Evolutionary causes and consequences of consistent individual variation in cooperative behaviour. *Philosophical Transactions of the Royal Society B: Biological Sciences*, **365**, 2751–2764.

Bergmüller, R., and Taborsky, M. (2007). Adaptive behavioural syndromes due to strategic niche specialization. *BMC Ecology*, **7**, 12.

Bergsma, R., Kanis, E., Knol, E. F., and Bijma, P. (2008). The contribution of social effects to heritable variation in finishing traits of domestic pigs (*Sus scrofa*). *Genetics*, **178**, 1559–1570.

Bernstein, I. S. (1964). Role of the dominant male rhesus monkey in response to external challenges to the group. *Journal of Comparative and Physiological Psychology*, **57**, 404–406.

Bernstein, I. S. (1981). Dominance: the baby and the bathwater. *Behavioral and Brain Sciences*, **4**, 419–429.

Bertram, S. M., and Gorelick, R. (2009). Quantifying and comparing mating systems using normalized mutual entropy. *Animal Behaviour*, **77**, 201–206.

Beshers, S. N., Huang, Z. Y., Oono, Y., and Robinson, G. E. (2001). Social inhibition and the regulation of temporal polyethism in honey bees. *Journal of Theoretical Biology*, **213**, 461–479.

Best, E. C., Seddon, J. M., Dwyer, R. G., and Goldizen, A. W. (2013). Social preference influences female community structure in a population of wild eastern grey kangaroos. *Animal Behaviour*, **86**, 1031–1040.

Best, P. B. (1979). Social organization in sperm whales, *Physeter macrocephalus*. In: Winn, H. E., and Olla, B. L., eds. *Behaviour of marine animals*, pp. 227–289. Plenum Press, New York.

Beukers, J. S., Jones, G. P., and Buckley, R. M. (1995). Use of implant microtags for studies on populations of small reef fish. *Marine Ecology Progress Series*, **125**, 61–66.

Beyer, K., Gozlan, R. E., and Copp, G. H. (2010). Social network properties within a fish assemblage invaded by a non-native sunbleak *Leucaspius delineatus*. *Ecological Modelling*, **221**, 2118–2122.

Bhadra, A., Jordán, F., Sumana, A., Deshpande, S. A., and Gadagkar, R. (2009). A comparative social network analysis of wasp colonies and classrooms: linking network structure to functioning. *Ecological Complexity*, **6**, 48–55.

Bijma, P., Muir, W. M., and van Arendonk, J. M. (2007). Multilevel selection, 1: Quantitative genetics of inheritance and response to selection. *Genetics*, **175**, 277–288.

Birkhead, T. R., and Pizzari, T. (2002). Postcopulatory sexual selection. *Nature Reviews Genetics*, **3**, 262–273.

Birkhead, T., Wimpenny, J., and Montgomerie, B. (2014). *Ten thousand birds: ornithology since Darwin*. Princeton University Press, Princeton, NJ.

Bjørnstad, O. N., Finkenstädt, B., and Grenfell, B. T. (2002). Dynamics of measles epidemics: estimating scaling of transmission rates using a time series SIR model. *Ecological Monographs*, **72**, 169–184.

Blomberg, S., and Shine, R. (2006). Reptiles. In: Sutherland, W. J., ed. *Ecological census techniques: a handbook*, 2nd edn. Cambridge University Press, Cambridge.

Blonder, B., and Dornhaus, A. (2011). Time-ordered networks reveal limitations to information flow in ant colonies. *PLoS ONE*,**6**, e20298.

Blonder, B., Wey, T. W., Dornhaus, A., James, R., and Sih, A. (2012). Temporal dynamics and network analysis. *Methods in Ecology and Evolution*, **3**, 958–972.

Bluff, L. A. and Rutz, C. (2008). A quick guide to video-tracking birds. *Biology Letters* **4**, 319–322.

Blumstein, D. T., Mennill, D. J., Clemins, P., Girod, L., Yao, K., Patricelli. G., et al. (2011). Acoustic monitoring in terrestrial environments using microphone arrays: applications, technological considerations and prospectus. *Journal of Applied Ecology*, **48**, 758–767.

Blumstein, D. T., Verneyre, L., and Daniel, J. C. (2004). Reliability and the adaptive utility of discrimination among alarm callers. *Proceedings of the Royal Society B: Biological Sciences*, **271**, 1851–1857.

Blumstein, D. T., Wey, T. W., and Tang, K. (2009). A test of the social cohesion hypothesis: interactive female marmots remain at home. *Proceedings of the Royal Society B: Biological Sciences*, **276**, 3007–3012.

Boatright-Horowitz, S. L., Horowitz, S. S., and Simmons, A. M. (2000). Patterns of vocal response in a bullfrog (*Rana catesbeiana*) chorus: preferential responding to far neighbors. *Ethology*, **106**, 701–712.

Bode, N.W. F.,Faria, J.J., Franks, D.W., Krause,J., and Wood, A J. (2010). How perceived threat increases synchronization in collectively moving animal groups. *Proceedings of the Royal Society B: Biological Sciences*, **277**, 3065–3070.

Bode, N. W. F., Franks, D. W., and Wood, A. J. (2011). Limited interactions in flocks: relating model simulations to empirical data. *Journal of the Royal Society Interface*, **8**, 301–304.

Bode, N. W. F., Franks, D. W., and Wood, A. J. (2012). Leading from the front? Social networks in navigating groups. *Behavioral Ecology and Sociobiology*, **66**, 835–843.

Bode, N. W. F., Franks, D. W., Wood, A. J., Piercy, J. J. B., Croft, D. P., and Codling, E. A. (2012). Distinguishing social from non-social navigation in moving animal groups. *American Naturalist*, **179**, 612–632.

Bode, N. W. F., Wood, A. J., and Franks, D. W. (2011a). Social networks and models for collective motion in animals. *Behavioral Ecology and Sociobiology*, **65**, 117–130.

Bode, N. W. F., Wood, A. J., and Franks, D. W. (2011b). The impact of social networks on animal collective motion. *Animal Behaviour*, **82**, 29–38.

Bode, N. W. F., Wood, A. J., and Franks, D. W. (2012). Social networks improve leaderless group navigation by facilitating long-distance communication. *Current Zoology*, **58**, 329–341.

Bogich, T. L., Funk, S., Malcolm, T. R., Chhun, N., Epstein, J. H., Chmura, A. A., et al. (2013). Using network theory to identify the causes of disease outbreaks of unknown origin. *Journal of the Royal Society Interface*, **10**, 20130127.

Bograd, S. J., Block, B. A., Costa, D. P., and Godley, B. J. (2010) Biologging technologies: new tools for conservation. Introduction. *Endangered Species Research* **10**, 1–7.

Böhm, M., Palphramand, K. L., Newton-Cross, G., Hutchings, M. R., and White, P. C. L. (2008). Dynamic interactions among badgers: implications for sociality and disease transmission. *Journal of Animal Ecology*, **77**, 735–745.

Bolger, D. T., Morrison, T. A., Vance, B., Lee, D., and Farid, H. (2012). A computer-assisted system for photographic mark–recapture analysis. *Methods in Ecology and Evolution*, **3**, 813–822.

Bonabeau, E., Dorigo, M., and Theraulaz, G. (1999). *Swarm intelligence: from natural to artificial systems, Volume 4*. Oxford University Press, New York.

Bonacich, P. (1987). Power and centrality: a family of measures. *American Journal of Sociology*, **92**, 1170–1182.

Bonter, D. N., and Bridge, E. S. (2011). Applications of radio frequency identification (RFID) in ornithological research: a review. *Journal of Field Ornithology*, **82**, 1–10.

Boogert, N. J., Reader, S. M., Hoppitt, W. J. E., and Laland, K. N. (2008). The origin and spread of innovations in starlings. *Animal Behaviour*, **75**, 1509–1518.

Boone, R. T., and Macy, M. W. (1999). Unlocking the doors of the prisoner's dilemma: dependence, selectivity, and cooperation. *Social Psychology Quarterly*, **62**, 32–52.

Booth, D. J. (1995). Juvenile groups in a coral-reef damselfish: density-dependent effects on individual fitness and population demography. *Ecology*, **76**, 91–106.

Borgatti, S. P. (2003). The key player problem. In:Breiger, R. C. K., and Pattison, P., eds. *Dynamic social network modeling and analysis: workshop summary and papers*, pp. 241–252. The National Academies Press, Washington, DC.

Borgatti. S. P. (2005). Centrality and network flow. *Social Networks*, **27**, 55–71.

Borgatti. S. P., and Foster, P. (2003). The network paradigm in organizational research: a review and typology. *Journal of Management*, **29**, 991–1013.

Borgatti. S. P., and Halgin, D. S. (2011). On network theory. *Organization Science*, **22**, 1168–1181.

Borgia, G., and Collis, K. (1989). Female choice for parasite-free male satin bowerbirds and the evolution of bright male plumage. *Behavioral Ecology and Sociobiology*, **25**, 445–454.

Barocas, A., Ilany, A., Koren, L., Kam, M., and Geffen, E. (2011). Variance in centrality within rock hyrax social networks predicts adult longevity. *PloS ONE*, **6**, e22375.

Borrel, V., Legendre, F., Dias de Amorim, M., Fdida, S. (2009). Simps: using sociology for personal mobility. *IEEE/ACM Transactions on Networking*, **17**, 831–842.

Botero, C. A., Pen, I., Komdeur, J., and Weissing, F. J. (2010). The evolution of individual variation in communication strategies. *Evolution*, **64**, 3123–3133.

Bouchard, S. (2001). Sex discrimination and roostmate recognition by olfactory cues in the African bats, *Mops condylurus* and *Chaerephon pumilus* (Chiroptera: Molossidae). *Journal of Zoology*, **254**, 109–117.

Bourke, A. F. G. (2011). *Principles of social evolution*. Oxford University Press, Oxford.

Bourke, A. F. G., and Franks, N. R. (1995). *Social evolution in ants*. Princeton University Press, Princeton, NJ.

Bower, J. L. (2000). *Acoustic interactions during naturally occurring territorial conflict in a song sparrow (*Melospiza melodia*) neighborhood*. PhD thesis, Cornell University, Ithaca, NY.

Bower, J. L. (2005). The occurrence and function of victory displays within communication networks. In: McGregor, P. K., ed. *Animal communication networks*, pp. 114–128. Cambridge University Press, Cambridge.

Bower, J. L., and Clark, C. W. (2005). A field test of the accuracy of a passive acoustic location system. *Bioacoustics*, **15**, 1–14.

Boyd, R., and Richerson, P. J. (1987). The evolution of ethnic markers. *Cultural Anthropology*, **2**, 65–79.

Boyland, N. K., James, R., Mlynski, D. T., Madden, J. R., and Croft, D. P. (2013). Spatial proximity loggers for recording animal social networks: consequences of interlogger variation in performance. *Behavioral Ecology and Sociobiology*, **67**, 1877–1890.

Bradbury, J. W., and Vehrencamp, S. L. (2000). Economic models of animal communication. *Animal Behaviour*, **59**, 259–268.

Bradbury, J. W., and Vehrencamp, S. L. (2011). Communication networks In: *Principles of animal communication*, 2nd edn, pp. 611–650. Sinauer Associates, Sunderland, MA.

Braun, A., Musse, S. R., de Oliveira, L. P. L., and Bodmann, B. E. J. (2003). Modeling individual behaviors in crowd simulation. In: *Proceedings of the 16th IEEE International Conference on Computer Animation and Social Agents*, pp. 143–148.

Brawn, J. D., and Samson, F. B. (1983). Winter behavior of tufted titmice. *The Wilson Bulletin*, **95**, 222–232.

Brent, L. J. N., Heilbronner, S. R., Horvath, J. E., Gonzalez-Martinez, J., Ruiz-Lambides, A., Robinson, A. G., et al. (2013). Genetic origins of social networks in rhesus macaques. *Scientific Reports*, **3**, 1042.

Brent, L. J. N., Lehmann, J., and Ramos-Fernández, G. (2011). Social network analysis in the study of nonhuman primates: a historical perspective. *American Journal of Primatology*, **73**, 1–11.

Brent, L. J. N., MacLarnon, A., Platt, M. L., and Semple, S. (2013). Seasonal changes in the structure of rhesus macaque social networks. *Behavioral Ecology and Sociobiology*, **67**, 349–359.

Brent, L. J. N., Semple, S., Dubuc, C., Heistermann, M., and MacLarnon, A. (2011). Social capital and physiological stress levels in free-ranging adult female rhesus macaques. *Behaviour*, **102**, 76–83.

Bridge, E. S., and Bonter, D. N. (2011). A low-cost radio frequency identification device for ornithological research. *Journal of Field Ornithology*, **82**, 52–59.

Brodie, E. D., III., Moore, A., and Janzen, F. (1995). Visualizing and quantifying natural selection. *Trends in Ecology and Evolution*, **10**, 313–318.

Brodin, T. (2009). Behavioral syndrome over the boundaries of life—carryovers from larvae to adult damselfly. *Behavioral Ecology*, **20**, 30–37.

Brotons, L. (2000). Individual food-hoarding decisions in a nonterritorial coal tit population: the role of social context. *Animal Behaviour*, **60**, 395–402.

Brown, C., Laland K., and Krause, J. (2011). *Fish cognition and behavior*. Wiley-Blackwell, Oxford.

Brown, J. L. (1983). Cooperation: a biologist's dilemma. *Advances in the Study of Behavior*, **13**, 1–37.

Bshary, R. (2003). The cleaner wrasse, *Labroides dimidiatus*, is a key organism for reef fish diversity at Ras Mohammed National Park, Egypt. *Journal of Animal Ecology*, **72**, 169–176.

Bshary, R., and Grutter, A. S. (2006). Image scoring and cooperation in a cleaner fish mutualism. *Nature*, **441**, 975–978.

Buffin, A., Denis, D., van Simaeys, G., Goldman, S., and Deneubourg, J.-L. (2009). Feeding and stocking up: radio-labelled food reveals exchange patterns in ants. *PLoS ONE*, **4**, e5919.

Bull, C. M., Godfrey, S. S., and Gordon, D. M. (2012). Social networks and the spread of *Salmonella* in a sleepy lizard population. *Molecular Ecology*, **21**, 4386–4392.

Bull, C. M., Griffin, C. L., and Johnston, G. R. (1999). Olfactory discrimination in scat-piling lizards. *Behavioral Ecology*, **10**, 136–140.

Bull, C. M., Griffin, C. L., Lanham, E. J., and Johnston, G. R. (2000). Recognition of pheromones from group members in a gregarious lizard, *Egernia stokesii*. *Journal of Herpetology*, **34**, 92–99.

Bumann, D., Krause, J., and Rubenstein, D. (1997). Mortality risk of spatial positions in animal groups: the danger of being in the front. *Behaviour*, **134**, 1063–1076.

Buscarino, A., Fortuna, L., Frasca, M., and Rizzo, A. (2006). Dynamical network interactions in distributed control of robots. *Chaos*, **16**, 015116.

Byrne, R. W., and Whiten, A. (1988). *Machiavellian intelligence: social expertise and the evolution of intellect in monkeys, apes, and humans*. Oxford University Press, Oxford.

Byrne, R. W., Whiten, A., and Henzi, S. P. (1989). Social relationships of mountain baboons: leadership and affiliation in a non-female-bonded primate. *American Journal of Primatology*, **18**, 191–207.

Caldwell, M. C., and Caldwell, D. K. (1965). Individualized whistle contours in bottlenose dolphins (*Tursiops truncatus*). *Nature*, **207**, 434–435.

Calsbeek, R., and Marnocha, E. (2006). Context dependent territory defense: the importance of habitat structure in *Anolis sagrei*. *Ethology*, **112**, 537–543.

Camazine, S., Deneubourg, J.-L., Franks, N. R., Sneyd, J., Theraulaz, G., and Bonabeau, B. (2003). *Self-organization in biological systems*. Princeton University Press, Princeton, NJ.

Cameron, E. Z., Setsaas, T. H., and Linklater, W. L. (2009). Social bonds between unrelated females increase reproductive success in feral horses. *Proceedings of the National Academy of Sciences*, **106**, 13850–13853.

Campbell, J., Kessler, B., Mayack, C., and Naug, D. (2010). Behavioral fever in infected honeybees: parasitic manipulation or coincidental benefits? *Parasitology*, **137**, 1487–1491.

Candolin, U., and Wong, B. B. M. (2008). Mate choice. In: Rocha, M. J., Arukwe, A., and Kapoor, B. G., eds. *Fish reproduction*, pp. 277–309. Science Publishers, Enfield, NH.

Cañon Jones, H. A., Hansen, L. A., Noble, C., Damsgård, B., Broom, D. M., and Pearce, G. P. (2010). Social network analysis of behavioural interactions influencing fin damage development in Atlantic salmon (Salmo salar) during feed-restriction. *Applied Animal Behaviour Science*, **127**, 139–151.

Cañon Jones, H. A., Noble, C., Damsgard, B., and Pearce, G. P. (2011). Social network analysis of the behavioural interactions that influence the development of fin damage in Atlantic salmon parr (*Salmo salar*) held at different stocking densities. *Applied Animal Behaviour Science*, **133**, 117–126.

Cañon Jones, H. A., Noble, C., Damsgard, B., and Pearce, G. P. (2012). Investigating the influence of predictable and unpredictable feed delivery schedules upon the behaviour and welfare of Atlantic salmon parr (*Salmo salar*) using social network analysis and fin damage. *Applied Animal Behaviour Science*, **138**, 132–140.

Cantor, M., Wedekin, L. L., Guimarães, P. R., Daura-Jorge, F. G., Rossi-Santos, M. R., and Simões-Lopes, P. C. (2012). Disentangling social networks from spatiotemporal dynamics: the temporal structure of a dolphin society. *Animal Behaviour*, **84**, 641–651.

Cantor, M., and Whitehead, H. (2013). The interplay between social networks and culture: theoretically and among whales and dolphins. *Philosophical Transactions of the Royal Society B: Biological Sciences*, **368**, 20120340.

Cao, T. T., and Dornhaus, A. (2013). Larger laboratory colonies consume proportionally less energy and have lower per capita brood production in *Temnothorax* ants. *Insectes Sociaux*, **60**, 1–5.

Carazo, P., Font, E., and Desfilis, E. (2007). Chemosensory assessment of rival competitive ability and scent-mark function in a lizard, *Podarcis hispanica*. *Animal Behaviour*, **74**, 895–902.

Carazo, P., Font, E., and Desfilis, E. (2008). Beyond 'nasty neighbours' and 'dear enemies'? Individual recognition by scent marks in a lizard (*Podarcis hispanica*). *Animal Behaviour*, **76**, 1953–1963.

Carazo, P., Font, E., and Desfilis, E. (2011). The role of scent marks in female choice of territories and refuges in a lizard (*Podarcis hispanica*). *Journal of Comparative Psychology*, **125**, 362–365.

Carrington, P. J., Scott, S., and Wasserman, S. (2005). *Models and methods in social network analysis*. Cambridge University Press, Cambridge.

Carter, A. J., Goldizen, A. W., and Tromp, S. A. (2010). Agamas exhibit behavioral syndromes: bolder males bask and feed more but may suffer higher predation. *Behavioral Ecology*, **21**, 655–661.

Carter, S. P., Delahay, R. J., Smith, G. C., Macdonald, D. W., Riordan, P., Etherington, T. R., et al. (2007) Culling-induced social perturbation in Eurasian badgers *Meles meles* and the management of TB in cattle: an analysis of a critical problem in applied ecology. *Proceedings of the Royal Society B: Biological Sciences*, **274**, 2769–2777.

Cassill, D. L., Stuy, A., and Buck, R. G. (1999). Emergent properties of food distribution among fire ant larvae. *Journal of Theoretical Biology*, **195**, 371–381.

Caswell, H. (2001) *Matrix population models: construction, analysis and interpretation*, 2nd edn. Sinauer Associates Inc., Sunderland, MA.

Chan, S., Fushing, H., Beisner, B. A., and McCowan, B. (2013). Joint modeling of multiple social networks to elucidate primate social dynamics: I. Maximum entropy principle and network-based interactions. *PLoS ONE*, **8**, e51903.

Chapais, B., and Mignault, C. (1991). Homosexual incest avoidance among females in captive Japanese macaques. *American Journal of Primatology*, **23**, 171–183.

Chapman, B. B., Hulthen, K., Blomqvist, D. R., Hansson, L. A., Nilsson, J. A., Brodersen, J., et al. (2011). To boldly go: individual differences in boldness influence migratory tendency. *Ecology Letters*, **14**, 871–876.

Chapman, B. B., Morrell, L. J., Benton, T. G., and Krause, J. (2008). Early interactions with adults mediate the development of predator defenses in guppies. *Behavioral Ecology*, **19**, 87–93.

Chapman, C. A. (1990). Association patterns of spider monkeys: the influence of ecology and sex on social organization. *Behavioral Ecology and Sociobiology*, **26**, 409–414.

Chapple, D. G., and Keogh, J. S. (2006). Group structure and stability in social aggregations of White's skink, *Egernia whitii*. *Ethology*, **112**, 247–257.

Charmantier, A., Buoro, M., Gimenez, O., and Weimerskirch, H. (2011). Heritability of short-scale natal dispersal in a large-scale foraging bird, the wandering albatross. *Journal of Evolutionary Bioliogy*, **24**, 1487–1496.

Charmantier, A., Keyser, A. J., and Promislow, D. E. (2007). First evidence for heritable variation in cooperative breeding behaviour. *Proceedings of the Royal Society B: Biological Sciences*, **274**, 1757–1761.

Chase, I. D. (1982). Behavioral sequences during dominance hierarchy formation in chickens. *Science*, **216**, 439–440.

Chaverri, G. (2010). Comparative social network analysis in a leaf-roosting bat. *Behavioral Ecology and Sociobiology*, **64**, 1619–1630.

Cheney, D. L. (1978a). Interactions of immature male and female baboons with adult females. *Animal Behaviour*, **26**, 389–408.

Cheney, D. L. (1978b). The play partners of immature baboons. *Animal Behaviour*, **26**, 1038–1050.

Cheney, D. L. (1987). Interactions and relationships between groups. In:Smuts, B. B., Cheney, D. L., Seyfarth, R. M., Wrangham, R. W., and Struhsaker, T. T. eds. *Primate societies*, pp. 267–281. University of Chicago Press, Chicago.

Cheney, D. L., and Seyfarth, R. (1990). *How monkeys see the world: inside the mind of another species*. University of Chicago Press, Chicago.

Chepko-Sade, B. D., Reitz, K. P., and Sade, D. S. (1989). Sociometrics of *Macaca mulatta* IV: network analysis of social structure of a pre-fision group. *Social Networks*, **11**, 293–314.

Childress, D., and Hartl, D. L. (1972). Sperm preference in *Drosophila melanogaster*. *Genetics*, **71**, 417–427.

Chinn, P. Z., Chvátalová, J., Dewdney, A. K., and Gibbs, N. E. (1982). The bandwidth problem for graphs and matrices—a survey. *Journal of Graph Theory*, **6**, 223–254.

Christakis, N. A., and Fowler, J. H. (2013). Friendship and natural selection. *arXiv*, 1308.5257.

Christley, R. M., Pinchbeck, G. L., Bowers, R. G., Clancy, D., French, N. P., Bennett, R., et al. (2005). Infections in social networks: using network analysis to identify high-risk individuals. *American Journal of Epidemiology*, **162**, 1024–1031.

Clapham, P. J. (2000). The humpback whale: seasonal feeding and breeding in a baleen whale. In: Mann, J., Connor, R. C., Tyack, P. L., and Whitehead, H., eds. *Cetacean societies: field studies of dolphins and whales*, pp. 173–196. University of Chicago Press, Chicago.

Clark, F. E. (2011). Space to choose: network analysis of social preferences in a captive chimpanzee community, and implications for management. *American Journal of Primatology*, **73**, 748–757.

Clay, C. A., Lehmer, E. M., Previtali, A., St Jeor, S., and Dearing, M. D. (2009). Contact heterogeneity in deer mice: implications for Sin Nombre virus transmission. *Proceedings of the Royal Society B: Biological Sciences*, **276**, 1305–1312.

Clutton-Brock, T. (2002). Breeding together: kin selection and mutualism in cooperative vertebrates. *Science*, **296**, 69–72.

Clutton-Brock, T. (2009). Cooperation between non-kin in animal societies. *Nature*, **462**, 51–57.

Clutton-Brock, T., and Sheldon, B. C. (2010). Individuals and populations: the role of long-term, individual-based studies of animals in ecology and evolutionary biology. *Trends in Ecology and Evolution*, **25**, 562–573.

Clutton-Brock, T. H., Albon, S. D., Gibson, R. M., and Guinness, F. E. (1979). The logical stag: adaptive aspects of fighting in red deer (*Cervus elaphus* L.). *Animal Behaviour*, **27**, 211–225.

Clutton-Brock, T. H., Brotherton, P. N. M., Russell, A. F., O'riain, M. J., Gaynor, D., Kansky, R., et al. (2001). Cooperation, control and concession in meerkat groups. *Science*, **291**, 478–481.

Clutton-Brock, T. H., Gaynor, D., Kansky, R., MacColl, A. D. C., McIlrath, G., Chadwick, P., et al. (1998). Costs of cooperative behaviour in suricates, *Suricata suricatta*. *Proceedings of the Royal Society B: Biological Sciences*, **265**, 185–190.

Clutton-Brock, T. H., and Harvey, P. H. (1976). Evolutionary rules and primate societies. In: Bateson, P. P. G., and Hinde, R. A., eds. *Growing points in ethology*, pp. 195–237. Cambridge University Press, Cambridge.

Clutton-Brock, T. H., and Parker, G. A. (1992). Potential reproductive rates and the operation of sexual selection. *Quarterly Review of Biology*, **67**, 437–456.

Cockburn, A. (1998). Evolution of helping behavior in cooperatively breeding birds. *Annual Review of Ecology and Systematics*, **29**, 141–177.

Codling, E. A., Pitchford, J. W., and Simpson, S. D. (2007). Group navigation and the 'many-wrongs principle' in models of animal movement. *Ecology*, **88**, 1864–1870.

Cohen, H., Geva, A. B., Matar, M. A., Zohar, J., and Kaplan, Z. (2008). Post-traumatic stress behavioural responses in inbred mouse strains: can genetic predisposition explain phenotypic vulnerability? *The International Journal of Neuropsychopharmacology*, **11**, 331–349.

Cole, S. W., Mendoza, S. P., and Capitanio, J. P. (2009). Social stress desensitizes lymphocytes to regulation by endogenous glucocorticoids: insights from in vivo cell trafficking dynamics in rhesus macaques. *Psychosomatic Medicine*, **71**, 591–597.

Coleing, A. (2009). The application of social network theory to animal behaviour. *Bioscience Horizons*, **2**, 32–43.

Collet, J., Richardson, D. S., Worley, K., and Pizzari, T. (2012). Sexual selection and the differential effect of polyandry. *Proceedings of the National Academy of Sciences*, **109**, 8641–8645.

Connor, R. C. (2000). Group living in whales and dolphins. In: Mann, J., Connor, R. C., Tyack, P. L., and Whitehead, H., eds. *Cetacean societies: field studies of dolphins and whales*, pp. 199–218. University of Chicago Press, Chicago.

Connor, R. C. (2001). Individual foraging specializations in marine mammals: culture and ecology. *Behavioral and Brain Sciences*, **24**, 329.

Connor, R. C. (2007). Dolphin social intelligence: complex alliance relationships in bottlenose dolphins and a consideration of selective environments for extreme brain size evolution in mammals. *Philosophical Transactions of the Royal Society B: Biological Sciences*, **362**, 587–602.

Connor, R. C. (2010). Cooperation beyond the dyad: on simple models and a complex society. *Philosophical Transactions of the Royal Society B: Biological Sciences*, **365**, 2687–2697.

Connor, R. C., Heithaus, M. R., and Barre, L. M. (1999). Superalliance of bottlenose dolphins. *Nature*, **397**, 571–572.

Connor, R. C., Heithaus, M. R., and Barre, L. M. (2001). Complex social structure, alliance stability and mating access in a bottlenose dolphin 'super-alliance'. *Proceedings of the Royal Society B: Biological Sciences*, **268**, 263–267.

Connor, R. C., Mann, J., Tyack, P. L., and Whitehead, H. (1998). Social evolution in toothed whales. *Trends in Ecology and Evolution*, **13**, 228–232.

Connor, R. C., and Smolker, R. S. (1985). Habituated dolphins (*Tursiops sp.*) in western Australia. *Journal of Mammalogy*, **66**, 398–400.

Connor, R. C., Smolker, R., and Bejder, L. (2006). Synchrony, social behaviour and alliance affiliation in Indian Ocean bottlenose dolphins, *Tursiops aduncus*. *Animal Behaviour*, **72**, 1371–1378.

Connor, R. C., Wells, R. S., Mann, J., and Read, A. J. (2000). The bottlenose dolphin: social relationships in a fission–fusion society. In: Mann, J., Connor, R. C., Tyack, P. L., and Whitehead, H., eds. *Cetacean societies: field studies of dolphins and whales*, pp. 91–126. University of Chicago Press, Chicago.

Conradt, L. (1998). Could asynchrony in activity between the sexes cause intersexual social segregation in ruminants? *Proceedings of the Royal Society B: Biological Sciences*, **265**, 1359–1363.

Conradt, L., Krause, J., Couzin, I. D., and Roper, T. J. (2009). 'Leading according to need' in self-organizing groups. *American Naturalist*, **173**, 304–312.

Conradt, L.,and Roper, T. J. (2000). Activity synchrony and social cohesion: a fission–fusion model. *Proceedings of the Royal Society B: Biological Sciences*, **267**, 2213–2218.

Consolini, L., Morbidi, F., Prattichizzo, D., and Tosques, M. (2008). Leader–follower formation control of nonholonomic mobile robots with input constraints. *Automatica*, **44**, 1343–1349.

Cook, K. S., and Emerson, R. M. (1978). Power, equity and commitment in exchange networks. *American Sociological Review*, **43**, 721–739.

Cooke, F., and Rockwell, R. F. (1988). Reproductive success in a lesser snow goose population. In: Clutton-Brock, T. H., ed. *Reproductive success*, pp. 237–250. University of Chicago Press, Chicago.

Cooke, S. J., Hinch, S. G., Lucas, M. C., and Lutcavage, M. (2013). Biotelemetry and biologging. In: Zale, A. V., Parrish, D. L., and Sutton, T. M., ed. *Fisheries techniques*, 3rd edn, pp. 819–881. American Fisheries Society, Bethesda.

Cords, M. (2002). Friendship among adult female blue monkeys (*Cercopithecus mitis*). *Behaviour*, **139**, 291–314.

Corner, L. A. L., Pfeiffer, D. U., de Lisle, G. W., Morris, R. S., and Buddle, B. M. (2002). Natural transmission of *Mycobacterium bovis* infection in captive brushtail possums (*Trichosurus vulpecula*). *New Zealand Veterinary Journal*, **50**, 154–162.

Corner, L. A. L., Pfeiffer, D. U., and Morris, R. S. (2003). Social-network analysis of *Mycobacterium bovis* transmission among captive brushtail possums (*Trichosurus vulpecula*). *Preventive Veterinary Medicine*, **59**, 147–167.

Cornwallis, C. K., and Uller, T. (2010). Towards an evolutionary ecology of sexual traits. *Trends in Ecology and Evolution*, **25**, 145–152.

Corradino, C. (1990). Proximity structure in a captive colony of Japanese monkeys (*Macaca fuscata fuscata*): an application of multidimensional scaling. *Primates*, **31**, 351–362.

Correa, C. M., Zanela, M. B., and Schmidt, V. (2010). Social behaviour in lactating dairy goats after regrouping. *Acta Scientiae Veterinariae*, **38**, 425–428.

Costenbader, E., and Valente, T. W. (2003). The stability of centrality measures when networks are sampled. *Social Networks*, **25**, 283–307.

Cote, J., and Clobert, J. (2007). Social personalities influence natal dispersal in a lizard. *Proceedings of the Royal Society B: Biological Sciences*, **274**, 383–390.

Cote, J., Fogarty, S., Brodin, T., Weinersmith, K., and Sih, A. (2011). Personality-dependent dispersal in the invasive mosquitofish: group composition matters. *Proceedings of the Royal Society B: Biological Sciences*, **278**, 1670–1678.

Cotter, S., and Kilner, R. (2010). Personal immunity versus social immunity. *Behavioral Ecology*, **21**, 663–668.

Coussi-Korbel, S., and Fragaszy, D.M. (1995). On the relation between social dynamics and social learning. *Animal Behaviour*, **50**, 1441–1453.

Couzin, I. D., James, R., Mawdsley, D., Croft, D. P., and Krause, J. (2006). Social organization and information transfer in schooling fishes. In: Brown, C., Laland K. N., and Krause, J., eds. *Fish cognition and behavior*, pp. 166–185. Blackwell Publishing, Oxford.

Couzin, I. D., and Krause, J. (2003). Self-organisation and collective behaviour of vertebrates. *Advances in the Study of Behaviour*, **32**, 1–75.

Couzin, I. D., Krause, J., Franks, N. R., and Levin, S. A. (2005). Effective leadership and decision-making in animal groups on the move. *Nature*, **433**, 513–516.

Couzin, I. D., Krause, J., James, R., Ruxton, G.D., and Franks, N.R. (2002). Collective memory and spatial sorting in animal groups. *Journal of Theoretical Biology*, **218**, 1–11.

Craft, M. E., and Caillaud, D. (2011). Network models: an underutilized tool in wildlife epidemiology? *Interdisciplinary Perspectives on Infectious Diseases*, **2011**, 676949.

Craft, M. E., Volz, E., Packer, C., and Meyers, L. A. (2011). Disease transmission in territorial populations: the small-world network of Serengeti lions. *Journal of the Royal Society Interface*, **8**, 776–786.

Crall, J. P., Stewart, C. V., Berger-Wolf, T. Y., Rubenstein, D. I., and Sundaresan, S. R. (2013). HotSpotter—pattern species instance recognition. In: *2013 IEEE Workshop On Applications Of Computer Vision*, pp. 230–237. Curran Associates, Red Hook, NY.

Crockford, C., Wittig, R. M., Seyfarth, R. M., and Cheney, D. L. (2007). Baboons eavesdrop to deduce mating opportunities. *Animal Behaviour*, **73**, 885–890.

Crofoot, M. C., Gilby, I. C., Wikelski, M. C., and Kays, R. W. (2008). Interaction location outweighs the competitive advantage of numerical superiority in *Cebus capucinus* intergroup contests. *Proceedings of the National Academy of Sciences*, **105**, 577–581.

Crofoot, M. C., Rubenstein, D. I., Maiya, A. S., and Berger-Wolf, T. Y. (2011). Aggression, grooming and group-level cooperation in white-faced capuchin monkeys (*Cebus capucinus*): insights from social networks. *American Journal of Primatology*, **73**, 821–833.

Croft, D. P., Arrowsmith, B. J., Bielby, J., Skinner, K., White, E., Couzin, I. D., et al. (2003). Mechanisms underlying shoal composition in the Trinidadian guppy (*Poecilia reticulata*). *Oikos*, **100**, 429–438.

Croft, D. P., Edenbrow, M., Darden, S. K., Ramnarine, I. W., van Oosterhout, C., and Cable, J. (2011). Effect of gyrodactylid ectoparasites on host behaviour and social network structure in guppies *Poecilia reticulata*. *Behavioral Ecology and Sociobiology*, **65**, 2219–2227.

Croft, D. P., Hamilton, P. B., Darden, S. K., Jacoby, D. M. P., James, R., Bettaney, E. M., et al. (2012). The role of relatedness in structuring the social network of a wild guppy population. *Oecologia*, **170**, 955–963.

Croft, D. P., James, R., and Krause, J. (2008). *Exploring animal social networks*. Princeton University Press, Princeton, NJ.

Croft, D. P., James, R., Thomas, P. O. R., Hathaway, C., Mawdsley, D., Laland K. N., et al. (2006). Social structure and co-operative interactions in a wild population of guppies (*Poecilia reticulata*). *Behavioral Ecology and Sociobiology*, **59**, 644–650.

Croft, D. P., James, R., Ward, A. J. W., Botham, M. S., Mawdsley, D., and Krause, J.. (2005). Assortative interactions and social networks in fish. *Oecologia*, **143**, 211–219.

Croft, D. P., Krause, J., Darden, S. K., Ramnarine, I. W., Faria, J. J., and James, R. (2009). Behavioural trait assortment in a social network: patterns and implications. *Behavioral Ecology and Sociobiology*, **63**, 1495–1503.

Croft, D. P., Krause, J., and James, R. (2004). Social networks in the guppy (*Poecilia reticulata*). *Proceedings of the Royal Society B: Biological Sciences*, **271**, S516–S519.

Croft, D. P., Madden, J. R., Franks, D. W., and James, R. (2011). Hypothesis testing in animal social networks. *Trends in Ecology and Evolution*, **26**, 502–507.

Crook, J. H., and Butterfield, P. A. (1970). Gender role in the social system of Quelea. In: Crook, J. H., ed. *Social behaviour in birds and mammals: essays on the social ethology of animals and man*, pp. 211–248. Academic Press, London.

Cross, P. C., Creech, T. G., Ebinger, M. R., Heisey, D. M., Irvine, K. M., and Scott, C. (2012). Wildlife contact analysis: emerging methods, questions, and challenges. *Behavioral Ecology and Sociobiology*, **66**, 1437–1447.

Cross, P. C., Creech, T. G., Ebinger, M. R., Manlove, K., Irvine, K., Henningsen, J., et al. (2013). Female elk contacts are neither frequency nor density dependent. *Ecology*, **94**, 2076–2086.

Cross, P. C., Johnson, P. L. F., Lloyd-Smith, J. O., and Getz, W. M. (2007). Utility of R_0 as a predictor of disease invasion in structured populations. *Journal of the Royal Society Interface*, **4**, 315–324.

Cross, P. C., Lloyd-Smith, J. O., Bowers, J. A., Hay, C. T., Hofmeyr, M., and Getz, W. M. (2004). Integrating association data and disease dynamics in a social ungulate: bovine tuberculosis in African buffalo. *Annales Zoologici Fennici*, **41**, 879–892.

Cross, P. C., Lloyd-Smith, J. O., and Getz, W. M. (2005). Disentangling association patterns in fission–fusion societies using African buffalo as an example. *Animal Behaviour*, **69**, 499–506.

Cyr, H., and Cyr, I. (2003). Temporal scaling of temperature variability from land to oceans. *Evolutionary Ecology Research*, **5**, 1183–1197.

Dabelsteen, T. (2005). Public, private or anonymous? Facilitating and countering eavesdropping. In: McGregor, P. K., ed. *Animal communication networks*, pp. 38–62. Cambridge University Press, Cambridge.

Dabelsteen, T., McGregor, P. K., Lampe, H. M., Langmore, N. E., and Holland, J. (1998). Quiet song in song birds: an overlooked phenomenon. *Bioacoustics*, **9**, 89–106.

Dall, S. R., Giraldeau, L. A., Olsson, O., McNamara, J. M., and Stephens, D. W. (2005). Information and its use by animals in evolutionary ecology. *Trends in Ecology and Evolution*, **20**, 187–193.

Danon, L., Ford, A. P., House, T., Jewell, C. P., Keeling, M. J., Roberts, G. O., et al. (2011). Networks and the epidemiology of infectious disease. *Interdisciplinary Perspectives in Infectious Diseases*, **2011**, 284909.

Darden, S. K., James, R., Ramnarine, I. W., and Croft, D. P. (2009). Social implications of the battle of the sexes: sexual harassment disrupts female sociality and social recognition. *Proceedings of the Royal Society B: Biological Sciences*, **276**, 2651–2656.

Darwin, C. (1859). *On the origin of species*. John Murray, London.

Darwin, C. (1871). *The descent of man and selection in relation to sex*. John Murray, London.

Daura-Jorge, F., Cantor, M., Ingram, S., Lusseau, D., and Simões-Lopes, P. (2012). The structure of a bottlenose dolphin society is coupled to a unique foraging cooperation with artisanal fishermen. *Biology Letters*, **8**, 702–705.

David, H. A. (1987). Ranking from unbalanced paired-comparison data. *Biometrika*, **74**, 432–436.

Dávid-Barrett, T., and Dunbar, R. I. M. (2012). Cooperation, behavioural synchrony and status in social networks. *Journal of Theoretical Biology*, **308**, 88–95.

Davies, N. B., and Hartley, I. R. (1996). Food patchiness, territory overlap and social systems: an experiment with dunnocks *Prunella modularis*. *Journal of Animal Ecology*, **65**, 837–846.

Davies, N. B., and Lundberg, A. (1984). Food distribution and a variable mating system in the dunnock, *Prunella modularis*. *Journal of Animal Ecology*, **53**, 895–912.

Davis, A. R., Corl, A., Surget-Groba, Y., and Sinervo, B. (2011). Convergent evolution of kin-based sociality in a lizard. *Proceedings of the Royal Society B: Biological Sciences*, **278**, 1507–1514.

Davis, K. M., and Burghardt, G. M. (2011). Turtles (*Pseudemys nelsoni*) lean about visual cues indicating food from experienced turtles. *Journal of Comparative Psychology*, **125**, 404–410.

Dawkins, R. (1976). *The selfish gene*. Oxford University Press, Oxford.

De, P., Singh, A. E., Wong, T., Yacoub, W., and Jolly, A. M. (2004). Sexual network analysis of a gonorrhoea outbreak. *Sexually Transmitted Infections*, **80**, 280–285.

de Silva, S., Ranjeewa, A. D., and Kryazhimskiy, S. (2011). The dynamics of social networks among female Asian elephants. *BMC Ecology*, **11**, 1375–1389.

de Vries, H. (1998). Finding a dominance order most consistent with a linear hierarchy: a new procedure and review. *Animal Behaviour*, **55**, 827–843.

DeBlois, E. M., and Rose, G. A. (1996). Cross-shoal variability in the feeding habits of migrating Atlantic cod (*Gadus morhua*). *Oecologia*, **108**, 192–196.

DeDeo, S., Krakaeur, D. C., and Flack, J. C. (2010). Inductive game theory and the dynamics of animal conflict. *PLoS Computational Biology*, **6**, e1000782.

Deecke, V. B., Barrett-Lennard, L., Spong, P., and Ford, J. (2010). The structure of stereotyped calls reflects kinship and social affiliation in resident killer whales (*Orcinus orca*). *Naturwissenschaften*, **97**, 513–518.

Deecke, V. B., Ford, J. K. B., and Slater, P. J. B. (2005). The vocal behaviour of mammal-eating killer whales: communicating with costly calls. *Animal Behaviour*, **69**, 395–405.

Deecke, V. B., Slater, P. J. B., and Ford, J. K. B. (2002). Selective habituation shapes acoustic predator recognition in harbour seals. *Nature*, **420**, 171–173.

DeNardo, D. F., and Licht, P. (1993). Effects of corticosterone on social behavior of male lizards. *Hormones and Behavior*, **27**, 184–199.

DeNardo, D. F., and Sinervo, B. (1994). Effects of steroid hormone interaction on activity and home-range size of male lizards. *Hormones and Behavior*, **28**, 273–287.

Dey, C. J., Reddon, A. R., O'Connor, C. M., and Balshine, S. (2013). Network structure is related to social conflict in a cooperatively breeding fish. *Animal Behaviour*, **85**, 395–402.

Dingemanse, N. J., Both, C., van Noordwijk, A. J., Rutten, A. L. and Drent, P. J. (2003). Natal dispersal and personalities in great tits (*Parus major*). *Proceedings of the Royal Society B: Biological Sciences*, **270**, 741–747.

Dingemanse, N. J., Dochtermann, N., and Nakagawa, S. (2012). Defining behavioural syndromes and the role of 'syndrome deviation' in understanding. *Behavioral Ecology and Sociobiology*, **66**, 1543–1548.

Dingemanse, N. J., and Wolf, M. (2010). Recent models for adaptive personality differences: a review. *Philosophical Transactions of the Royal Society B: Biological Sciences*, **365**, 3947–3958.

Diniz-Filho, J. A. F., and Bini, L. M. (2012). Thirty-five years of spatial autocorrelation analysis in population genetics: an essay in honour of Robert Sokal (1926–2012). *Biological Journal of the Linnean Society*, **107**, 721–736.

Doligez, B., Daniel, G., Warin, P., Pärt, T., Gustafsson, L., and Réale, D. (2012). Estimation and comparison of heritability and parent–offspring resemblance in dispersal probability from capture–recapture data using different methods: the Collared Flycatcher as a case study. *Journal of Ornithology*, **152**, 539–554.

Donohue, K. (2003). The influence of neighbor relatedness on multilevel selection in the Great Lakes sea rocket. *American Naturalist*, **162**, 77–92.

Donohue, K. (2004). Density-dependent multilevel selection in the Great lakes sea rocket. *Ecology*, **85**, 180–191.

Double, M. C., Peakall, R., Beck, N. R., and Cockburn, A. (2005). Dispersal, philopatry, and infidelity: dissecting local genetic structure in superb fairy-wrens (*Malurus cyaneus*). *Evolution*, **59**, 625–635.

Dow, M. M., and deWaal, F. B. M. (1989). Assignment methods for the analysis of network subgroup interactions. *Social Networks*, **11**, 237–255.

Drent, P. J. (2003). *The functional ethology of territoriality in the great tit (*Parus major *L.)*. University of Groningen, Groningen.

Drewe, J. A. (2010). Who infects whom? Social networks and tuberculosis transmission in wild meerkats. *Proceedings of the Royal Society B: Biological Sciences*, **277**, 633–642.

Drewe, J. A., Eames, K. T. D., Madden, J. R., and Pearce, G. P. (2011). Integrating contact network structure into tuberculosis epidemiology in meerkats in South Africa: implications for control. *Preventive Veterinary Medicine*, **101**, 113–120.

Drewe, J. A., Madden, J. R., and Pearce, G. P. (2009). The social network structure of a wild meerkat population: 1. Inter-group interactions. *Behavioral Ecology and Sociobiology*, **63**, 1295–1306.

Drewe, J. A., O'Connor, H., Weber, N., McDonald, R. A., and Delahay, R. J. (2013). Patterns of direct and indirect contact among cattle and badgers naturally infected with tuberculosis. *Epidemiology and Infection*, **141**, 1467–1475.

Drews, C. (1993). The concept and definition of dominance in animal behaviour. *Behaviour*, **125**, 283–313.

Dube, C., Ribble, C., Kelton, D., and McNab, B. (2011). Introduction to network analysis and its implications for animal disease modelling. *Revue scientifique et technique (International Office of Epizootics)*, **30**, 425–436.

Duckett, P. E., Morgan, M. H., and Stow, A. J. (2012). Tree-dwelling populations of the skink *Egernia striolata* aggregate in groups of close kin. *Copeia*, **2012**, 130–134.

Duffield, G. A., and Bull, C. M. (2002). Stable social aggregations in an Australian lizard, *Egernia stokesii*. *Naturwissenschaften*, **89**, 424–427.

Dufour, V., Sueur, C., Whiten, A., and Buchanan-Smith, H. M. (2011). The impact of moving to a novel environment on social networks, activity and wellbeing in two new world primates. *American Journal of Primatology*, **73**, 802–811.

Dugatkin, L. A. (1991). Dynamics of the tit-for-tat strategy during predator inspection in the guppy (*Poecilia reticulata*). *Behavioral Ecology and Sociobiology*, **29**, 127–132.

Dugatkin, L. A. (1992). Sexual selection and imitation: females copy the mate choice of others. *American Naturalist*, **139**, 1384–1389.

Dugatkin, L. A. (1997). *Cooperation among animals: an evolutionary perspective*. Oxford University Press, Oxford.

Dunbar, R. I. M. (1988). *Primate social systems*. Cornell University Press, Ithaca, NY.

Dunbar, R. I. M. (1991). Functional significance of social grooming in primates. *Folia Primatologica*, **57**, 121–131.

Dunnet, G. M., Ollason, J. C., and Anderson, A. (1979). A 28-year study of breeding fulmars *Fulmarus glacialis* in Orkney. *Ibis*, **121**, 293–300.

Dynel, M. (2011). Revisiting Goffman's postulates on participant statuses in verbal interaction. *Language and Linguistics Compass*, **5**, 454–465.

Dzieweczynski, T. L., Sullivan, K. R., Forrette, L. M., and Hebert, O. L. (2012). Repeated recent aggressive encounters do not affect behavioral consistency in male Siamese fighting fish. *Ethology*, **118**, 351–359.

Eberhard, W. (1996). *Female control: sexual selection by cryptic female choice*. Princeton University Press, Princeton, NJ.

Ebrahimi, M., and Bull, C. M. (2012). Food supplementation reduces post-release dispersal during simulated translocations of the endangered pygmy bluetongue lizard *Tiliqua adelaidensis*. *Endangered Species Research*, **18**, 169–178.

Ebrahimi, M., Fenner, A. L., and Bull, C. M. (2012). Lizard behaviour suggests a new design for artificial burrows. *Wildlife Research*, **39**, 295–300.

Edenbrow, M., and Croft, D. P. (2011). Behavioural types and life history strategies during ontogeny in the mangrove killifish, *Kryptolebias marmoratus*. *Animal Behaviour*, **82**, 731–741.

Edenbrow, M., and Croft, D. P. (2013). Environmental and genetic effects shape the development of personality traits in the mangrove killifish *Kryptolebias marmoratus*. *Oikos*, **122**, 667–681.

Edenbrow, M., Darden, S. K., Ramnarine, I. W., Evans, J. P., James, R., and Croft, D. P. (2011). Environmental effects on social interaction networks and male reproductive behaviour in guppies, *Poecilia reticulata*. *Animal Behaviour*, **81**, 551–558.

Efferson, C., Lalive, R., and Fehr, E. (2008). The coevolution of cultural groups and ingroup favoritism. *Science*, **321**, 1844–1849.

Ehman, K. D., and Scott, M. E. (2001). Urinary odour preferences of MHC-congenic female mice, *Mus domesticus*: implications for kin recognition and detection of parasitized males. *Animal Behaviour*, **62**, 781–789.

Ekman, J. (1989). Ecology of non-breeding social systems of Parus. *The Wilson Bulletin*, **101**, 263–288.

Eldakar, O. T., Wilson, D. S., Dlugos, M. J., and Pepper, J. W. (2010). The role of multilevel selection in the evolution of sexual conflict in the water strider *Aquarius remigis*. *Evolution*, **64**, 3183–3189.

Elena, V. P., Thomas, C. G., Patricia, G. P., and Paul, F. D. (1999). Patch size and composition of social groups in wintering tufted titmice. *The Auk*, **116**, 1152–1155.

Ellegren, H. (2008). Sequencing goes 454 and takes large-scale genomics into the wild. *Molecular Ecology*, **17**, 1629–1631.

Ellen, E. D., Visscher, J., van Arendonk, J. A. M., and Bijma, P. (2008). Survival of laying hens: genetic parameters for direct and associative effects in three purebred layer lines. *Poultry Science*, **87**, 233–239.

Elo, A. E. (1978). *The rating of chess players, past and present*. Arco, New York.

Emlen, S. T. (1994). Benefits, constraints, and the evolution of the family. *Trends in Ecology and Evolution*, **9**, 282–285.

Emlen, S. T., and Oring, L. W. (1977). Ecology, sexual selection, and the evolution of mating systems. *Science*, **197**, 215–223.

Emmering, Q. C., and Schmidt, K. A. (2011). Nesting songbirds assess spatial heterogeneity of predatory chipmunks by eavesdropping on their vocalizations. *Journal of Animal Ecology*, **18**, 1305–1312.

Enoksson, B. (1988). Age and sex-related differences in dominance and foraging behaviour of nuthatches *Sitta europaea*. *Animal Behaviour*, **36**, 231–238.

Erwin, N., and Erwin, J. (1976). Social density and aggression in captive groups of pigtail monkeys (*Macaca nemestrina*). *Applied Animal Ethology*, **2**, 265–269.

Estevez, I., Keeling, L. J., and Newberry, R. C. (2003). Decreasing aggressions with increasing group size in young domestic fowl. *Applied Animal Behaviour Science*, **84**, 213–218.

Ewbank, R., Meese, G. B., and Cox, J. E. (1974). Individual recognition and the dominance hierarchy in the domesticated pig: the role of sight. *Animal Behaviour*, **22**, 473–480.

Ezenwa, V. O. (2004). Host social behavior and parasitic infection: a multifactorial approach. *Behavioral Ecology*, **15**, 446–454.

Faria, J. J., Dyer, J. R. G., Clement, R. O., Couzin, I. D., Holt, N., Ward, A. J. W., et al. (2010). A novel method for investigating the collective behaviour of fish: introducing 'Robofish'. *Behavioral Ecology and Sociobiology*, **64**, 1211–1218.

Farine, D. R., Garroway, C. J., and Sheldon, B. C. (2012). Social network analysis of mixed-species flocks: exploring the structure and evolution of interspecific social behaviour. *Animal Behaviour*, **84**, 1271–1277.

Farine, D. R., and Milburn, P. J. (2013). Social organisation of thornbill-dominated mixed-species flocks using social network analysis. *Behavioral Ecology and Sociobiology*, **67**, 321–330.

Farwell, M., and McLaughlin, R. L. (2009). Alternative foraging tactics and risk taking in brook charr (*Salvelinus fontinalis*). *Behavioral Ecology*, **20**, 913–921.

Faust, K. (2010). A puzzle concerning triads in social networks: graph constraints and the triad census. *Social Networks*, **32**, 221–233.

Fedigan, L. M. (1972). Roles and activities of male geladas (*Theropithecus gelada*). *Behaviour*, **41**, 82–90.

Fehl, K., van der Post, D. J., and Semmann, D. (2011). Co-evolution of behaviour and social network structure promotes human cooperation. *Ecology Letters*, **14**, 546–551.

Feigenbaum, C., and Naug, D. (2010). The influence of social hunger on food distribution and its implications for disease transmission in a honeybee colony. *Insectes Sociaux*, **57**, 217–222.

Fenner, A. L., and Bull, C. M. (2010). The use of scats as social signals in a solitary, endangered scincid lizard, *Tiliqua adelaidensis*. *Wildlife Research*, **37**, 582–587.

Fenner, A. L., and Bull, C. M. (2011). Central-place territorial defence in a burrow-dwelling skink: aggressive responses to conspecific models in pygmy bluetongue lizards. *Journal of Zoology*, **283**, 45–51.

Fenner, A. L., Godfrey, S. S., and Bull, C. M. (2011). Using social networks to deduce whether residents or dispersers spread parasites in a lizard population. *Journal of Animal Ecology*, **80**, 835–843.

Fenner, A. L., Pavey, C. R., and Bull, C. M. (2012). Behavioural observations and use of burrow systems by an endangered Australian arid-zone lizard, Slater's skink (*Liopholis slateri*). *Australian Journal of Zoology*, **60**, 127–132.

Fenton, M. B. (2003). Eavesdropping on the echolocation and social calls of bats. *Mammal Review*, **33**, 193–204.

Fernald, R. D., and Maruska, K. P. (2012). Social information changes the brain. *Proceedings of the National Academy of Sciences*, **109**, 17194–17199.

Ferrari, M. J., Bansal, S., Meyers, L. A., and Bjørnstad, O. N. (2006). Network frailty and the geometry of herd immunity. *Proceedings of the Royal Society B: Biological Sciences*, **273**, 2743–2748.

Festa-Bianchet, M. (1989). Individual differences, parasites, and the costs of reproduction for bighorn ewes (*Ovis canadensis*). *The Journal of Animal Ecology*, **58**, 785–795.

Fewell, J. H. (2003). Social insect networks. *Science*, **301**, 1867–1870.

Fichtenberg, C. M., Muth, S. Q., Brown, B., Padian, N. S., Glass, T. A., and Ellen, J. M. (2009). Sexual network position and risk of sexually transmitted infections. *Sexually Transmitted Infections*, **85**, 493–498.

Flack, J. C. (2012). Multiple time-scales and the developmental dynamics of social systems. *Philosophical Transactions of the Royal Society B: Biological Sciences*, **367**, 1802–1810.

Flack, J. C., and de Waal, F., B. M. (2007). Context modulates signal meaning in primate communication. *Proceedings of the National Academy of Science*, **104**, 1581–1586.

Flack, J. C., de Waal, F. B. M., and Krakauer, D. C. (2005). Social structure, robustness, and policing cost in a cognitively sophisticated species. *The American Naturalist*, **165**, E126–E139.

Flack, J. C., Girvan, M., de Waal, F. B. M., and Krakauer, D. C. (2006). Policing stabilizes construction of social niches in primates. *Nature*, **439**, 426–429.

Flack, J. C., and Krakaeur, D. C. (2006). Encoding power in communication networks. *The American Naturalist*, **168**, E87–E102.

Flack, J. C., Krakauer, D. C., and de Waal, F. B. M. (2005). Robustness mechanisms in primate societies: a perturbation study. *Proceedings of the Royal Society B: Biological Sciences*, **272**, 1091–1099.

Fleagle, J. G., Janson, C. H., and Reed, K., eds. (1999). *Primate communities*. Cambridge University Press, Cambridge.

Flesch, J. S., Duncan, M. G., Pascoe, J. H., and Mulley, R. C. (2009). A simple method of attaching GPS tracking devices to free-ranging lace monitors (*Varanus varius*). *Herpetological Conservation and Biology*, **4**, 411–414.

Fletcher, J. A., and Doebeli, M. (2009). A simple and general explanation for the evolution of altruism. *Proceedings of the Royal Society B: Biological Sciences*, **276**, 13–19.

Font, E., Barbosa, D., Sampedro, C., and Carazo, P. (2012). Social behavior, chemical communication and adult neurogenesis: studies of scent mark function in *Podarcis* wall lizards. *General and Comparative Endocrinology*, **177**, 9–17.

Foote, A. D. (2008). Mortality rate acceleration and post-reproductive lifespan in matrilineal whale species. *Biology Letters*, **4**, 189–191.

Foote, J. R., Fitzsimmons, L. P., Mennill, D. J., and Ratcliffe, L. M. (2010). Black-capped chickadee dawn choruses are interactive communication networks. *Behaviour*, **147**, 1219–1248.

Ford, J. K. B. (1991). Vocal traditions among resident killer whales (*Orcinus orca*) in coastal waters of British Columbia. *Canadian Journal of Zoology*, **69**, 1454–1483.

Ford, J. K. B. (2002a). Dialects. In: Perrin, W. F., Wursig, B., and Thewissen, J. G. M., eds. *The encyclopedia of marine mammals* pp. 322–323. Academic Press, New York.

Ford, J. K. B. (2002b). Killer whales. In: Perrin, W. F., Wursig, B., Thewissen, J. G. M., eds. *The encyclopedia of marine mammals*, pp. 669–676. Academic Press, New York.

Ford, J. K. B., and Ellis, G. M. (2006). Selective foraging by fish-eating killer whales, *Orcinus orca*, in British Columbia. *Marine Ecology Progress Series*, **316**, 185–199.

Formica, V. A., Augat, M. E., Barnard, M. E., Butterfield, R. E., Wood, C. W., and Brodie, E. D., III. (2010). Using home range estimates to construct social networks for species with indirect behavioral interactions. *Behavioral Ecology and Sociobiology*. **64**, 1199–1208.

Formica, V. A., McGlothlin J. W., Wood, C. W., Augat, M. E., Butterfield, R. E., Barnard, M. E., et al. (2011). Phenotypic assortment mediates the effect of social selection in a wild beetle population. *Evolution*, **65**, 2771–2781.

Formica, V. A., Wood, C. W., Larsen, W. B., Butterfield, R. E., Augat, M. E., Hougen, H. Y., et al. (2012). Fitness consequences of social network position in a wild population of forked fungus beetles (*Bolitotherus cornutus*). *Journal of Evolutionary Biology*. **25**, 130–137.

Fortin, M.-J., and Dale, M. (2005). *Spatial analysis: a guide for ecologists*. Cambridge University Press, Cambridge.

Fortunato, S. (2010). Community detection in graphs. *Physics Reports*, **486**, 75–174.

Foster, E. A., Franks, D. W., Morrell, L. J., Balcomb, K. C., Parsons, K. M., van Ginneken, A., et al. (2012). Social network correlates of food availability in an endangered population of killer whales, *Orcinus orca*. *Animal Behaviour*, **83**, 731–736.

Foster, K. R., Wenseleers, T., and Ratnieks, F. L. W. (2006). Kin selection is the key to altruism. *Trends in Ecology and Evolution*, **21**, 57–60.

Fowler, J. H., and Christakis, N. A. (2010). Cooperative behavior cascades in human social networks. *Proceedings of the National Academy of Sciences*, **107**, 5334–5338.

Fowler, J. H., Dawes, C. T., and Christakis, N. A. (2009). Model of genetic variation in human social networks. *Proceedings of the National Academy of Sciences*, **106**, 1720–1724.

Franks, D. W., James, R., Noble, J., and Ruxton, G. D. (2009). A foundation for developing a methodology for social network sampling. *Behavioral Ecology and Sociobiology*, **63**, 1079–1088.

Franks, D. W., Ruxton, G. D., and James, R. (2010). Sampling animal association networks with the gambit of the group. *Behavioral Ecology and Sociobiology*, **64**, 493–503.

Franz, M., and Nunn, C. L. (2009). Network-based diffusion analysis: a new method for detecting social learning. *Proceedings of the Royal Society B: Biological Sciences*, **276**, 1829–1836.

Franz, M., and Nunn, C. L. (2010). Investigating the impact of observation errors on the statistical performance of network-based diffusion analysis. *Learning and Behavior*, **38**, 235–242.

Freeberg, T. M., Dunbar, R. I. M., and Ord, T. J. (2012). Social complexity as a proximate and ultimate factor in communicative complexity. *Philosophical Transactions of the Royal Society B: Biological Sciences*, **367**, 1785–1801.

Freeman, L. C. (1979). Centrality in social networks: conceptual clarification. *Social Networks*, 1, 215–239.

Fréon, P., Coetzee, J. C., van der Lingen, C. D., Connell, A. D., O'Donoghue, S. H., Roberts, M. J., et al. (2010). A review and tests of hypotheses about causes of the KwaZulu–Natal sardine run. *African Journal of Marine Science*, **32**, 449–479.

Frère, C. H., Krützen, M., Mann, J., Connor, R.C., Bejder, L., and Sherwin, W.B. (2010). Social and genetic interactions drive fitness variation in a free-living dolphin population. *Proceedings of the National Academy of Sciences*, **107**, 19949–19954.

Fridman, N., and Kaminka, G. A. (2007). Towards a cognitive model of crowd behavior based on social comparison theory. *Proceedings of the National Conference on Artificial Intelligence*, **22**, 731–737.

Fripp, D., Owen, C., Quintana-Rizzo, E., Shapiro, A., Buckstaff, K., Jankowski, K., et al. (2005). Bottlenose dolphin (*Tursiops truncatus*) calves appear to model their signature whistles on the signature whistles of community members. *Animal Cognition*, **8**, 17–26.

Frost, A. J., Winrow-Giffen, A., Ashley, P. J., and Sneddon, L. U. (2007). Plasticity in animal personality traits: does prior experience alter the degree of boldness? *Proceedings of the Royal Society B: Biological Sciences*, **274**, 333–339.

Frost, C. L., and Bergmann, P. J. (2012). Spatial distribution and habitat utilization of the zebra-tailed lizard (*Callisaurus draconoides*). *Journal of Herpetology*, **46**, 203–208.

Fruteau, C., Voelkl, B., van Damme, E., and Noë, R. (2009). Supply and demand determine the market value of food providers in vervet monkeys. *Proceedings of the National Academy of Sciences*, 106, 12007–12012.

Fuller, S. J., Bull, C. M., Murray, K., and Spencer, R. J. (2005). Clustering of related individuals in a population of the Australian lizard, *Egernia frerei*. *Molecular Ecology*, **14**, 1207–1213.

Furuichi, T. (1985). Inter-male associations in a wild Japanese macaque troop on Yakushima Island, Japan. *Primates*, **26**, 219–237.

Fushing, H., Jordàb, Ò., Beisnerd, B., and McCowan, B. (2013). Computing systemic risk using multiple behavioral and keystone networks: the emergence of a crisis in primate societies and banks. *International Journal of Forecasting*, **30**, 797–806.

Galeano, J., Pastor, J. M., and Iriondo, J. M. (2009). Short communication: weighted-interaction nestedness estimator (WINE): a new estimator to calculate over frequency matrices. *Environmental Modelling and Software*, **24**, 1342–1346.

Galef, B. G., Jr. (1988). Imitation in animals: history, definition and interpretation of the data from the psychological laboratory. In: Galef, B. G., Jr, and Zentall, T. R., eds. *Social learning: psychological and biological perspectives*, pp. 3–28. Erlbaum, Hillsdale.

Gamerman, D. (1997). *Markov chain Monte Carlo*. Chapman and Hall, London.

Gamerman, D., and Lopes, H. F. (2006). *Markov chain Monte Carlo: stochastic simulation for Bayesian inference*, 2nd edn. Chapman and Hall/CRC Press, Boca Raton, FL.

Gammell, M. P., de Vries, H., Jennings, D. J., Carlin, C. M., and Hayden, T. J. (2003). David's score: a more appropriate dominance ranking method than Clutton-Brock et al.'s index. *Animal Behaviour*, **66**, 601–605.

Gardner, A., and West, S. A. (2007). Social evolution: the decline and fall of genetic kin recognition. *Current Biology*, **17**, R810–R812.

Gardner, A., and West, S. A. (2009). Greenbeards. *Evolution*, **64**, 25–38.

Gardner, M. G., Bull, C. M., Cooper, S. J. B., and Duffield, G. A. (2001). Genetic evidence for a family structure in stable social aggregations of the Australian lizard *Egernia stokesii*. *Molecular Ecology*, **10**, 175–183.

Gardner, M. G., Bull, C. M., Fenner, A. L., Murray, K., and Donnellan, S. C. (2007). Consistent social structure within aggregations of the Australian lizard *Egernia stokesii* across seven disconnected rocky outcrops. *Journal of Ethology*, **25**, 263–270.

Garland, E. C., Goldizen, A. W., Rekdahl, M. L., Constantine, R., Garrigue, C., Hauser, N. D., et al. (2011). Dynamic horizontal cultural transmission of humpback whale song at the ocean basin scale. *Current Biology*, **21**, 687–691.

Garroway, C. J., Bowman, J., and Wilson, P. J. (2012). Complex social structure of southern flying squirrels is related to spatial proximity but not kinship. *Behavioral Ecology and Sociobiology*, **67**, 113–122.

Garvin, M. R., Saitoh, K., and Gharrett, A. J. (2010). Application of single nucleotide polymorphisms to non-model species: a technical review. *Molecular Ecology Resources*, **10**, 915–934.

Gasparini, C., Serena, G., and Pilastro, A. (2013). Do unattractive friends make you look better? Context-dependent male mating preferences in the guppy. *Proceedings of the Royal Society B: Biological Sciences*, **280**, 20123072.

Gavier-Widen, D., Duff, J. P., and Meredith, A., eds. (2012). *Infectious diseases of wild mammals and birds in europe*. Wiley-Blackwell, Oxford.

Gazda, S. K., Connor, R. C., Edgar, R. K., and Cox, F. (2005). A division of labour with role specialization in group-hunting bottlenose dolphins (*Tursiops truncatus*) off Cedar Key, Florida. *Proceedings of the Royal Society B: Biological Sciences*, **272**, 135–140.

Gelman, A. (2006). Prior distributions for variance parameters in hierarchical models. *Bayesian Analysis*, **1**, 515–533.

Gelman, A., Carlin, J. B., Stern, H. S., and Rubin, D. B. (2004). *Bayesian data analysis*. Chapman and Hall/CRC Press, Boca Raton, FL.

George, J. C., Bada, J., Zeh, J., Scott, L., Brown, S. E., O'Hara, T., et al. (1999). Age and growth estimates of bowhead whales (*Balaena mysticetus*) via aspartic acid racemization. *Canadian Journal of Zoology*, **77**, 571–580.

Gero, S., Bejder, L., Whitehead, H., Mann, J., and Connor, R. C. (2005). Behaviourally specific preferred associations in bottlenose dolphins, *Tursiops* spp. *Canadian Journal of Zoology*, **83**, 1566–1573.

Gero, S., Engelhaupt, D., Rendell, L., and Whitehead, H. (2009). Who cares? Between-group variation in alloparental caregiving in sperm whales. *Behavioral Ecology*, **20**, 838–843.

Gero, S., Engelhaupt, D., and Whitehead, H. (2008). Heterogeneous associations within a sperm whale unit reflect pairwise relatedness. *Behavioral Ecology and Sociobiology*, **63**, 143–151.

Gero, S., Gordon, J., and Whitehead, H. (2013). Calves as social hubs: dynamics of the social network within sperm whale units. *Proceedings of the Royal Society B: Biological Sciences*, **280**, 20131113.

Gero, S., Milligan, M., Rinaldi, C., Francis, P., Gordon, J., Carlson, C., et al. (2013). Behavior and social structure of the sperm whales of Dominica, West Indies. *Marine Mammal Science*, **30**, 905–922.

Getz, W. M., Lloyd-Smith, J. O., Cross, P. C., Bar-David, S., Johnson, P. L., Porco, T. C., et al. (2006). Modeling the invasion and spread of contagious diseases in heterogeneous populations. *DIMACS: Series in Discrete Mathematics and Theoretical Computer Science*, **71**, 113.

Ghani, A. C., Donnelly, C. A., and Garnett, G. P. (1998). Sampling biases and missing data in explorations of sexual partner networks for the spread of sexually transmitted diseases. *Statistics in Medicine*, **17**, 2079–2097.

Gibbons, J. W., and Andrews, K. M. (2004). PIT tagging: simple technology at its best. *BioScience*, **54**, 447–454.

Girvan, M., and Newman, M. E. J. (2002). Community structure in social and biological networks. *Proceedings of the National Academy of Sciences*, **99**, 7821–7826.

Godfrey, S. S. (2013). Networks and the ecology of parasite transmission: a framework for wildlife parasitology. *International Journal for Parasitology: Parasites and Wildlife*, **2**, 235–245.

Godfrey, S. S., Bradley, J. K., Sih, A., and Bull, C. M. (2012). Lovers and fighters in sleepy lizard land: where do aggressive males fit in a social network? *Animal Behaviour*, **83**, 209–215.

Godfrey, S. S., Bull, C. M., James, R., and Murray, K. (2009). Network structure and parasite transmission in a group living lizard, the gidgee skink, *Egernia stokesii*. *Behavioral Ecology and Sociobiology*, **63**, 1045–1056.

Godfrey, S. S., Moore, J. A., Nelson, N. J., and Bull, M. (2010). Social network structure and parasite infection patterns in a territorial reptile, the tuatara (*Sphenodon punctatus*). *International Journal for Parasitology*, **40**, 1575–1585.

Godfrey, S. S., Sih, A., and Bull, C. M. (2013). The response of a sleepy lizard social network to altered ecological conditions. *Animal Behaviour*, **86**, 763–772.

Godin, J.-G. J., and Dugatkin, L. A. (1996). Female mating preference for bold males in the guppy, *Poecilia reticulata*. *Proceedings of the National Academy of Sciences*, **93**, 10262–10267.

Gonzalez, A., Rayfield, B., and Lindo, Z. (2011). The disentangled bank: how loss of habitat fragments and disassembles ecological networks. *American Journal of Botany*, **98**, 503–516.

Goodale, E., Beauchamp, G., Magrath, R. D., Nieh, J. C., and Ruxton, G. D. (2010). Interspecific information transfer influences animal community structure. *Trends in Ecology and Evolution*, **25**, 354–361.

Goodall, J. (1968). *The behaviour of free-living chimpanzees in the Gombe Stream Reserve*. Baillière, Tindall and Cassell, London.

Goodall, J. (1986). *The chimpanzees of Gombe: patterns of behavior*. Belknap Press of Harvard University Press, Cambridge, MA.

Gordon, D. M., Goodwin, B. C., and Trainor, L. E. H. (1992). A parallel distributed model of the behaviour of ant colonies. *Journal of Theoretical Biology*, **156**, 293–307.

Gordon, D. M. (1996). The organization of work in social insect colonies. *Nature*, **380**, 121–124.

Gordon, D. M. (2010). *Ant encounters: interaction networks and colony behavior*. Princeton University Press, Princeton, NJ.

Gosden, T. P., and Svensson, E. I. (2008). Spatial and temporal dynamics in a geographic sexual selection mosaic. *Evolution*, **62**, 845–856.

Gosling, L. M. (1986). Evolution of the mating strategies in male antelopes. In: Rubenstein, D.I., and Wrangham, R.W., eds. *Ecological aspects of social evolution: birds and mammals*, pp. 244–281. Princeton University Press, Princeton, NJ.

Gowans, S., Würsig, B., and Karczmarski, L. (2007). The social structure and strategies of delphinids: predictions based on an ecological framework. *Advances in Marine Biology*, **53**, 195–294.

Grabowska-Zhang, A. M., Wilkin, T. A., and Sheldon, B. C. (2012). Effects of neighbor familiarity on reproductive success in the great tit (*Parus major*). *Behavioral Ecology*, **23**, 322–333.

Grafe, T. U. (2005). Anuran choruses as communication networks. In: McGregor, P. K., ed. *Animal communication networks*, pp. 277–299. Cambridge University Press, Cambridge.

Grandin, T. (2001). Cattle vocalizations are associated with handling and equipment problems at beef slaughter plants. *Applied Animal Behaviour Science*, **71**, 191–201.

Grandin, T. (2003). Transferring results of behavioral research to industry to improve animal welfare on the farm, ranch and the slaughter plant. *Applied Animal Behaviour Science*, **81**, 215–228.

Granovetter, M. S. (1973). The strength of weak ties. *American Journal of Sociology*, **78**, 1360–1380.

Grant, P. R. (1999). *Ecology and evolution of Darwin's finches*. Princeton University Press, Princeton, NJ.

Graves, B. M., and Duvall, D. (1987). An experimental study of aggregation and thermoregulation in prairie rattlesnakes (*Crotalus viridis viridis*). *Herpetologica*, **43**, 259–264.

Grear, D. A., Perkins, S. E., and Hudson, P. J. (2009). Does elevated testosterone result in increased exposure and transmission of parasites? *Ecology Letters*, **12**, 528–537.

Greig, E. I., and Pruett-Jones, S. (2010). Danger may enhance communication: predator calls alert females to male displays. *Behavioral Ecology*, **21**, 1360–1366.

Greiner, B., and Levati, M. V. (2005). Indirect reciprocity in cyclical networks: an experimental study. *Journal of Economic Psychology*, **26**, 711–731.

Griesser, M., Barnaby, J., Schneider, N. A., Figenschau, N., Wright, J., Griffith, S. C., et al. (2009). Influence of winter ranging behaviour on the social organization of a cooperatively breeding bird species, the apostlebird. *Ethology*, **115**, 888–896.

Griffin, R. H., and Nunn, C. L. (2012). Community structure and the spread of infectious disease in primate social networks. *Evolutionary Ecology*, **26**, 779–800.

Griffiths, S. W., and Magurran, A. E. (1999). Schooling decisions in guppies (*Poecilia reticulata*) are based on familiarity rather than kin recognition by phenotype matching. *Behavioral Ecology and Sociobiology*, **45**, 437–443.

Griffiths, S. W., and Ward, A. (2011). Social recognition of conspecifics. In: Brown, C., Laland K. N., and Krause, J., eds. *Fish cognition and behavior*, 2nd edn. Wiley-Blackwell, Oxford.

Groothuis, T. G. G., and Trillmich, F. (2011). Unfolding personalities: the importance of studying ontogeny. *Developmental Psychobiology*, **53**, 641–655.

Grosenick, L., Clement, T. S., and Fernald, R. D. (2007). Fish can infer social rank by observation alone. *Nature*, **445**, 429–432.

Groves, C. P. (2005). Order primates. In:Wilson, D. E., and Reeder, D. M., eds. *Mammal species of the world*, pp. 111–184. Johns Hopkins University Press, Baltimore.

Gueron, S.,Levin, S. A., and Rubenstein, D. I. (1996). The dynamics of herds: from individuals to aggregations. *Journal of Theoretical Biology*, **182**, 85–98.

Guilford, T. C., Meade, J., Freeman, R., Biro, D., Evans, T.,Bonadonna, F., et al. (2008). GPS tracking of the foraging movements of Manx Shearwaters *Puffinus puffinus* breeding on Skomer Island, Wales. *Ibis*, **150**, 462–473.

Guimarães, P. R., Jr, de Menezes, M. A., Baird, R. W., Lusseau, D., Guimarães, P., and dos Reis, S. F. (2007). Vulnerability of a killer whale social network to disease outbreaks. *Physical Review E*, **76**, 042901.

Gupta, S., Anderson, R. M., and May, R. M. (1989). Networks of sexual contacts: implications for the pattern of spread of HIV. *AIDS*, **3**, 807–818.

Gustafson, P., Hossain, S., and Macnab Ying, C. (2006). Conservative prior distributions for variance parameters in hierarchical models. *Canadian Journal of Statistics*, **34**, 377–390.

Guttridge, T. L., Gruber, S. H., DiBattista, J. D., Feldheim, K. A., Croft, D. P., Krause, S., et al. (2011). Assortative interactions and leadership in a free-ranging population of juvenile lemon sharks, *Negaprion brevirostris*. *Marine Ecology Progress Series*, **423**, 235–245.

Guttridge, T. L., Gruber, S. H., Krause, J., and Sims, D.W. (2010). Novel acoustic technology for studying free-ranging shark social behaviour by recording individuals' interactions. *PLoS ONE*, **5**, e9324.

Gyuris, E., Fero, O., and Barta, Z. (2012). Personality traits across ontogeny in firebugs, *Pyrrhocoris apterus*. *Animal Behaviour*, **84**, 103–109.

Habiba. (2013). *Finding critical individuals in dynamic networks*. PhD dissertation, University of Illinois at Chicago, Chicago.

Hagen, I. J., and Bull, C. M. (2011). Home ranges in the trees: radiotelemetry of the prehensile tailed skink, *Corucia zebrata*. *Journal of Herpetology*, **45**, 36–39.

Hall, M. L., and Magrath, R. D. (2007). Temporal coordination signals coalition quality. *Current Biology*, **17**, R406–407.

Hamede, R. K., Bashford, J., Jones, M., and McCallum, H. (2012). Simulating devil facial tumour disease outbreaks across empirically derived contact networks. *Journal of Applied Ecology*, **49**, 447–456.

Hamede, R. K., Bashford, J., McCallum, H., and Jones, M. (2009). Contact networks in a wild Tasmanian devil (*Sarcophilus harrisii*) population: using social network analysis to reveal seasonal variability in social behaviour and its implications for transmission of devil facial tumour disease. *Ecology Letters*, **12**, 1147–1157.

Hamilton, I. M., and Taborsky, M. (2005). Contingent movement and cooperation evolve under generalized reciprocity. *Proceedings of the Royal Society B: Biological Sciences*, **272**, 2259–2267.

Hamilton, W. D. (1963). Evolution of altruistic behavior. *American Naturalist*, **97**, 354–356.

Hamilton, W. D. (1964a). Genetical evolution of social behaviour. I. *Journal of Theoretical Biology*, **7**, 1–16.

Hamilton, W. D. (1964b). Genetical evolution of social behaviour. II. *Journal of Theoretical Biology*, **7**, 17–52.

Hamilton, W. D. (1971). Selection of selfish and altruistic behavior in some extreme models. In: Eisenberg, J. F., and Dillon, W. S., eds. *Man and beast: comparative social behavior*. Smithsonian Press, Washington, DC.

Hamilton, W. D., and Zuk, M. (1982). Heritable true fitness and bright birds: a role for parasites. *Science*, **218**, 384–387.

Hammerstein, P. (2002). Why is reciprocity so rare in social animals? A protestant appeal. In:Hammerstein, P., ed. *Genetic and cultural evolution of cooperation*. MIT Press, Cambridge, MA.

Hampson, K.,Dushoff, J., Cleaveland S., Haydon, D. T., and Kaare, M. (2009). Transmission dynamics and prospects for the elimination of canine rabies. *PLoS Biology*, **7**, e1000053.

Han, J. D.,Dupuy, D., Bertin, N., Cusick, M. E., and Vidal, M. (2005). Effect of sampling on topology predictions of protein–protein interaction networks. *Nature Biotechnology*, **23**, 839–844.

Handcock, R. N., Swain, D. L., Bishop-Hurley, G. J.,Patison, K. P.,Wark, T., Valencia, P., et al. (2009). Monitoring animal behaviour and environmental interactions using wireless sensor networks, GPS collars and satellite remote sensing. *Sensors*, **9**, 3586–3603.

Hanneman, R. A., and Riddle, M. (2005). *Introduction to social network methods*. <http://faculty.ucr.edu/~hanneman/>.

Harcourt, J. L., Sweetman, G., Manica, A., and Johnstone, R. A. (2010). Pairs of fish resolve conflicts over coordinated movement by taking turns. *Current Biology*, **20**, 156–160.

Harris, M. (1970). Territory limiting the size of the breeding population of the oystercatcher (*Haematopus ostralegus*)—a removal experiment. *The Journal of Animal Ecology*, **39**, 707–713.

Hartmann, E.,Christensen, J. W., and Keeling, L. J. (2009). Social interactions of unfamiliar horses during paired encounters: effect of pre-exposure on aggression level and so risk of injury. *Applied Animal Behaviour Science*, **121**, 214–221.

Hasegawa, T., and Hiraiwa, M. (1980). Social interactions of orphans observed in a free-ranging troop of Japanese macaques. *Folia Primatologica*, **33**, 129–158.

Hassall, S. A.,Ward, W. R., and Murray, R. D. (1993). Effects of lameness on the behaviour of cows during the summer. *Veterinary Record*, **132**, 578–580.

Hatchwell, B. J. (2010). Cryptic kin selection: kin structure in vertebrate populations and opportunities for kin-directed cooperation. *Ethology*, **116**, 203–216.

Hatchwell, B. J., Sharp, S. P., Beckerman, A. P., and Meade, J. (2013). Ecological and demographic correlates of helping behaviour in a cooperatively breeding bird. *Journal of Animal Ecology*, **82**, 486-494.

Heatwole, H., and Taylor, J. (1987). *Ecology of reptiles*. Surrey Beatty and Sons, Chipping Norton, Australia.

Hedrick, A. V., and Kortet, R. (2012). Sex differences in the repeatability of boldness over metamorphosis. *Behavioral Ecology and Sociobiology*, **66**, 1–6.

Heisler, I., and Damuth, J. (1987). A method for analyzing selection in hierarchically structured populations. *American Naturalist*, **130**, 582–602.

Helbing, D., Farkas, I., and Vicsek, T. (2000). Simulating dynamical features of escape panic. *Nature*, **407**, 487–490.

Helms, A. M., De Moraes, C. M., Tooker, J. F., and Mescher, M. C. (2013). Exposure of *Solidago altissima* plants to volatile emissions of an insect antagonist (*Eurosta solidaginis*) deters subsequent herbivory. *Proceedings of the National Academy of Sciences*, **110**, 199–204.

Hemelrijk, C. K. (2000). Towards the integration of social dominance and spatial structure. *Animal Behaviour*, **59**, 1035–1048.

Hemelrijk, C. K., and Kunz, H. (2005). Density distribution and size sorting in fish schools: an individual-based model. *Behavioral Ecology*, **16**, 178–187.

Hemelrijk, C. K., Wantia, J., and Gygax, L. (2005). The construction of dominance order: comparing performance of five methods using an individual-based model. *Behaviour*, **142**, 1037–1058.

Henry, L., Bourguet, C., Coulon, M., Aubry, C., and Hausberger, M. (2013). Sharing mates and nest boxes is associated with female friendship in European starlings, *Sturnus vulgaris*. *Journal of Comparative Psychology*, **127**, 1–13.

Henzi, P. S., Barrett, L., Gaynor, D., Greeff, J., Weingrill, T., and Hill, R. A. (2003). Effect of resource competition on the long-term allocation of grooming by female baboons: evaluating Seyfarth's model. *Animal Behaviour*, **66**, 931–938.

Henzi, P. S., Lusseau, D., Weingrill, T., van Schaik, C. P., and Barrett, L. (2009). Cyclicity in the structure of female baboon social networks. *Behavioral Ecology and Sociobiology*, **63**, 1015–1021.

Herbert-Read, J. E., Krause, S., Morrell, L. J., Schaerf, T. M., Krause, J., and Ward, A. J. W. (2013). The role of individuality in collective group movement. *Proceedings of the Royal Society B: Biological Sciences*, **280**, 20122564.

Herbert-Read, J. E., Perna, A., Mann, R. P., Schaerf, T. M., Sumpter, D. J., and Ward, A. J. W. (2011). Inferring the rules of interaction of shoaling fish. *Proceedings of the National Academy of Sciences*, **108**, 18726–18731.

Hews, D. K., Date, P., Hara, E., and Castellano, M. J. (2011). Field presentation of male secretions alters social display in *Sceloporus virgatus* but not *S. undulatus* lizards. *Behavioral Ecology and Sociobiology*, **65**, 1403–1410.

Heyes, C. M. (1994). Social learning in animals: categories and mechanisms. *Biological Reviews*, **69**, 207–231.

Hiby, L., and Lovell, P. (1990). Individual recognition of cetaceans: use of photo-identification and other techniques to estimate population parameters. *Report of the International Whaling Commission*, **Special Issue 12**, 42–43.

Hill, R. A., Bentley, R. A., and Dunbar, R. I. M. (2008). Network scaling reveals consistent fractal pattern in hierarchical mammalian societies. *Biology Letters*, **4**, 748–751.

Hillgarth, N. (1990). Parasites and female choice in the ring-necked pheasant. *American Zoologist*, **30**, 227–233.

Hinde, R. A. (1976). Interactions, relationships and social structure. *Man*, **11**, 1–17.

Hinde, R. A. (1983). A conceptual framework. In: Hinde, R. A., ed. *Primate social relationships*, pp. 1–7. Blackwell Scientific Publications, Oxford.

Hirsch, B. T., Prange, S., Hauver, S. A., and Gehrt, S. D. (2013). Genetic relatedness does not predict racoon social network structure. *Animal Behaviour*, **85**, 463–470.

Hoare, D. J., Krause, J., Ruxton, G. D., and Godin J.-G. J. (2000). The social organisation of free-ranging fish shoals. *Oikos*, **89**, 546–554.

Hobson, E. A.,Avery, M. L., and Wright, T. F. (2013). An analytical framework for quantifying and testing patterns of temporal dynamics in social networks. *Animal Behaviour*, **85**, 83–96.

Hock, K., and Fefferman, N. H. (2012). Social organization patterns can lower disease risk without associated disease avoidance or immunity. *Ecological Complexity*, **12**, 34–42.

Hoelzel, A. R. (1991). Killer whale predation on marine mammals at Punta Norte, Argentina; food sharing, provisioning and foraging strategy. *Behavioral Ecology and Sociobiology*, **29**, 197–204.

Hoem, S. A., Melis, C., Linnell, J. D. C., and Andersen, R. (2007). Fighting behaviour in territorial male roe deer *Capreolus capreolus*: the effects of antler size and residence. *European Journal of Wildlife Research*, **53**, 1–8.

Hoff, P. D. (2009). *A first course in Bayesian statistical methods*. Springer, New York.

Holekamp, K. E., Smith, J. E., Strelioff, C. C., Van Horn, R. C., and Watts, H. E. (2012). Society, demography and genetic structure in the spotted hyena. *Molecular Ecology*, **21**, 613–632.

Holland P. W., and Leinhardt, S. (1976). Local structure in social networks. *Sociological Methodology*, **7**, 1–45.

Holmberg, J., Norman, B., and Arzoumanian, Z. (2009). Estimating population size, structure, and residency time for whale sharks *Rhincodon typus* through collaborative photo-identification. *Endangered Species Reserarch*, **7**, 39–53.

Hoppit, W., Boogert, N. J., and Laland K. N. (2010). Detecting social transmission in networks. *Journal of Theoretical Biology*, **263**, 544–555.

Hoppitt, W., Kandler, A., Kendal, J. R., and Laland, K. N. (2010). The effect of task structure on diffusion dynamics: implications for diffusion curve and network-based analyses. *Learning and Behavior*, **38**, 243–251.

Hoppitt, W., and Laland K. N. (2011). Detecting social learning using networks: a user's guide. *American Journal of Primatology*, **73**, 834–844.

Hoppitt, W., and Laland K. N. (2013). *Social learning*. Princeton University Press, Princeton, NJ.

Horn, A. G., and McGregor, P. K. (2013). Influence and information in communication networks. In: Stegmann, U., ed. *Animal communication theory: information and influence*, pp. 43–62. Cambridge University Press, Cambridge.

Hou, C., Kaspari, M., Vander Zanden, H. B., and Gillooly, J. F. (2010). Energetic basis of colonial living in social insects. *Proceedings of the National Academy of Sciences*, **107**, 3634–3638.

Houde, A. E., and Torio, A. J. (1992). Effect of parasite infection on male colour pattern and female choice in guppies. *Behavioral Ecology*, **3**, 346–351.

Hu, J., and Hong, Y. (2007). Leader-following coordination of multi-agent systems with coupling time delays. *Physica A: Statistical Mechanics and its Applications*, **374**, 853–863.

Huang, Z. Y., and Robinson, G. E. (1992). Honeybee colony integration: worker–worker interactions mediate hormonally regulated plasticity in division of labor. *Proceedings of the National Academy of Sciences*, **89**, 11726–11729.

Hughes, D. P., Border, J., and Thomas, F. (2012). *Host manipulation by parasites*. Oxford University Press, Oxford.

Hughes, N. K., Kelly, J. L., and Banks, P. B. (2012). Dangerous liaisons: the predation risks of receiving social signals. *Ecology Letters*, **15**, 1326–1339.

Hughes, N. K., Korpimäki, E., and Banks, P. B. (2010). The predation risks of interspecific eavesdropping: weasel–vole interactions. *Oikos*, **119**, 1210–1216.

Hunt, J., Breuker, C. J., Sadowski, J. A., and Moore, A. J. (2009). Male–male competition, female mate choice and their interaction: determining total sexual selection. *Journal of Evolutionary Biology*, **22**, 13–26.

Husak, J. F., Irschick, D. J., Meyers, J. J., Laılvaux, S. P., and Moore, I. T. (2007). Hormones, sexual signals, and performance of green anole lizards (*Anolis carolinensis*). *Hormones and Behavior*, **52**, 360–367.

Hutchinson, J., and Gigerenzer, G. (2005). Simple heuristics and rules of thumb: where psychologists and behavioural biologists might meet. *Behavioural Processes*, **69**, 97–124.

Huth, A., and Wissel, C. (1992). The simulation of the movement of fish schools. *Journal of Theoretical Biology*, **156**, 365–385.

Ibanez, A., Lopez, P., and Martin, J. (2012). Discrimination of conspecifics' chemicals may allow Spanish terrapins to find better partners and avoid competitors. *Animal Behaviour*, **83**, 1107–1113.

Ioannou, C. C., Guttal, V., and Couzin, I. D. (2012). Predatory fish select for coordinated collective motion in virtual prey. *Science*, **337**, 1212.

Ioannou, C. C., Payne, M., and Krause, J. (2008). Ecological consequences of the bold–shy continuum—the effect of predator boldness on prey risk. *Oecologia*, **157**, 177–182.

Isbell, L. A., Pruetz, J. D., Lewis, M., and Young, T. P. (1999). Rank differences in ecological behavior: a comparative study of patas monkeys (*Erythrocebus patas*) and vervets (*Cercopithecus aethiops*). *International Journal of Primatology*, **20**, 257–272.

Ivkovich, T. V., Filatova, O. A., Burdin, A. M., Sato, H., and Hoyt, E. (2010). The social organization of resident-type killer whales (*Orcinus orca*) in Avacha Gulf, Northwestern Pacific, as revealed through association patterns and acoustic similarity. *Mammalian Biology-Zeitschrift für Säugetierkunde*, **75**, 198–210.

Jacobs, A., and Petit, O. (2011). Social network modeling: a powerful tool for the study of group scale phenomena in primates. *American Journal of Primatology*, **73**, 741–747.

Jacoby, D. M. P., Brooks, E. J., Croft, D. P., and Sims, D. W. (2012). Developing a deeper understanding of animal movements and spatial dynamics through novel application of network analyses. *Methods in Ecology and Evolution*, **3**, 574–583.

Jacoby, D. M. P., Busawon, D. S., and Sims, D. W. (2010). Sex and social networking: the influence of male presence on social structure of female shark groups. *Behavioral Ecology*, **21**, 808–818.

Jacoby, D. M. P., Croft, D. P., and Sims, D. W. (2012). Social behaviour in sharks and rays: analysis, patterns and implications for conservation. *Fish and Fisheries*, **13**, 399–417.

Jacomy, M., Heymann, S., Venturini, T., and Bastian, M. (2012). *ForceAtlas2, a continuous graph layout algorithm for handy network visualization*.<http://bit.ly/1jWUO98>.

Jadbabaie, A., Lin, J., and Morse, A. S. (2003). Coordination of groups of mobile autonomous agents using nearest neighbor rules. *IEEE Transactions on Automatic Control*, **48**, 988–1001.

James, A., Pitchford, J. W., and Plank, M. J. (2012). Disentangling nestedness from models of ecological complexity. *Nature*, **487**, 227–230.

James, R., Croft, D. P., and Krause, J. (2009). Potential banana skins in animal social network analysis. *Behavioral Ecology and Sociobiology*, **63**, 989–997.

Jameson, K. A., Appleby, M. C., and Freeman, L. C. (1999). Finding an appropriate order for a hierarchy based on probabilistic dominance. *Animal Behaviour*, **57**, 991–998.

Janik, V. M., and Slater, P. B. (1997). Vocal learning in mammals. *Advances in the Study of Behavior*, **26**, 59–99.

Jansen, V. A. A., and van Baalen, M. (2006). Altruism through beard chromodynamics. *Nature*, **440**, 663–666.

Jarman, P. (1974). The social organisation of antelope in relation to their ecology. *Behaviour*, **48**, 215–267.

Jaynes, E. T. (2003). *Probability theory: the logic of science*. Cambridge University Press, Cambridge.

Jeanne, R. L. (1986). The organization of work in *Polybia occidentalis*: costs and benefits of specialization in a social wasp. *Behavioral Ecology and Sociobiology*, **19**, 333–341.

Jeanne, R. L. (1988). *Interindividual behavioral variability in social insects*. Westview Press, Boulder, CO.

Jeanne, R. L. (1999). Group size, productivity, and information flow in social wasps. In: Detrain, C., Deneubourg, J. L., and Pasteels, J. M., eds. *Information processing in social insects*, pp. 3–30. Birkhauser, Basel.

Jeanson, R. (2012). Long-term dynamics in proximity networks in ants. *Animal Behaviour*, **83**, 915–923.

Johnson, J. S., Kropczynski, J. N., Lacki, M. J., and Langlois, G. D. (2012). Social networks of Rafinesque's big-eared bats (*Corynorhinus rafinesquii*) in bottomland hardwood forests. *Journal of Mammalogy*, **93**, 1545–1558.

Johnson, M. A., Revell, L. J., and Losos, J. B. (2009). Behavioral convergence and adaptive radiation: effects of habitat use on territorial behavior in *Anolis* lizards. *Evolution*, **64**, 1151–1159.

Johnson, M. P., and Tyack, P. L. (2003). A.digital acoustic recording tag for measuring the response of wild marine mammals to sound. *IEEE Journal of Oceanic Engineering*, **28**, 3–12.

Jolles, J. W., King, A. J., Manica, A., and Thornton, A. (2013). Heterogeneous structure in mixed-species corvid flocks in flight. *Animal Behaviour*, **85**, 743–750.

Jones, D. L., Jones, R. L., and Ratnam, R. (2014). Calling dynamics and call synchronization in a local group of unison bout callers. *Journal of Comparative Physiology A*, **200**, 93–107.

Joppa, L. N., and Pimm, S. (2010). On nestedness in ecological networks. *Evolutionary Ecology Research*, **12**, 35–46.

Julio-Pieper, M., O'Mahony, C. M., Clarke, G., Bravo, J. A., Dinan, T. G., and Cryan, J. F. (2012). Chronic stress-induced alterations in mouse colonic, 5-HT and defecation responses are strain dependent. *Stress*, **15**, 218–226.

Kalinoski, R. (1975). Intra- and interspecific aggression in house finches and house sparrows. *Condor*, **77**, 375–384.

Kao, R. R., Danon, L., Green, D. M., and Kiss, I. Z. (2006). Demographic structure and pathogen dynamics on the network of livestock movements in Great Britain. *Proceedings of the Royal Society B: Biological Sciences*, **273**, 1999–2007.

Kasper, C., and Voelkl, B. (2009). A social network analysis of primate groups. *Primates*, **50**, 343–356.

Kasumovic, M. M., Bruce, M. J., Andrade, M. C. B., and Herberstein, M. E. (2008). Spatial and temporal demographic variation drives within-season fluctuations in sexual selection. *Evolution*, **62**, 2316–2325.

Katz, Y., Tunstrom, K., Ioannou, C. C., Huepe, C., and Couzin, I. D. (2011). Inferring the structure and dynamics of interactions in schooling fish. *Proceedings of the National Academy of Sciences*, **108**, 18720–18725.

Kaufmann, J. H. (1983). On the definitions and functions of dominance and territoriality. *Biological Reviews*, **58**, 1–20.

Kauffman, S. A. (1993). *The origins of order: self-organization and selection in evolution*. Oxford University Press, New York.

Kavaliers, M., Choleris, E., Agmo, A., Braun, W. J., Colwell, D. D., Muglia, L. J., et al. (2006). Inadvertant social information and the avoidance of parasitised male mice: a role for oxytocin. *Proceedings of the National Academy of Sciences*, **103**, 4293–4298.

Kavaliers, M., and Colwell, D. D. (1995). Discrimination by female mice between the odours of parasitised and non-parasitised males. *Proceedings of the Royal Society B: Biological Sciences*, **261**, 31–35.

Kavaliers, M., Colwell, D. D., and Choleris, E. (1998). Parasitised female mice display reduced aversive responses to the odours of infected males. *Proceedings of the Royal Society B: Biological Sciences*, **265**, 1111–1118.

Kavaliers, M., Colwell, D. D., Ossenkopp, K.-P., and Perrot-Sinal, T. S. (1997). Altered responses to female odors in parasitized male mice: neuromodulatory mechanisms and relations to female choice. *Behavioral Ecology and Sociobiology*, **40**, 373–384.

Kawamura, S. (1958). Matriarchal social ranks in the Minoo-B troop: a study of the rank system of Japanese monkeys. *Primates*, **1**, 149–156.

Kays, R., Tilak, S., Crofoot, M., Fountain, T., Obando, D., Ortega, A., et al. (2011). Tracking animal location and activity with an automated radio telemetry system in a tropical rainforest. *The Computer Journal*, **54**, 1931–1948.

Kazem, A. J. N., and Aureli, F. (2005). Redirection of aggression: multiparty signalling in a network? In: McGregor, P. K., ed. *Animal communication networks*, pp. 191–218. Cambridge University Press, Cambridge.

Kearnes, M. (2009). *Male harassment influences female feral horse (Equus caballus) movement on Shackleford Banks, NC*. Senior thesis, Princeton University, Princeton, NJ.

Keeling, L. J., and Gonyou, H. (2001). *Social behaviour in farm animals*. CABI Publishing, Wallingford, UK.

Keeling, M. J. (1999). The effects of local spatial structure on epidemiological invasions. *Proceedings of the Royal Society B: Biological Sciences*, **266**, 859–867.

Keller, L., and Ross, K. G. (1998). Selfish genes: a green beard in the red fire ant. *Nature*, **394**, 573–575.

Kelly, J. L., Morrell, L. J., Inskip, C., Krause, J., and Croft, D. P. (2011). Predation risk shapes social networks in fission–fusion populations. *PLoS ONE*, **6**, e24280.

Kendal, R. L., Custance, D. M., Kendal, J. R., Vale, G., Stoinski, T. S., Rakotomalala, N. L., et al. (2010). Evidence for social learning in wild lemurs (*Lemur catta*). *Learning and Behavior*, **38**, 220–234.

Kendal, R. L., Galef, B. G., and van Schaik, C. P. (2010).Social learning research outside the laboratory: how and why? *Learning and Behavior*, **38**, 187–194.

Kennedy, C. E. J., Endler, J. A., Poynton, S. L., and McMinn, H. (1987). Parasite load predicts mate choice in guppies. *Behavioral Ecology and Sociobiology*, **21**, 291–295.

Kerr, G. D., Bull, C. M., and Cottrell, G. R. (2004). Use of an 'on board' datalogger to determine lizard activity patterns, body temperature and microhabitat use for extended periods in the field. *Wildlife Research*, **31**, 171–176.

Kerr, G. D., Bull, C. M., and Mackay, D. (2004). Human disturbance and stride frequency in the sleepy lizard (*Tiliqua rugosa*): implications for behavioral studies. *Journal of Herpetology*, **38**, 519–526.

Kerth, G. (2008). Causes and consequences of sociality in bats. *Bioscience*, **58**, 737–746.

Kerth, G., Perony, N., and Schweitzer, F. (2011) Bats are able to maintain long-term social relationships despite the high fission–fusion dynamics of their groups. *Proceedings of the Royal Society B: Biological Sciences*, **278**, 2761–2767.

Kiester, A. R. (1979). Conspecifics as cues: a mechanism for habitat selection in the Panamanian grass anole (*Anolis auratus*). *Behavioral Ecology and Sociobiology*, **5**, 323–330.

Kimura, T., Ohashi, M., Okada, R., and Ikeno, H. (2011). A new approach for the simultaneous tracking of multiple honeybees for analysis of hive behavior. *Apidologie*, **42**, 607–617.

King, A. J., Clark, F. E., and Cowlishaw, G. (2011). The dining etiquette of desert baboons: the roles of social bonds, kinship, and dominance in co-feeding networks. *American Journal of Primatology*, **73**, 768–774.

King, R., Morgan, B. J. T., Giminez, O., and Brooks, S. P. (2010). *Bayesian analysis for population ecology*. Chapman and Hall/CRC, Boca Raton, FL.

King, S. L., and Janik, V. M. (2013). Bottlenose dolphins can use learned vocal labels to address each other. *Proceedings of the National Academy of Sciences*, **110**, 13216–13221.

King, S. L., Sayigh, L. S., Wells, R. S., Fellner, W., and Janik, V. M. (2013). Vocal copying of individually distinctive signature whistles in bottlenose dolphins. *Proceedings of the Royal Society B: Biological Sciences*, **280**, 20130053.

Kingsolver, J. G., Hoekstra, H. E., Hoekstra, J. M., Berrigan, D.,Vignieri, S. N., Hill, C. E., et al.. (2001). The strength of phenotypic selection in natural populations. *American Naturalist*, **157**, 245–261.

Kingsolver, J. G., and Pfennig, D. W. (2007). Patterns and power of phenotypic selection in nature. *Bioscience*, **57**, 561–572.

Kiss, I. Z., Green, D. M., and Kao, R. R. (2006). The network of sheep movements within Great Britain: network properties and their implications for infectious disease spread. *Journal of the Royal Society Interface*, **3**, 669–677.

Kitchen-Wheeler, A. M. (2010). Visual identification of individual manta ray (*Manta alfredi*) in the Maldives Islands, Western Indian Ocean. *Marine Biology Research*, **6**, 351–363.

Klefoth, T., Skov, C., Krause, J., and Arlinghaus, R. (2012). Assessing the genetic basis of boldness in fish: the role of ecological context and risk-stimuli. *Behavioral Ecology and Sociobiology*, **66**, 547–559.

Kleinberg, J. M. (1999). Authoritative sources in a hyperlinked environment. *Journal of the ACM*, **46**, 604–632.

Klingel, H. (1977). Observations on social organization and behaviour of African and Asiatic wild asses (*Equus africanus* and *E. hemionus*). *Zeitschrift für Tierpsychologie*, **44**, 323–331.

Klovdahl, A. S. (1985). Social networks and the spread of infectious diseases: the AIDS example. *Social Science and Medicine*, **21**, 1203–1216.

Kluijver, H. N. (1951). *The population ecology of the great tit*, Parus m. major *L*. Brill, Leiden.

Klump, G. M., Kretzschamar, E., and Curio, E. (1986). The hearing of an avian predator and its avian prey. *Behavioral Ecology and Sociobiology*, **18**, 317–323.

Köenig, M., Tessone, C., and Zenou, Y. (2012). Nestedness in networks: a theoretical model and some applications. Centre for Economic Policy Research Discussion Paper No. DP8807. <http://ssrn.com/abstract=2013798>.

Komers, P. E. (1997). Behavioural plasticity in variable environments. *Canadian Journal of Zoology*, **75**, 161–169.

Kondrashov, A. S., and Shpak, M. (1998). On the origin of species by means of assortative mating. *Proceedings of the Royal Society B: Biological Sciences*, **265**, 2273–2278.

Koops, M. A. (2004). Reliability and the value of information. *Animal Behaviour*, **67**, 103–111.

Kopps, A. M., and Sherwin, W. B. (2012). Modelling the emergence and stability of a vertically transmitted cultural trait in bottlenose dolphins. *Animal Behaviour*, **84**, 1347–1362.

Koprivnikar, J., Gibson, C. H., and Redfern, J. C. (2011). Infectious personalities: behavioural syndromes and disease risk in larval amphibians. *Proceedings of the Royal Society B: Biological Sciences*, **279**, 1544–1550.

Kraaijeveld, K., and Dickinson, J. (2001). Family-based winter territoriality in western bluebirds, *Sialia mexicana*: the structure and dynamics of winter groups. *Animal Behaviour*, **61**, 109–117.

Kraus, S. D., and Hatch, J. J. (2001). Mating strategies in the North Atlantic right whale (*Eubalaena glacialis*). *Journal of Cetacean Research and Management*, **Special Issue 2**, 237–244.

Krause, J., Butlin, R., Peuhkuri, N., and Pritchard, V. L. (2000). The social organisation of fish shoals: a test of the predictive power of laboratory experiments for the field. *Biological Reviews*, **75**, 477–501.

Krause, J., Croft, D. P., and James, R. (2007). Social network theory in the behavioural sciences: potential applications. *Behavioral Ecology and Sociobiology*, **62**, 15–27.

Krause, J., James, R., and Croft, D. P. (2010). Personality in the context of social networks. *Philosophical Transactions of the Royal Society B: Biological Sciences*, **365**, 4099–4106.

Krause, J., Krause, S., Arlinghaus, R., Psorakis, I., Roberts, S., and Rutz, C. (2013). Reality mining of animal social systems. *Trends in Ecology and Evolution*, **28**, 541–551.

Krause, J., Lusseau, D., and James, R. (2009). Animal social networks: an introduction. *Behavioral Ecology and Sociobiology*, **63**, 967–973.

Krause, J., and Ruxton, G. D. (2002). *Living in groups*. Oxford University Press, Oxford.

Krause, J., and Ruxton, G. D. (2010). Important topics in group living. In: Székely, T., Moore, A. J., and Komdeur, J., eds. *Social behavior: genes, ecology and evolution*, pp. 203–225. Cambridge University Press, Cambridge.

Krause, J., Ruxton, G. D., and Rubenstein, D. I. (1998). Is there an influence of group size on predator hunting success? *Journal of Fish Biology*, **52**, 494–501.

Krause, J., Wilson, A. D. M., and Croft, D. P. (2011). New technology facilitates the study of social networks. *Trends in Ecology and Evolution*, **26**, 5–6.

Krause, J., Winfield, A. F. T., and Deneubourg, J. L. (2011). Interactive robots in experimental biology. *Trends in Ecology and Evolution*, **26**, 369–375.

Krause, S., Mattner, L., James, R., Guttridge, T. L., Corcoran, M. J., Gruber, S. H., et al. (2009). Social network analysis and valid Markov chain Monte Carlo tests of null models. *Behavioral Ecology and Sociobiology*, **63**, 1089–1096.

Krebs, J. R. (1982). Territorial defense in the great tit (*Parus major*): do residents always win? *Behavioral Ecology and Sociobiology*, **11**, 185–194.

Kroodsma, D. E. (2004). The diversity and plasticity of birdsong. In: Marler, P. and Slabbekoorn, H., eds., *Nature's music*, pp. 108–131. Elsevier Academic Press, San Diego.

Krutzen, M., Mann, J., Heithaus, M.R., Connor, R. C., Bejder, L., and Sherwin, W.B. (2005). Cultural transmission of tool use in bottlenose dolphins. *Proceedings of the National Academy of Sciences*, **102**, 8939–8943.

Kruuk, L., and Hadfield, J. (2007). How to separate genetic and environmental causes of similarity between relatives. *Journal of Evolutionary Biology*, **20**, 1890–1903.

Kühl, H. S., and Burghardt, T. (2013). Animal biometrics: quantifying and detecting phenotypic appearance. *Trends in Ecology and Evolution*, **28**, 432–441.

Kukielka, E., Barasona, J. A., Cowie, C .E., Drewe, J. A., Gortazar, C., Cotarelo, I., et al. (2013). Spatial and temporal interactions between livestock and wildlife in South Central Spain assessed by camera traps. *Preventive Veterinary Medicine*, **112**, 213–221.

Kummer, H. (1957). Soziales Verhalten einer Mantelpavianen-Gruppe. *Schweizerische Zeitschrift fuer Psychologie*, **Beiheft 33**, 1–91.

Kummer, H. (1968). *Social organization of hamadryas baboons*. University of Chicago Press, Chicago.

Kummer, H. (1995). *In the quest of the sacred baboon*. Princeton University Press, Princeton, NJ.

Kurvers, R. H. J. M., Eijkelenkamp, B., van Oers, K., van Lith, B., van Wieren, S. E., Ydenberg, R. C., et al. (2009). Personality differences explain leadership in barnacle geese. *Animal Behaviour*, **78**, 447–453.

Kurvers, R. H. J. M., Krause, J., Croft, D. P., Wilson, A. D. M., and Wolf, M. (2014). Ecological and evolutionary consequences of social networks: emerging topics. *Trends in Ecology and Evolution*, **29**, 326–335.

Kvarnemo, C., and Simmons, L. W. (2013). Polyandry as a mediator of sexual selection before and after mating. *Philosophical Transactions of the Royal Society B: Biological Sciences*, **368**, 20120042.

Lachlan, R. F., Crooks, L., and Laland, K. N. (1998). Who follows whom? Shoaling preferences and social learning of foraging information in guppies. *Animal Behaviour*, **56**, 181–190.

Lack, D. (1964). A long-term study of the great tit (*Parus major*). *Journal of Animal Ecology*, **33**, 159–173.

Laiolo, P., and Obeso, J. R. (2012). Multilevel selection and neighbourhood effects from individual to metapopulation in a wild passerine. *PLoS ONE*, **7**, e38526.

Laland, K. N., and Galef, B. G. Jr. (2009). *The question of animal culture*. Harvard University Press, Cambridge, MA.

Laland K. N., and Williams, K. (1998). Social transmission of maladaptive information in the guppy. *Behavioral Ecology*, **9**, 493–499.

Lambert, T. D., Kays, R. W., Jansen, P. A., Aliaga-Rossel, E., and Wikelski, M. (2009). Nocturnal activity by the primarily diurnal Central American agouti (*Dasyprocta punctata*) in relation to environmental conditions, resource abundance and predation risk. *Journal of Tropical Ecology*, **25**, 211–215.

Lande, R., and Arnold, S. J. (1983). The measurement of selection on correlated characters. *Evolution*, **37**, 1210–1226.

Laws, R. M., Parker, I. S. C., and Johnstone, R. C. B. (1975). *Elephants and their habitats: the ecology of elephants in North Bunyoro, Uganda*. Clarendon Press, Oxford.

Le Galliard, J. F., Ferrière, R., and Dieckmann, U. (2003). The adaptive dynamics of altruism in spatially heterogeneous populations. *Evolution*, **57**, 1–17.

Lea, A. J., Blumstein, D. T., Wey, T. W., and Martin, J. G. (2010). Heritable victimization and the benefits of agonistic relationships. *Proceedings of the National Academy of Sciences*, **107**, 21587–21592.

Leblond, C., and Reebs, S. G. (2006). Individual leadership and boldness in shoals of golden shiners (*Notemigonus crysoleucas*). *Behaviour*, **143**, 1263–1280.

Lee, Peter. (1989). *Bayesian statistics: an introduction*. Hodder Arnold, London.

Legendre, P., and Legendre, L. (2012). *Numerical ecology*, 3rd English edn. Elsevier, Amsterdam.

Lehmann, J., Andrews, K., and Dunbar, R. I. M. (2010). Social networks and social complexity in female-bonded primates. *Proceedings of the British Academy*, **158**, 57–82.

Lehmann, J., and Dunbar, R. I. M. Network cohesion, group size and neocortex size in female-bonded Old World primates. *Proceedings of the Royal Society B: Biological Sciences*, **276**, 4417–4422.

Lehmann, J., and Ross, C. (2011). Baboon (*Papio anubis*) social complexity—a network approach. *American Journal of Primatology*, **73**, 775–789.

Lehmann, L., and Boesch, C. (2009). Sociality of the dispersing sex: the nature of social bonds in West African female chimpanzees, *Pan troglodytes*. *Animal Behaviour*, 77, 377–387.

Leigh, J. (2010). The group selection controversy. *Journal of Evolutionary Biology*, **23**, 6–19.

Lemasson, B. H., Anderson, J. J., and Goodwin, R. A. (2009). Collective motion in animal groups from a neurobiological perspective: the adaptive benefits of dynamic sensory loads and selective attention. *Journal of Theoretical Biology*, **261**, 501–510.

Lesaffre, E., and Lawson, A. B. (2012). *Bayesian biostatistics*. John Wiley and Sons Ltd, Chichester.

Leu, S. T., Bashford, J., Kappeler, P. M., and Bull, C. M. (2010). Association networks reveal social organization in the sleepy lizard. *Animal Behaviour*, **79**, 217–225.

Leu, S. T., Kappeler, P. M., and Bull, C. M. (2010). Refuge sharing network predicts ectoparasite load in a lizard. *Behavioral Ecology and Sociobiology*, **64**, 1495–1503.

Leu, S. T., Kappeler, P. M., and Bull, C. M. (2011). The influence of refuge sharing on social behaviour in the lizard *Tiliqua rugosa*. *Behavioral Ecology and Sociobiology*, **65**, 837–847.

Lieberman, E., Hauert, C., and Nowak, M. A. (2005). Evolutionary dynamics on graphs. *Nature*, **433**, 312–316.

Liker, A., Bókony, V., Kulcsár, A., Tóth, Z., Szabó, K., Kaholek, B., et al. (2009). Genetic relatedness in wintering groups of house sparrows (*Passer domesticus*). *Molecular Ecology*, **18**, 4696–4706.

Linklater, W. L., Cameron, E. Z., Minot, E. O., and Stafford, K. J. (1999). Stallion harassment and the mating system of horses. *Animal Behaviour*, **58**, 295–306.

Lion, S., and van Baalen, M. (2008). Self-structuring in spatial evolutionary ecology. *Ecology Letters*, **11**, 277–295.

Liu, Y., Passino, K. M., and Polycarpou, M. M. (2003). Stability analysis of M-dimensional asynchronous swarms with a fixed communication topology. *IEEE Transactions on Automatic Control*, **48**, 76–95.

Lloyd-Smith, J. O., Schreiber, S. J., Kopp, P. E., and Getz, W. M. (2005). Superspreading and the effect of individual variation on disease emergence, *Nature*, **438**, 355–359.

Lopes, P. C., Adelman, J., Wingfield, J. C., and Bentley, G. E. (2012). Social context modulates sickness behavior. *Behavioral Ecology and Sociobiology*, **66**, 1421–1428.

Lopez, P., Hawlena, D., Polo, V., Amo, L., and Martin, J. (2005). Sources of individual shy–bold variations in antipredator behaviour of male Iberian rock lizards. *Animal Behaviour*, **69**, 1–9.

López-Riquelme, G. O., Malo, E. A., Cruz-López, L., and Fanjul-Moles, M. L. (2006). Antennal olfactory sensitivity in response to task-related odours of three castes of the ant *Atta mexicana* (hymenoptera: formicidae). *Physiological Entomology*, **31**, 353–360.

Lusseau, D. (2003). The emergent properties of a dolphin social network. *Proceedings of the Royal Society B: Biological Sciences*, **270**, 186–188.

Lusseau, D. (2007a). Evidence for social role in a dolphin social network. *Evolutionary Ecology*, **21**, 357–366.

Lusseau, D. (2007b). Why are male social relationships complex in the Doubtful Sound bottlenose dolphin population? *PLoS ONE*, **2**, e348.

Lusseau, D., Barrett, L., and Henzi, S. P. (2011). Formalising the multidimensional nature of social networks. arXiv preprint arXiv:1101.3735.

Lusseau, D., Barrett, L., and Henzi, S. P. (2012). Formalising the multidimensional nature of social networks. *Philosophical Transactions of the Royal Society B: Biological Sciences*, **367**, 2108–2118.

Lusseau, D., and Conradt, L. (2009). The emergence of unshared consensus decisions in bottlenose dolphins. *Behavioral Ecology and Sociobiology*, **63**, 1067–1077.

Lusseau, D., and Newman, M. E. J. (2004). Identifying the role that animals play in their social networks. *Proceedings of the Royal Society B: Biological Sciences*, **271**, S477–S481.

Lusseau, D., Whitehead, H., and Gero, S. (2008). Incorporating uncertainty into the study of animal social networks. *Animal Behaviour*, **75**, 1809–1815.

Lusseau, D., Wilson, B., Hammond, P. S., Grellier, K., Durban, J. W., Parsons, K. M., et al. (2006). Quantifying the influence of sociality on population structure in bottlenose dolphins. *Journal of Animal Ecology*, **75**, 14–24.

Luttbeg, B., and Trussell, G. C. (2013). How the informational environment shapes how prey estimate predation risk and the resulting indirect effects of predators. *American Naturalist*, **181**, 182–194.

Macdonald, S., Schuelke, O., and Ostner, J. (2014). The structure and stability of female social relationships in Assamese macaques (*Macaca assamensis*). Manuscript in preparation.

Macedo, R. H., Manica, L., and Dias, R. I. (2012). Conspicuous sexual signals in a socially monogamous passerine: the case of neotropical blue-black grassquits. *Journal of Ornithology*, **153**, S15–S22.

MacGregor, L. H., Cumming, G. S., and Hockey, P. A. R. (2011). Understanding pathogen transmission dynamics in waterbird communities: at what scale should interactions be studied? *South African Journal of Science*, **107**, 56–65.

MacIntosh, A. J. J., Jacobs, A., Garcia, C., Shimizu, K., Mouri, K., Huffman, M. A., et al. (2012) Monkeys in the middle: parasite transmission through the social network of a wild primate. *PLoS ONE*, **7**, e51144.

Madden, J. R. (2008). Do bowerbirds exhibit cultures? *Animal Cognition*, **11**, 1–12.

Madden, J. R., Drewe, J. A., Pearce, G. P., and Clutton-Brock, T. H. (2009). The social network structure of a wild meerkat population: 2. Intragroup interactions. *Behavioral Ecology and Sociobiology*, **64**, 81–95.

Madden, J. R., Nielsen, J. F., Clutton-Brock, T. H. (2012). Do networks of social interactions reflect patterns of kinship? *Current Zoology*, **58**, 319–328.

Madosky, J. M., Rubenstein, D. I., Howard, J. J., and Stuska, S. (2010). The effects of immunocontraception on harem fidelity in a feral horse (*Equus caballus*) population. *Applied Animal Behaviour Science*, **128**, 50–56.

Magnhagen, C., Braithwaite, V. A., Forsgren, E., and Kapoor, B. G., eds. (2008). *Fish behaviour*. Science Publishers, Enfield, NH.

Magrath, R. D., and Bennett, T. H. (2012). A micro-geography of fear: learning to eavesdrop on alarm calls of neighbouring heterospecifics. *Proceedings of the Royal Society B: Biological Sciences*, **279**, 902–909.

Magrath, R. D., Pitcher, B. J., and Gardner, J. L. (2009). An avian eavesdropping network: alarm signal reliability and heterospecific response. *Behavioral Ecology*, **20**, 745–752.

Magurran, A. E. (2005). *Evolutionary biology: the Trinidadian guppy*. Oxford University Press, Oxford.

Main, A. R., and Bull, C. M. (1996). Mother–offspring recognition in two Australian lizards, *Tiliqua rugosa* and *Egernia stokesii*. *Animal Behaviour*, **52**, 193–200.

Makagon, M. M., McCowan, B., and Mench, J. A. (2012). How can social network analysis contribute to social behavior research in applied ethology? *Applied Animal Behaviour Science*, **138**, 152–161.

Mann, J., Connor, R. C., Tyack, P. L., and Whitehead, H. (2000). *Cetacean societies: field studies of dolphins and whales*. University of Chicago Press, Chicago.

Mann, J., Sargeant, B. L., Watson-Capps, J. J., Gibson, Q. A., Heithaus, M. R., Connor, R. C., et al. (2008). Why do dolphins carry sponges? *PLoS ONE*, **3**, e3868.

Mann, J., Stanton, M. A., Patterson, E. M., Bienenstock, E. J., and Singh, L. O. (2012). Social networks reveal cultural behaviour in tool-using dolphins. *Nature Communications*, **3**, 980.

Marino, L., Connor, R. C., Fordyce, R. E., Herman, L. M., Hof, P. R., Lefebvre, L., et al. (2007). Cetaceans have complex brains for complex cognition. *PLoS Biology*, **5**, e139.

Marler, P. (1955). Characteristics of some animal calls. *Nature*, **176**, 6–8.

Marsh, M. K., Hutchings, M. R., McLeod, S. R., and White, P. C. L. (2011). Spatial and temporal heterogeneities in the contact behaviour of rabbits. *Behavioral Ecology and Sociobiology*. **65**, 183–195.

Martin, J., and Lopez, P. (1999). When to come out from a refuge: risk-sensitive and state-dependent decisions in an alpine lizard. *Behavioral Ecology*, **10**, 487–492.

Mason, R. T., and Parker, M. R. (2010). Social behavior and pheromonal communication in reptiles. *Journal of Comparative Physiology*, **196**, 729–749.

Matessi, G., Matos, R. J., and Dabelsteen, T. (2008). Communication in social networks of territorial animals: networking at different levels in birds and other systems. In: d'Ettorre, P., and Hughes, D. P., eds. *Sociobiology of communication: an interdisciplinary perspective*, pp. 33–53. Oxford University Press, Oxford.

Matessi, G., Matos, R. J., Peake, T. M., McGregor, P. K., and Dabelsteen, T. (2010). Effects of social environment and personality on communication in male Siamese fighting fish in an artificial network. *Animal Behaviour*, **79**, 43–49.

Matessi, G., McGregor, P. K., Peake, T. M., and Dabelsteen, T. (2005). Do male birds intercept and use rival courtship calls to adjust paternity protection behaviours? *Behaviour*, **142**, 507–542.

Matos, R. J., and Schlupp, I. (2005). Performing in front of an audience: signallers and the social environment. In: McGregor, P. K., ed. *Animal communication networks*, pp. 63–83. Cambridge University Press, Cambridge.

Matsuda, I., Zhang, P., Swedell, L., Mori, U., Tuuga, A., Bernard, H., and Sueur, C. (2012). Comparisons of intraunit relationships in nonhuman primates living in multilevel social systems. *International Journal of Primatology*, **33**, 1038–1053.

May, R. M. (2006). Network structure and the biology of populations. *Trends in Ecology and Evolution*, **21**, 394–399.

May, R. M. (2013). Networks and webs in ecosystems and financial systems. *Philosophical Transactions of the Royal Society A: Mathematical, Physical and Engineering Sciences*, **371**, 20120376.

May, R. M., Levin, S. A., and Sugihara, G. (2008). Complex systems: ecology for bankers. *Nature*, **451**, 893–895.

Mayberry, D., Woolnough, A., Twigg, L., Martin, G., Lowe, T., and McLeod, S. (2010). *Preliminary investigations for measuring animal contact rates: final report*. Wildlife and Exotic Disease Preparedness Program, Department of Agriculture, Fisheries and Forestry, Canberra.

Maynard Smith, J. (1982). *Evolution and the theory of games*. Cambridge University Press, Cambridge.

Maynard Smith, J. (1983). Game theory and the evolution of cooperation. In: Bendall, D. S., ed. *Evolution from molecules to men*, pp. 445–456. Cambridge University Press, Cambridge.

Maynard Smith, J., and Harper, D. (2003). *Animal signals*. Oxford University Press, Oxford.

Maynard Smith, J., and Szathmáry, E. R. (1995). *The major transitions in evolution*. W. H. Freeman Spektrum, Oxford.

McAlpin, S., Duckett, P. E., and Stow, A. J. (2011). Lizards cooperatively tunnel to construct a long-term home for family members. *PLoS ONE*, **6**, e19041.

McCallum, H., Barlow, N., and Hone, J. (2001). How should pathogen transmission be modelled? *Trends in Ecology and Evolution*, **16**, 295–300.

McCallum, H., Jones, M., Hawkins, C., Hamede, R., Lachish, S., Sinn, D. L., et al. (2009). Transmission dynamics of Tasmanian devil facial tumor disease may lead to disease-induced extinction. *Ecology*, **90**, 3379–3392.

McCarthy, M. (2007). *Bayesian methods for ecology*. Cambridge University Press, Cambridge.

McCleery, R. H., Pettifor, R. A., Armbruster, P., Meyer, K., Sheldon, B. C. and Perrins, C. M. (2004). Components of variance underlying fitness in a natural population of the great tit *Parus major*. *American Naturalist*, **164**, E62–E72.

McComb, K., Moss, C., Sayialel, S., and Baker, L. (2000). Unusually extensive networks of vocal recognition in African elephants. *Animal Behaviour*, **59**, 1103–1109.

McComb, K., and Semple, S. (2005). Coevolution of vocal communication and sociality in primates. *Biology Letters*, **1**, 381–385.

McCowan, B., Anderson, K., Heagarty, A., and Cameron, A. (2008). Utility of social network analysis for primate behavioral management and well-being. *Applied Animal Behaviour Science*, **109**, 396–405.

McCowan, B., Beisner, B. A., Capitanio, J. P., Jackson, M. E., Cameron, A. N., Seil, S., et al. (2011). Network stability is a balancing act of personality, power, and conflict dynamics in rhesus macaque societies. *PLoS ONE*, **6**, e22350.

McDiarmid, R. W., Foster, M. S., Guyer, C., Gibbons, J. W., and Chernoff, N. (2012). *Reptile biodiversity: standard methods for inventory and monitoring*. University of California Press, Berkeley.

McDonald, D. B. (2007). Predicting fate from early connectivity in a social network. *Proceedings of the National Academy of Sciences*, **104**, 10910–10914.

McDonald, D. B. (2009). Young-boy networks without kin clusters in a lek-mating manakin. *Behavioral Ecology and Sociobiology*, **63**, 1029–1034.

McDonald, D. B., and Shizuka, D. (2013). Comparative transitive and temporal orderliness in dominance networks. *Behavioral Ecology*, **24**, 511–520.

McDonald, G. C., James, R., Krause, J., and Pizzari, T. (2013). Sexual networks: measuring sexual selection in structured, polyandrous populations. *Philosophical Transactions of the Royal Society B: Biological Sciences*, **368**, 20120356.

McElreath, R., Boyd, R., and Richerson, P. J. (2003). Shared norms and the evolution of ethnic markers. *Current Anthropology*, **44**, 122–129.

McGregor, P. K. (2004). Communication. In: Bolhuis, J. J., and Giraldeau, L.-A., eds. *The behavior of animals: mechanisms, function and evolution*, pp. 226–250. Blackwell Scientific Publications, Oxford.

McGregor, P. K., ed. (2005). *Animal communication networks*. Cambridge University Press, Cambridge.

McGregor, P. K. (2009). Communication networks and eavesdropping in animals. In: Squire, L. R., ed. *Encyclopedia of neuroscience, Volume 2*, pp. 1179–1184. Academic Press, Oxford.

McGregor, P. K., and Dabelsteen, T. (1996). Communication networks. In,Kroodsma, D. E. and Miller, E. H., eds. *Ecology and evolution of acoustic communication in birds*, pp. 409–425. Cornell University Press, Ithaca, NY.

McGregor, P. K., Horn, A. G., Leonard, M. L., and Thomsen, F. (2013). Anthropogenic noise and conservation. In: Brumm, H., ed. *Animal communication and noise*, pp. 409–440. Springer Verlag, Berlin.

McGregor, P. K., Otter, K. A., and Peake, T. M. (2000). Communication networks: receiver and signaller perspectives. In:Espmark, Y., Amundsen, T. and Rosenqvist, G., eds. *Animal signals: signalling and signal design in animal communication*, pp. 329–340. Tapir Academic Press, Trondheim.

McGregor, P. K., and Peake, T. M. (2000). Communication networks: social environments for receiving and signalling behaviour. *Acta Ethologica*, **2**, 71–81.

McNamara, K. B., Brown, R. L., Elgar, M. A., and Jones, T. M. (2008). Paternity costs from polyandry compensated by increased fecundity in the hide beetle. *Behavioral Ecology and Sociobiology*, **19**, 433–440.

Mehlman, P., and Chapais, B. (1988). Differential effects of kinship, dominance, and the mating season on female allogrooming in a captive group of *Macaca fuscata*. *Primates*, **29**, 195–217.

Melleti, M., Delgado, M. M., Penteriani, V., Mirabile, M., and Boitani, L. (2010). Spatial properties of a forest buffalo herd and individual positioning as a response to environmental cues and social behavior. *Journal of Ethology*, **28**, 421–428.

Mennill, D. J., Doucet, S. M., Ward, K. A. A., Maynard, D. F., Otis, B., and Burt, J. M. (2012). A novel digital telemetry system for tracking wild animals: a field test for studying mate choice in a lekking tropical bird. *Methods in Ecology and Evolution*, **3**, 663–672.

Mennill, D. J., Ratcliffe, L. M., and Boag, P. T. (2002). Female eavesdropping on male song contests in songbirds. *Science*, **296**, 873.

Mersch, D. P., Crespi, A., and Keller, L. (2013). Tracking individuals shows spatial fidelity is a key regulator of ant social organization. *Science*, **340**, 1090–1093.

Mesnick, S. L. (2001). Genetic relatedness in sperm whales: evidence and culture implications. *Behavioral and Brain Sciences*, **24**, 346–347.

Metcalfe, N. B., and Thomson, B. C. (1995). Fish recognize and prefer to shoal with poor competitors. *Proceedings of the Royal Society B: Biological Sciences*, **259**, 207–210.

Michael, D. R., Cunningham, R. B., and Lindenmayer, D. B. (2010). The social elite: habitat heterogeneity, complexity and quality in granite inselbergs influence patterns of aggregation in *Egernia striolata* (Lygosominae: Scincidae). *Austral Ecology*, **35**, 862–870.

Michler, S., Nicolaus, M.,Ubels, R., van der Velde, M., Komdeur, J., Both, C., and Tinbergen, J. M. (2011). Sex-specific effects of the local social environment on juvenile post-fledging dispersal in great tits. *Behavioral Ecology and Sociobiology*, **65**, 1975–1986.

Milinski, M. (1987). Tit for tat in sticklebacks and the evolution of cooperation. *Nature*, **325**, 433–435.

Milinski, M., Semmann, D., Bakker, T. C. M., and Krambeck, H. J. (2001). Cooperation through indirect reciprocity: image scoring or standing strategy? *Proceedings of the Royal Society B: Biological Sciences*, **268**, 2495–2501.

Milinski, M., and Wedekind, C. (1998). Working memory constrains human cooperation in the Prisoner's Dilemma. *Proceedings of the National Academy of Sciences*, **95**, 13755–13758.

Miller, J. L., King, A. P., and West, M. J. (2008). Female social networks influence male vocal development in brown-headed cowbirds, *Molothrus ater*. *Animal Behaviour*, **76**, 931–941.

Milo, R., Shen-Orr, S., Itzkovitz, S., Kashtan, N., Chklovskii, D., and Alon, U. (2002). Network motifs:

simple building blocks of complex networks. *Science*, **298**, 824–827.

Moller, A. P., Dufva, R., and Allander, K. (1993). Parasites and the evolution of host social behavior. *Advances in the Study of Behavior*, **22**, 65–102.

Moore, S. L., and Wilson, K. (2002). Parasites as a viability cost of sexual selection in natural populations of mammals. *Science*, **297**, 2015–2018.

Mooring, M. S., and Hart, B. L. (1992). Animal grouping for protection from parasites: selfish herd and encounter-dilution effects. *Behaviour*, **123**, 173–193.

Morand-Ferron, J., and Quinn, J. L. (2011). Larger groups of passerines are more efficient problem solvers in the wild. *Proceedings of the National Academy of Sciences*, **108**, 15898–15903.

Moreau, M., Arrufat, P., Latil, G., and Jeanson, R. (2011). Use of radio-tagging to map spatial organization and social interactions in insects. *The Journal of Experimental Biology*, **214**, 17–21.

Morrell, L. J., Croft, D. P., Dyer, J. R. G., Chapman, B. B., Kelley, J. L., Laland K. N., and Krause, J. (2008). Association patterns and foraging behaviour in natural and artificial guppy shoals. *Animal Behaviour*, **76**, 855–864.

Morrel, L. J., and Romey, W. L. (2008). Optimal individual positions within animal groups. *Behavioral Ecology*, **19**, 909–919.

Moss, C. J., and Poole, J. H. (1983). Relationships and social structure in African elephants. In: Hinde, R. A., ed. *Primate social relationships: a n integrated approach*, pp. 315–325. Blackwell Scientific, Oxford.

Mourier, J., Vercelloni, J., and Planes, S. (2012). Evidence of social communities in a spatially structure network of a free-ranging shark species. *Animal Behaviour*, **83**, 389–401.

Moussaïd, M., Perozo, N., Garnier, S., Helbing, D., and Theraulaz, G. (2010). The walking behaviour of pedestrian social groups and its impact on crowd dynamics. *PLoS ONE*, **5**, e10047.

Moyle, P. B., and Cech, J. J., Jr. (2004). *Fishes: an introduction to ichthyology*. Prentice Hall, Englewood Cliffs, NJ.

Mundinger, P. C. (1982). Microgeographic and macrogeographic variation in the acquired vocalizations of birds. In: Kroodsma, D. E. and Miller, E. H., eds. *Acoustic communication in birds, Volume 2*, pp. 147–208. Academic Press, New York.

Musolesi, M., Hailes, S., and Mascolo, C. (2004). An ad hoc mobility model founded on social network theory. In: *Proceedings of the 7th ACM international symposium on modeling, analysis and simulation of wireless and mobile systems*, pp. 20–24. Association for Computing Machinery, New York.

Musse, S. R., and Thalmann, D. (1997). A model of human crowd behavior: group inter-relationship and collision detection analysis. *Computer Animation and Simulation*, **97**, 39–51.

Myers, J. P. (1983). Space, time and the pattern of individual associations in a group-living species: sanderlings have no friends. *Behavioral Ecology and Sociobiology*, **12**, 129–134.

Naguib, M., Amrhein, V., and Kunc, H. P. (2004). Effects of territorial intrusions on eavesdropping neighbors: communication networks in nightingales. *Behavioral Ecology*, **15**, 1011–1015.

Nagy, M., Ákos, Z., Biro, D., and Vicsek, T. (2010). Hierarchical group dynamics in pigeon flocks. *Nature*, **464**, 890–893.

Nagy, M., Vásárhely, G., Pettit, B., Roberts-Mariani, I., Vicsek, T., and Biro, D. (2013). Context-dependent hierarchies in pigeons. *Proceedings of the National Academy of Sciences*, **110**, 13049–13054.

Napier, J. R., and Napier, P. H. (1997). *The natural history of the primates*. MIT Press, Cambridge, MA.

National Agricultural Statistics Service. (2011). *Cattle death loss*. United States Department of Agriculture, Washington, DC.

National Research Council. (2011). *Guide for the care and use of laboratory animals*, 8th edn. The National Academies Press, Washington, DC.

Naug, D. (2008). Structure of the social network and its influence on transmission dynamics in a honeybee colony. *Behavioral Ecology and Sociobiology*, **62**, 1719–1725.

Naug, D. (2009). Structure and resilience of the social network in an insect colony as a function of colony size. *Behavioral Ecology and Sociobiology*, **63**, 1023–1028.

Naug, D., and Camazine, S. (2002). The role of colony organization on pathogen transmission in social insects. *Journal of Theoretical Biology*, **215**, 427–439.

Naug, D., and Gadagkar, R. (1999). Flexible division of labor mediated by social interactions in an insect colony—a simulation model. *Journal of Theoretical Biology*, **197**, 123–133.

Naug, D., and Gibbs, A. (2009). Behavioral changes mediated by hunger in honeybees infected with *Nosema ceranae*. *Apidologie*, **40**, 595–599.

Nettle, D. (1999). Language variation and the evolution of societies. In: Dunbar, R. I. M., Knight, C., and Power, C., eds. *The evolution of culture*, pp. 214–227. Rutgers University Press, Piscataway, NJ.

Neumann, C., Duboscq, J., Dubuc, C., Ginting, A., Irwan, A. M., Agil, M., et al. (2011). Assessing dominance hierarchies: validation and advantages of progressive evaluation with Elo-rating. *Animal Behaviour*, **82**, 911–921.

Newman, M. E. J. (2002a). Assortative mixing in networks. *Physical Review Letters*, **89**, 208701.

Newman, M. E. J. (2002b). The spread of epidemic disease on networks. *Physical Review E*, **66**, 016128.

Newman, M. E. J. (2003). Mixing patterns in networks. *Physical Review E*, **67**, 026126.

Newman, M. E. J. (2006). Modularity and community structure in networks. *Proceedings of the National Academy of Sciences*, **103**, 8577–8582.

Newman, M. E. J. (2010). *Networks: an introduction*. Oxford University Press, Oxford.

Newman, M. E. J., and Girvan, M. (2004). Finding and evaluating community structure in networks. *Physical Review E*, **69**, 026113.

Newton, I. (1985). Lifetime reproductive output of female sparrowhawks. *The Journal of Animal Ecology*, **54**, 241–253.

Nicolaus, M., Bouwman, K. M., and Dingemanse, N. J. (2008). Effect of PIT tags on the survival and recruitment of great tits *Parus major*. *Ardea*, **96**, 286–292.

Nieh, J. C. (1999). Stingless-bee communication. *American Science*, **87**, 428–435.

Nieh, J. C., Barreto, L. S., Contrera, F. A. L., and Imperatriz-Fonseca, V. L. (2004). Olfactory eavesdropping by a competitively foraging stingless bee, *Trigona spinipes*. *Proceedings of the Royal Society B: Biological Sciences*, **271**, 1633–1640.

Nilsson, J.-A., and Smith, H. G. (1985). Early fledgling mortality and the timing of juvenile dispersal in the marsh tit *Parus palustris*. *Ornis Scandinavica*, **16**, 293–298.

Noad, M. J., Cato, D. H., Bryden, M. M., Jenner, M. N., and Jenner, K. C. S. (2000). Cultural revolution in whale songs. *Nature*, **408**, 537.

Nowak, M. A. (2006). Five rules for the evolution of cooperation. *Science*, **314**, 1560–1563.

Nowak, M. A., Bonhoeffer, S., and May, R. M. (1994). Spatial games and the maintenance of cooperation. *Proceedings of the National Academy of Sciences*, **91**, 4877–4881.

Nowak, M. A., and May, R. M. (1992). Evolutionary games and spatial chaos. *Nature*, **359**, 826–829.

Nowak, M. A., and Roch, S. (2007). Upstream reciprocity and the evolution of gratitude. *Proceedings of the Royal Society B: Biological Sciences*, **274**, 605–610.

Nowak, M. A., and Sigmund, K. (1992). Tit for tat in heterogeneous populations. *Nature*, **355**, 250–253.

Nowak, M., and Sigmund, K. (1993). A strategy of win–stay, lose–shift that outperforms tit-for-tat in the Prisoner's Dilemma game. *Nature*, **364**, 56–58.

Nowak, M. A., and Sigmund, K. (1998). Evolution of indirect reciprocity by image scoring. *Nature*, **393**, 573–577.

Nowak, M. A., and Sigmund, K. (2005). Evolution of indirect reciprocity. *Nature*, **437**, 1291–1298.

Nowak, M. A., Tarnita, C. E., and Antal, T. (2010). Evolutionary dynamics in structured populations. *Philosophical Transactions of the Royal Society B: Biological Sciences*, **365**, 19–30.

Nuñez, C. M. V., Adelman, J. S., and Rubenstein, D. I. (2010). Immunocontraception in wild horses (*Equus caballus*) extends reproductive cycling beyond the normal breeding season. *PLoS ONE*, **5**, e13635.

Nuñez, C. M., Adelman, J. S., and Rubenstein, D. I. (2014). Sociality increases juvenile survival after a catastrophic event in the feral horse (*Equus caballus*). *Behavioral Ecology*, doi:10.1093/beheco/aru163

Nunn, C., L. (2012). Primate disease ecology in comparative and theoretical perspective. *American Journal of Primatology*, **74**, 497–509.

Nunney, L. (1985). Group selection altruism and structured-deme models. *American Naturalist*, **126**, 212–230.

Oates-O'Brien, R. S., Farver, T. B., Anderson-Vicino, K. C., McCowan, B., and Lerche, N. W. (2010). Predictors of matrilineal overthrows in large captive breeding groups of rhesus macaques (*Macaca mulatta*). *Journal of the American Association for Laboratory Animal Science*, **49**, 196–201.

O'Connor, D., and Shine, R. (2003). Lizards in 'nuclear families': a novel reptilian social system in *Egernia saxatilis* (Scincidae). *Molecular Ecology*, **12**, 743–752.

O'Connor, D., and Shine, R. (2006). Kin discrimination in the social lizard *Egernia saxatilis* (Scincidae). *Behavioral Ecology*, **17**, 206–211.

O'Donnell, S., and Bulova, S. J. (2007). Worker connectivity: a simulation model of variation in worker communication and its effects on task performance. *Insectes Sociaux*, **54**, 211–218.

Oh, K. P., and Badyaev, A. V. (2006). Adaptive genetic complementarity in mate choice coexists with selection for elaborate sexual traits. *Proceedings of the Royal Society B: Biological Sciences*, **273**, 1913–1919.

Oh, K. P., and Badyaev, A. V. (2010). Structure of social networks in a passerine bird: consequences for sexual selection and the evolution of mating strategies. *American Naturalist*, **176**, E80–E89.

Ohtsuki, H., Hauert, C., Lieberman, E., and Nowak, M. A. (2006). A simple rule for the evolution of cooperation on graphs and social networks. *Nature*, **441**, 502–505.

Ohtsuki, H., and Nowak, M. A. (2007). Direct reciprocity on graphs. *Journal of Theoretical Biology*, **247**, 462–470.

Oi, T. (1988). Sociological study on the troop fission of wild Japanese monkeys (*Macaca fuscata yakui*) on Yakushima Island. *Primates*, **29**, 1–19.

Okasha, S. (2004a). Multilevel selection and the partitioning of covariance : a.comparison of three approaches. *Evolution*. **58**, 486–494.

Okasha, S. (2004b). Multi-level selection, covariance and contextual analysis. *British Journal for the Philosophy of Science*. **55**, 481–504.

Olesen, J. M., Bascompte, J., Dupont, Y. L., and Jordano, P. (2007). The modularity of pollination networks. *Proceedings of the National Academy of Sciences*, **104**, 19891–19896.

Opsahl, T., and Panzarasa, P. (2009). Clustering in weighted networks. *Social Networks*, **31**, 155–163.

Orbell, J. M., Schwartz-Shea, P., and Simmons, R. T. (1984). Do cooperators exit more readily than defectors? *The American Political Science Review*, **78**, 147–162.

Ord, T. J., Blumstein, D. T., and Evans, C. S. (2002). Ecology and signal evolution in lizards. *Biological Journal of the Linnean Society*, **77**, 127–148.

Ord, T. J., and Stamps, J. A. (2008). Alert signals enhance animal communication in 'noisy' environments. *Proceedings of the National Academy of Sciences*, **105**, 18830–18835.

Oster, G. F., and Wilson, E. O. (1978). *Caste and ecology in the social insects*. Princeton University Press, Princeton, NJ.

Osterwalder, K., Klingenbock, A., and Shine, R. (2004). Field studies on a social lizard: home range and social organization in an Australian skink, *Egernia major*. *Austral Ecology*, **29**, 241–249.

Otte, D. (1974). Effects and functions in the evolution of signalling systems. *Annual Review of Ecology and Systematics*, **5**, 385–417.

Otter, K., McGregor, P. K., Terry, A. M. R., Burford, F. R. L., Peake, T. M., and Dabelsteen, T. (1999). Do female great tits (*Parus major*) assess males by eavesdropping? A field study using interactive song playback. *Proceedings of the Royal Society B: Biological Sciences*, **266**, 1305–1309.

Otterstatter, M. C., and Thomson, J. D. (2007). Contact networks and transmission of an intestinal pathogen in bumble bee (*Bombus impatiens*) colonies. *Oecologia*, **154**, 411–421.

Pacala, S., Gordon, D. M., and Godfray, H. C. J. (1996). Effects of social group size on information transfer and task allocation. *Evolutionary Ecology*, **10**, 127–165.

Pacheco, J. M., Traulsen, A., and Nowak, M. A. (2006). Coevolution of strategy and structure in complex networks with dynamical linking. *Physical Review Letters*, **97**, 258103.

Packer, C. (1977). Reciprocal altruism in *Papio anubis*. *Nature*, **265**, 441–443.

Packer, C. (1986). The ecology of sociality in felids. In: Rubenstein, D. I., and Wrangham, R. W., eds. *Ecological aspects of social evolution*, pp. 429–451. Princeton University Press, Princeton, NJ.

Page, R. E., and Mitchell, S. D. (1991). Self organization and adaptation in insect societies. In: Fine, A., Forbes, M., and Wessels, L., eds. *PSA: proceedings of the biennial meeting of the Philosophy of Science Association, 1990, Volume 2*, pp. 289–298. Philosophy of Science Association, East Lansing, MI.

Pagel, M., and Dawkins, M. S. (1997). Peck orders and group size in laying hens: 'future contracts' for non-aggression. *Behavioural Processes*, **20**, 13–25.

Pankiw, T., Waddington, K., and Page, R. (2001). Modulation of sucrose response thresholds in honey bees (*Apis mellifera* L.): influence of genotype, feeding, and foraging experience. *Journal of Comparative Physiology A*, **187**, 293–301.

Parker, G. A. (1970). Sperm competition and its evolutionary consequences in insects. *Biological Reviews*, **45**, 525–567.

Parker, G. A. (1998). Sperm competition and the evolution of ejaculates: towards a theory base. In: Birkhead, T. R., and Møller, A. P., eds. *Sperm competition and sexual selection*, pp.3–54. Academic Press, London.

Parker, G. A., and Birkhead, T. R. (2013). Polyandry: the history of a revolution. *Philosophical Transactions of the Royal Society B: Biological Sciences*, **368**, 20120335.

Parker, G. A., and Pizzari, T. (2010). Sperm competition and ejaculate economics. *Biological Reviews*, **85**, 897–934.

Parra, G. J., Corkeron, P. J., and Arnold, P. (2011). Grouping and fission–fusion dynamics in Australian snubfin and Indo-Pacific humpback dolphins. *Animal Behaviour*, **82**, 1423–1433.

Parsons, J., and Baptista, L. F. (1980). Crown color and dominance in the white-crowned sparrow. *Auk*, **97**, 807–815.

Pascalis, O., and Bachevalier, J. (1998). Face recognition in primates: a cross-species study. *Behavioural Processes*, **43**, 87–96.

Patriquin, K. J., Leonard, M. L., Broders, H. G., and Garroway, C. J. (2010). Do social networks of female northern long-eared bats vary with reproductive period and age? *Behavioral Ecology and Sociobiology*, **64**, 899–913.

Patzner, R. A. (2008). Reproductive strategies of fish. In: Rocha, M. J., Arukwe, A., and Kapoor, B. G., eds. *Fish reproduction*, pp. 311–350. Science Publishers, Enfield, NH.

Payne, R., and Webb, D. (1971). Orientation by means of long-range acoustic signaling in baleen whales. *Annals of the New York Academy of Sciences*, **188**, 110–141.

Payne, R. S., and Guinee, L. N. (1983). Humpback whale (*Megaptera novaeangliae*) songs as indicators of 'stocks'. In: Payne, R. S., ed. *Communication and behaviour of whales*, pp. 333–358. Westview Press, Boulder, CO.

Payne, R. S., and McVay, S. (1971). Songs of humpback whales. *Science*, **173**, 587–597.

Peake, T. M. (2005). Eavesdropping in communication networks. In: McGregor, P. K., ed. *Animal communication networks*, pp. 13–37. Cambridge University Press, Cambridge.

Peake, T. M., Terry, A. M. R., McGregor, P. K., and Dabelsteen, T. (2002). Do great tits assess rivals by combining direct experience with information gathered by eavesdropping? *Proceedings of the Royal Society B: Biological Sciences*, **269**, 1925–1929.

Penn, D., Schneider, G., White, K., Slev, P., and Potts, W. (1998). Influenza infection neutralizes the attractiveness of male odour to female mice (*Mus musculus*). *Ethology*, **104**, 685–694.

Pennisi, E. (2009). *On the origin of cooperation*. *Science*, **325**, 1196–1199.

Perc, M., and Szolnoki, A. (2010). Coevolutionary games—a mini review. *BioSystems*, **99**, 109–125.

Perkins, S. E., Cagnacci, F., Stradiotto, A., Arnoldi, D., and Hudson, P. J. (2009). Comparison of social networks derived from ecological data: implications for inferring infectious disease dynamics. *Journal of Animal Ecology*, **78**, 1015–1022.

Perkins, S. E., Cattadori, I. M., Tagliapietra, V., and Hudson, P. J. (2003). Empirical evidence for key hosts in persistence of a tick-borne disease. *International Journal for Parasitology*, **9**, 909–917.

Perkins, T. A., Scott, T. W., Le Menach, A., and Smith, D. L. (2013). Heterogeneity, mixing, and the spatial scales of mosquito-borne pathogen transmission. *PLoS Computational Biology*, **9**, e1003327.

Pernetta, A. P., Reading, C. J., and Allen, J. A. (2009). Chemoreception and kin discrimination by neonate smooth snakes, *Coronella austriaca*. *Animal Behaviour*, **77**, 363–368.

Perreault, C. (2010). A note on reconstructing animal social networks from independent small-group observations. *Animal Behaviour*, **80**, 551–562.

Perry, G., Wallace, M. C., Perry, D., Curzer, H., and Muhlberger, P. (2011). Toe clipping of amphibians and reptiles: science, ethics and the law. *Journal of Herpetology*, **45**, 547–555.

Perry, S., Baker, M., Fedigan, L., Gros-Louis, J., Jack, K., MacKinnon, K. C, et al. (2003). Social conventions in wild white-faced capuchin monkeys. *Current Anthropology*, **44**, 241–268.

Petit, O., and Bon, R. (2010). Decision-making processes: the case of collective movements. *Behavioural Processes*, **84**, 635–647.

Petit, S., Waudby, H. P., Walker, A. T., Zanker, R., and Rau, G. (2012). A non-mutilating method for marking small wild mammals and reptiles. *Australian Journal of Zoology*, **60**, 64–71.

Pfeiffer, T., Rutte, C., Killingback, T., Taborsky, M., and Bonhoeffer, S. (2005). Evolution of cooperation by generalized reciprocity. *Proceedings of the Royal Society B: Biological Sciences*, **272**, 1115–1120.

Pike, T. W., Samanta, M., Lindstroem, J., and Royle, N. J. (2008). Behavioural phenotype affects social interactions in an animal network. *Proceedings of the Royal Society B: Biological Sciences*, **275**, 2515–2520.

Pinter-Wollman, N., Hobson, E. A., Smith, J. E., Edelman, A. J., Shizuka, D., de Silva, S., et al. (2014). The dynamics of animal social networks: analytical, conceptual, and theoretical advances. *Behavioral Ecology*, **25**, 242–255.

Pinter-Wollman, N., Wollman R., Guertz, A., Holmes, S., and Gordon, D. M. (2011). The effect of individual variation on the structure and function of interaction networks in harvester ants. *Journal of Royal Society Interface*, **8**, 1562–1573.

Pitcher, T. J., Green, D. A., and Magurran, A. E. (1986). Dicing with death: predator inspection behaviour in minnow shoals. *Journal of Fish Biology*, **28**, 439–448.

Pitman, R. L., Ballance, L. T., Mesnick, S. L., and Chivers, S. J. (2001). Killer whale predation on sperm whales: observations and implications. *Marine Mammal Science*, **17**, 494–507.

Piyapong, C., Butlin, R. K., Faria, J. J., Wang, J., and Krause, J. (2011). Kin assortment in juvenile shoals in wild guppy populations. *Heredity*, **106**, 747–756.

Pizzari, T., and Gardner, A. (2012). The sociobiology of sex: inclusive fitness consequences of inter-sexual interactions. *Philosophical Transactions of the Royal Society B: Biological Sciences*, **367**, 2314–2323.

Planqué, R., Britton, N. F., and Slabbekoorn, H. (2014). On the maintenance of bird song dialects. *Journal of Mathematical Biology*, **68**, 505–531.

Ploog, D. W., Blitz, J., and Ploog, F. (1963). Studies on social and sexual behavior of the squirrel monkey (*Saimiri sciureus*). *Folia Primatologica*, **1**, 29–66.

Plowright, R. C., and Plowright, C. M. S. (1988). Elitism in social insects: a positive feedback model. In: Jeanne, R. L., ed. *Interindividual behavioral variability in social insects*, pp. 419–431. Westview, Boulder, CO.

Poelman, E. H., Bruinsma, M., Zhu, F., Weldegergis, B. T., Boursault, A. E., Jongema, Y., et al. (2012). Hyperparasitoids use herbivore-induced plant volatiles to locate their parasitoid host. *PLoS Biology*, **10**, e1001435.

Poole, J. H., Tyack, P. L., Stoeger-Horwath, A. S., and Watwood, S. (2005). Animal behaviour: elephants are capable of vocal learning. *Nature*, **434**, 455–456.

Pope, D. S. (2005). Waving in a crowd: fiddler crabs signal in networks. In: McGregor, P. K., ed. *Animal communication networks*, pp. 252–276. Cambridge University Press, Cambridge.

Porphyre, T., McKenzie, J., and Stevenson, M. A. (2011). Contact patterns as a risk factor for bovine tuberculosis infection in a free-living adult brushtail possum *Trichosurus vulpecula* population. *Preventive Veterinary Medicine*, **100**, 221– 230.

Poulin, R. (1996). Sexual inequalities in helminth infections: a cost of being a male? *American Naturalist*, **147**, 287.

Preston, B. T., Stevenson, I. R., Pemberton, J., and Wilson, K. (2001). Dominant rams lose out by sperm depletion. *Nature*, **409**, 681–682.

Price-Rees, S. J., and Shine, R. (2011). A backpack method for attaching GPS. transmitters to bluetongue lizards (*Tiliqua*, Scincidae). *Herpetological Conservation and Biology*, **6**, 142–148.

Psorakis, I., Roberts, S. J., Rezek, I., and Sheldon, B. C. (2012). Inferring social network structure in ecological systems from spatio-temporal data streams. *Journal of the Royal Society Interface*, **9**, 3055–3066.

Psorakis, I., Voelkl, B., Garroway, C. J., Radersma, R., Aplin, L. M., Crates, R. A., et al. (2014). Inferring social structure from temporal data. Manuscript submitted for publication.

Putman, R. J. (1995). Ethical considerations and animal welfare in ecological field studies. *Biodiversity and Conservation*, **4**, 903–915.

Qi, Y., Noble, D. W. A., Fu, J., and Whiting, M. J. (2012). Spatial and social organization in a burrow-dwelling lizard (*Phrynocephalus vlangalii*) from China. *PLoS ONE*, **7**, e41130.

Qiu, F., and Hu, X. (2010). Modeling group structures in pedestrian crowd simulation. *Simulation Modelling Practice and Theory*, **18**, 190–205.

Quera, V., Beltran, F. S., and Dolado, R. (2010). Flocking behaviour: agent-based simulation and hierarchical leadership. *Journal of Artificial Societies and Social Simulation*, **13**, 8.

Quick, N. J., and Janik, V. M. (2012). Bottlenose dolphins exchange signature whistles when meeting at sea. *Proceedings of the Royal Society B: Biological Sciences*, **279**, 2539–2545.

Radersma, R., Garroway, C. J., Santure, A. W., Farine, D. R., Slate, J. and Sheldon, B. C. (2014). Social and spatial effects on the genetic structure of a wild bird population. Manuscript in preparation.

Radostits, O. M., Blood, D. C., and Gay, C. C. (1994). *Veterinary medicine: a textbook of the diseases of cattle, sheep, goats, pigs and horses*, 8th edn. Ballière Tindall, London.

Raihani, N., and Bshary, R. (2011). Resolving the iterated prisoner's dilemma: theory and reality. *Journal of Evolutionary Biology*, **24**, 1628–1639.

Ramos-Fernández, G., Boyer, D., Aureli, F., and Vick, L. G. (2009). Association networks in spider monkeys (*Ateles geoffroyi*). *Behavioral Ecology and Sociobiology*, **63**, 999–1013.

Rand, D. G., Arbesman, S., and Christakis, N. A. (2011). Dynamic social networks promote cooperation in experiments with humans. *Proceedings of the National Academy of Sciences*, **108**, 19193–19198.

Rankin, D. J., and Taborsky, M. (2009). Assortment and the evolution of generalized reciprocity. *Evolution*, **63**, 1913–1922.

Ratnieks, F. L. W., and Anderson, C. (1999). Task partitioning in insect societies. *Insectes Sociaux*, **46**, 95–108.

Raussi, S., Boissy, A., Delval, E., Pradel, P., Kaihilahti, J., and Veissier, I. (2005). Does repeated regrouping alter the social behaviour of heifers? *Applied Animal Behaviour Science*, **93**, 1–12.

Rea, E., Laflèche, J., Stalker, S., Guarda, B. K., Shapiro, H., Johnson, I., et al. (2007). Duration and distance of exposure are important predictors of transmission among community contacts of Ontario SARS cases. *Epidemiology and Infection*, **11**, 914–921.

Réale, D., Garant, D., Humphries, M. M., Bergeron, P., Careau, V., and Montiglio, P. O. (2010). Personality and the emergence of the pace-of-life syndrome concept at the population level. *Philosophical Transactions of the Royal Society B: Biological Sciences*, **365**, 4051–4063.

Réale, D., Reader, S. M., Sol, D., McDougall, P. T., and Dingemanse, N. J. (2007) Integrating animal temperament within ecology and evolution. *Biological Reviews*, **82**, 291–318.

Rendell, L., and Gero, S. (2014). The behavioral ecologist's essential social networks cookbook—comment on Pinter-Wollman et al. *Behavioral Ecology*, **25**, 257–258.

Rendell, L., Mesnick, S. L., Dalebout, M. L., Burtenshaw, J., and Whitehead, H. (2012). Can genetic differences explain vocal dialect variation in sperm whales, *Physeter macrocephalus*? *Behavior Genetics*, **42**, 332–343.

Rendell, L., and Whitehead, H. (2001). Culture in whales and dolphins. *Behavioral and Brain Sciences*, **24**, 309–324.

Rendell, L. E., and Whitehead, H. (2003). Vocal clans in sperm whales (*Physeter macrocephalus*). *Proceedings of the Royal Society B: Biological Sciences*, **270**, 225–231.

Reynolds, J. D., Goodwin, N. B., and Freckleton, R. P. (2002). Evolutionary transitions in parental care and live bearing in vertebrates. *Philosophical Transactions of the Royal Society B: Biological Sciences*, **357**, 269–281.

Reynolds, R. G., and Fitzpatrick, B. M. (2007). Assortative mating in poison-dart frogs based on an ecologically important trait. *Evolution*, **61**, 2253–2259.

Reynolds, V. (1970). Roles and role change in monkey society: the consort relationship of rhesus monkeys. *Man*, **5**, 449–465.

Riesch, R., Barrett-Lennard, L. G., Ellis, G. M., Ford, J. K. B., and Deecke, V. B. (2012). Cultural traditions and the evolution of reproductive isolation: ecological speciation in killer whales? *Biological Journal of the Linnean Society*, **106**, 1–17.

Riolo, R. L., Cohen, M. D., and Axelrod, R. (2001). Evolution of cooperation without reciprocity. *Nature*, **414**, 441–443.

Rivier, C., and Rivest, S. (1991). Effect of stress on the activity of the hypothalami–pituitary-–gonadal axis: peripheral and central mechanisms. *Biology of Reproduction*, **45**, 523–532.

Robins, G., Pattison, P., and Elliott, P. (2001). Network models for social influence processes. *Psychometrika*, **66**, 161–189.

Robins, G., Pattison, P., Kalish, Y., and Lusher, D. (2007). An introduction to exponential random graph (p*) models for social networks. *Social Networks*, **29**, 173–191.

Robson, S. K., and Traniello, J. F. A. (1999). Key individuals and the organization of labor in ants. In: Detrain, C., Deneubourg, J. L., and Pasteels, J. M., eds. *Information processing in social insects*, pp. 239–259. Birkhauser, Basel.

Rocha, M. J., Arukwe, A., and Kapoor, B. G. (2008). *Fish reproduction*. Science Publishers, Enfield, NH.

Rollins, L. A., Browning, L. E., Holleley, C. E., Savage, J. L., Russell, A. F., and Griffith, S. C. (2012). Building genetic networks using relatedness information: a novel approach for the estimation of dispersal and characterization of group structure in social animals. *Molecular Ecology*, **21**, 1727–1740.

Roney, J. R., and Maestripieri, D. (2003). Social development and affiliation. In: Maestripieri, D., ed. *Primate psychology*, pp. 171–204. Harvard University Press, Cambridge, MA.

Ross, K. G. (2001). Molecular ecology of social behaviour: analyses of breeding systems and genetic structure. *Molecular Ecology*, **10**, 265–284.

Rowell, T. E. (1974). The concept of social dominance. *Behavioral Biology*, **11**, 131–154.

Royle, N. J., Pike, T. W., Heeb, P., Richner, H., and Koelliker, M. (2012). Offspring social network structure predicts fitness in families. *Proceedings of the Royal Society B: Biological Sciences*, **279**, 4914–4922.

Rubenstein, D. I. (1986). Ecology of sociality in horses and zebras. In: Rubenstein, D. I., and Wrangham, R. W., eds. *Ecological aspects of social evolution*, pp. 282–302. Princeton University Press, Princeton, NJ.

Rubenstein, D. I. (1994). The ecology of female social behavior in horses, zebras, and asses. In: Jarman, P., and Rossiter, A., eds. *Animal societies: individuals, interactions, and organization*, pp. 13–28. Kyoto University Press, Kyoto.

Rubenstein, D. I., and Hack, M. (2004). Natural and sexual selection and the evolution of multi-level societies: insights from zebras with comparisons to primates. In: Kappeler, P., and van Schaik, C. P., eds. *Sexual selection in primates: new and comparative perspectives*, pp. 266–279. Cambridge University Press, Cambridge.

Rubenstein, D. I., and Nuñez, C. M. (2009). Sociality and reproductive skew in horses and zebras.In Hager, R., and Jones, C. B., eds. *Reproductive skew in vertebrates: proximate and ultimate causes*, pp. 196–226. Cambridge University Press, Cambridge.

Rubenstein, D. I., Sundaresan, S., Fischhoff, I., and Saltz, D. (2007). Social networks in wild asses: comparing patterns and processes among populations. In: Stubbe, A., Kaczensky, P., Wesche, K., Samjaa, R., and Stubbe, M., eds. *Exploration into the biological resources of Mongolia, Volume 10*, pp. 159–176. Martin-Luther-University Halle-Wittenberg, Halle (Saale).

Rubenstein, D. I., and Wrangham, R. W., eds. (1986). *Ecological aspects of social evolution: birds and mammals*. Princeton University Press, Princeton, NJ.

Ruckstuhl, K., and Neuhaus, P. (2005). *Sexual segregation in vertebrates: ecology of the two sexes*. Cambridge University Press, Cambridge.

Russell, S. T., Kelley, J. L., Graves, J. A., Magurran, A. E. (2004). Kin structure and shoal composition dynamics in the guppy, *Poecilia reticulata*. *Oikos*, **106**, 520–526.

Rutte, C., and Taborsky, M. (2007). Generalized reciprocity in rats. *PLoS Biology*, **5**, e196.

Rutz, C., Bluff, L. A., Weir, A. A., and Kacelnik, A. (2007). Video cameras on wild birds. *Science*, **318**, 765.

Rutz, C., Burns, Z. T., James, R., Ismar, S. M. H., Burt, J., Otis, B., et al. (2012). Automated mapping of social networks in wild birds. *Current Biology*, **22**, R669–R671.

Rutz, C., and Hays, G. C. (2009). New frontiers in biologging science. *Biology Letters*, **5**, 289–292.

Ryder, T. B., Blake, J. G., Parker, P. G., and Loiselle, B. A. (2011). The composition, stability, and kinship of reproductive coalitions in a lekking bird. *Behavioral Ecology*, **22**, 282–290.

Ryder, T. B., McDonald, D. B., Blake, J. G., Parker, P. G., and Loiselle, B. A. (2008). Social networks in the lek-mating wire-tailed manakin (*Pipra filicauda*). *Proceedings of the Royal Society B: Biological Sciences*, **275**, 1367–1374.

Saavedra, S., Stouffer, D. B., Uzzi, B., and Bascompte, J. (2011). Strong contributors to network persistence are the most vulnerable to extinction. *Nature*, **478**, 233–235.

Sacchi, R., Scali, S., Pellitteri-Rosa, D., Pupin, F., Gentilli, A., Tettamanti, S., et al. (2010). Photographic identification in reptiles: a matter of scales. *Amphibia-Reptilia*, **31**, 489–502.

Sachs, J. L., Mueller, U. G., Wilcox, T. P., and Bull, J. J. (2004). The evolution of cooperation. *Quarterly Review of Biology*, **79**, 135–160.

Sade, D. S. (1965). Some aspects of parent–off spring and sibling relations in a group of rhesus monkeys, with a discussion of grooming. *American Journal of Physical Anthropology*, **23**, 1–18.

Sade, D. S. (1967). Determinants of dominance in a group of free-ranging rhesus monkeys. In: Stuart, A. A., ed. *Social communication among primates*. University of Chicago Press, Chicago.

Sade, D. S. (1972). Sociometrics of *Macaca mulatta*, I. Linkages and cliques in grooming matrices. *Folia Primatologica*, **18**, 196–223.

Sade, D. S. (1989). Sociometrics of *Macaca mulatta* III: n-path centrality in grooming networks. *Social Networks*, **11**, 273–292.

Sade, D. S. Altmann, M., Loy, J., Hausfater, G., and Breuggeman, J. A. (1988). Sociometrics of *Macaca mulatta*: II. Decoupling centrality and dominance in rhesus monkey social networks. *American Journal of Physical Anthropology*, **77**, 409–425.

Şahin, E., Labella, T. H., Trianni, V., Deneubourg, J. L., Rasse, P., Floreano, D., et al. (2002). SWARM-BOT: Pattern formation in a swarm of self-assembling mobile robots. In: *Proceedings of the IEEE International Conference on Systems, Man and Cybernetics, Hammamet, Tunisia*, pp. 6–11.

Sakaluk, S. K., and Belwood, J. J. (1984). Gecko phonotaxis to cricket calling song: a case of satellite predation. *Animal Behaviour*, **32**, 659–662.

Salathé, M., Bengtsson, L., Bodnar, T. J., Brewer, D. D., Brownstein, J. S., Buckee, C., et al. (2012). Digital epidemiology. *PLoS Computational Biology*, **8**, e1002616.

Salathé, M., and Jones, J. H. (2010). Dynamics and control of diseases in networks with community structure. *PLoS Computational Biology*, **6**, e1000736.

Santos, F. C., and Pacheco, J. M. (2006). A new route to the evolution of cooperation. *Journal of Evolutionary Biology*, **19**, 726–733.

Santos, F. C., Pacheco, J. M., and Lenaerts, T. (2006a) Cooperation prevails when individuals adjust their social ties. *PLoS Computational Biology*, **2**, 1284–1291.

Santos, F. C., Pacheco, J. M., and Lenaerts, T. (2006b) Evolutionary dynamics of social dilemmas in structured heterogeneous populations. *Proceedings of the National Academy of Sciences*, **103**, 3490–3494.

Santos, F. C., Rodrigues, J. F., and Pacheco, J. M. (2006). Graph topology plays a determinant role in the evolution of cooperation. *Proceedings of the Royal Society B: Biological Sciences*, **273**, 51–55.

Sapolsky, R. M. (2005). The influence of social hierarchy on primate health. *Science*, **308**, 648–652.

Sapolsky, R. M., and Share, L. J. (2004). A pacific culture among wild baboons: its emergence and transmission. *PLoS Biology*, **2**, E106.

Sargeant, B. L., and Mann, J. (2009). From social learning to culture: intrapopulation variation in bottlenose dolphins. In: Laland K. N., Galef B. G., Jr,eds. *The question of animal culture*, pp. 152–173. Harvard University Press, Cambridge, MA.

Šárová, R., Špinka, M., Panamá, J. L. A., Šimeáek, P. (2010). Graded leadership by dominant animals in a herd of female beef cattle on pasture. *Animal Behaviour*, **79**, 1037–1045.

Sartori, C., and Mantovani, R. (2013). Indirect genetic effects and the genetic bases of social dominance: evidence from cattle. *Heredity*, **110**, 3–9.

Saul, Z. M., and Filkov, V. (2007). Exploring biological network structure using exponential random graph models. *Bioinformatics*, **23**, 2604–2611.

Sayigh, L. S., Tyack, P. L., Wells, R. S., and Scott, M. D. (1990). Signature whistles of free-ranging bottlenose dolphins, *Tursiops truncatus*: stability and mother–offspring comparisons. *Behavioral Ecology and Sociobiology*, **26**, 247–260.

Schalk, G., and Forbes, M. R. (1997). Male biases in parasitism of mammals: effects of study type, host age, and parasite taxon. *Oikos*, **78**, 67–74.

Schartl, M., Wilde, B., Schlupp, I., and Parzefall, J. (1995). Evolutionary origin of a parthenoform, the Amazon molly *Poecilia formosa*, on the basis of a molecular genealogy. *Evolution*, **49**, 827–835.

Scheid, C. (2004). Diploma thesis, Université de Strasbourg, Strasbourg.

Scherr, H., Bowman, J., and Abraham, K. F. (2010). Migration and winter movements of double-crested cormorants breeding in Georgian Bay, Ontario. *Waterbirds*, **33**, 451–460.

Schlupp, I., Marler, C., and Ryan, M. J. (1994). Benefit to male sailfin mollies of mating with heterospecific females. *Science*, **263**, 373–374.

Schluter, D. (2000). *The ecology of adaptive radiation*. Oxford University Press, Oxford.

Schmid, V. S., and de Vries, H. (2013). Finding a dominance order most consistent with a linear hierarchy: an improved algorithm for the I&SI method. *Animal Behaviour*, **86**, 1097–1105.

Schmid-Hempel, P. (1998). *Parasites in social insects*. Princeton University Press, Princeton, NJ.

Schmidt, K. A., Dall, S. R., and Van Gils, J. A. (2009). The ecology of information: an overview on the ecological significance of making informed decisions. *Oikos*, **119**, 304–316.

Schmidt-Nielsen, K. (1972). *How animals work*. Cambridge University Press. Cambridge.

Schmolke, S. A., Li, Y. Z., and Gonyou, H. W. (2004). Effects of group size on social behavior following regrouping of growing-finishing pigs. *Applied Animal Behaviour Science*, **88**, 27–38.

Schneider, G., and Krueger, K. (2012). Third-party interventions keep social partners from exchanging affiliative interactions with others. *Animal Behaviour*, **83**, 377–387.

Scholl, J., and Naug, D. (2011). Olfactory discrimination of age-specific hydrocarbons generates behavioral segregation in a honeybee colony. *Behavioral Ecology and Sociobiology*, **65**, 1967–1973.

Schroeder, J., Cleasby, I. R., Nakagawa, S., Ockendon, N., and Burke, T. (2011). No evidence for adverse effects on fitness of fitting passive integrated transponders (PITs) in wild house sparrows *Passer domesticus*. *Journal of Avian Biology*, **42**, 271–275.

Schulz, T. M., Whitehead, H., Gero, S., and Rendell, L. (2008). Overlapping and matching of codas in vocal interactions between sperm whales: insights into communication function. *Animal Behaviour*, **76**, 1977–1988.

Schulz, T. M., Whitehead, H., Gero, S., and Rendell, L. (2011). Individual vocal production in a sperm whale (*Physeter macrocephalus*) social unit. *Marine Mammal Science*, **27**, 149–166.

Schürch, R., and Heg, D. (2010a). Life history and behavioral type in the highly social cichlid *Neolamprologus pulcher*. *Behavioral Ecology*, **21**, 588–598.

Schürch, R., and Heg, D. (2010b). Variation in helper type affects group stability and reproductive decisions in a cooperative breeder. *Ethology*, **116**, 257–269.

Schürch, R., Rothenberger, S., and Heg, D. (2010). The building-up of social relationships: behavioural types, social networks and cooperative breeding in a cichlid.

Philosophical Transactions of the Royal Society B: Biological Sciences, **365**, 4089–4098.

Scott, J. (2000). *Social network analysis: a handbook*, 2nd edn. SAGE Publications, London.

Searcy, W. A., and Nowicki, S. (2010). *The evolution of animal communication: reliability and deception in signaling systems*. Princeton University Press, Princeton, NJ.

Sedgewick, R. (2002). *Algorithms in C, Part 5: graph algorithms*, 3rd edn. Addison-Wesley, Boston, MA.

Seeley, T. D. (1982). Adaptive significance of the age polyethism schedule in honeybee colonies. *Behavioral Ecology and Sociobiology*, **11**, 287–293.

Seeley, T. D. (1989). Social foraging in honey bees: how nectar foragers assess their colony nutritional status. *Behavioral Ecology and Sociobiology*, **24**, 181–199.

Seinen, I., and Schram, A. (2006). Social status and group norms: indirect reciprocity in a repeated helping experiment. *European Economic Review*, **50**, 581–602.

Sendova-Franks, A. B., Hayward, R. K., Wulf, B., Klimek, T., James, R., Planqué, R., et al. (2010). Emergency networking: famine relief in ant colonies. *Animal Behaviour*, **79**, 473–485.

Seppänen, J. -T., Forsman, J. T., Mönkkönen, M., and Thomson, R. L. (2007). Social information use is a process across time, space, and ecology, reaching heterospecifics. *Ecology*, **88**, 1622–1633.

Seyfarth, R. M. (1980). The distribution of grooming and related behaviours among adult female vervet monkeys. *Animal Behaviour*, **28**, 798–813.

Seyfarth, R. M., and Cheney, D. L. (1984). Grooming, alliances and reciprocal altruism in vervet monkeys. *Nature*, **308**, 541–543.

Shah, B., Shine, R., Hudson, S., and Kearney, M. (2003). Sociality in lizards: why do thick-tailed geckos (*Nephrurus milii*) aggregate? *Behaviour*, **140**, 1039–1052.

Sherman, P. W., Lacey, E. A., Reeve, H. K., and Keller, L. (1995). The eusociality continuum. *Behavioral Ecology*, **6**, 102–108.

Shev, A., Hsieh, F., Beisner, B., and McCowan, B. (2012). Using Markov chain Monte Carlo (MCMC) to visualize and test the linearity assumption of the Bradley–Terry class of models. *Animal Behaviour*, **84**, 1523–1531.

Shik, J. (2010). The metabolic costs of building ant colonies from variably sized subunits. *Behavioral Ecology and Sociobiology*, **64**, 1981–1990.

Shine, R., Shine, T., Shine, J. M., and Shine, B. G. (2005). Synchrony in capture dates suggests cryptic social organization in sea snakes (*Emydocephalus annulatus*, Hydrophiidae). *Austral Ecology*, **30**, 805–811.

Shizuka, D., and McDonald, D. B. (2012). A social network perspective on measurements of dominance hierarchies. *Animal Behaviour*, **83**, 925–934.

Shoham, D. A., Tong, L., Lamberson, P. J., Auchincloss, A. H., Zhang, J., Dugas, L., et al. (2012). An actor-based model of social network influence on adolescent body size, screen time, and playing sports. *PLoS ONE*, **7**, e39795.

Shuster, S., and Wade, M. J. (2003). *Mating systems and strategies*. Princeton University Press, Princeton, NJ.

Shutt, K., MacLarnon, A., Heistermann, M., and Semple, S. (2007). Grooming in barbary macaques: better to give than to receive? *Biology Letters*, **3**, 231–233.

Siepielski, A. M., DiBattista, J. D., Evans, J. A., and Carlson, S. M. (2011). Differences in the temporal dynamics of phenotypic selection among fitness components in the wild. *Proceedings of the Royal Society B: Biological Sciences*, **278**, 1572–1580.

Sih, A., Bell, A., and Johnson, J. C. (2004). Behavioral syndromes: an ecological and evolutionary overview. *Trends in Ecology and Evolution*, **19**, 372–378.

Sih, A., Bell, A. M., Johnson, J. C., and Ziemba, R. E. (2004). Behavioral syndromes: an integrative overview. *The Quarterly Review of Biology*, **79**, 241–277.

Sih, A., Cote, J., Evans, M., Fogarty, S., and Pruitt, J. (2012). Ecological implications of behavioural syndromes. *Ecology Letters*, **15**, 278–289.

Sih, A., Hanser, S. F., and McHugh, K. A. (2009). Social network theory: new insights and issues for behavioral ecologists. *Behavioral Ecology and Sociobiology*, **63**, 975–988.

Sih, A., and Watters, J. V. (2005). The mix matters: behavioural types and group dynamics in water striders. *Behaviour*, **142**, 1417–1431.

Silk, J. B. (2007). The adaptive value of sociality in mammalian groups. *Philosophical Transactions of the Royal Society B: Biological Sciences*, **362**, 539–559.

Silk, J. B., Altmann, J., and Alberts, S. C. (2006). Social relationships among adult female baboons (*Papio cynocephalus*) I. Variation in the strength of social bonds. *Behavioral Ecology and Sociobiology*, **61**, 183–195.

Silk, J. B., Beehner, J. C., Bergman, T. J., Crockford, C., Engh, A. L., Moscovice, L. R., et al. (2009). The benefits of social capital: close social bonds among female baboons enhance offspring survival. *Proceedings of the Royal Society B: Biological Sciences*, **276**, 3099–3104.

Simões-Lopes, P. C., Fabian, M. E., and Menegheti, J. O. (1998). Dolphin interactions with the mullet artisinal fishing on southern Brazil: a qualitative and quantitative approach. *Revista Brasileira de Zoologia*, **15**, 709–726.

Simon, V. B. (2011). Communication signal rates predict interaction outcome in the brown anole lizard, *Anolis sagrei*. *Copeia*, **2011**, 38–45.

Simons, A. M. (2004). Many wrongs: the advantage of group navigation. *Trends in Ecology and Evolution*, **19**, 453–455.

Sinn, D. L., Gosling, S. D., and Moltschaniwskyj, N. A. (2008). Development of shy/bold behaviour in squid: context-specific phenotypes associated with developmental plasticity. *Animal Behaviour*, **75**, 433–442.

Širović, A., Hildebrand J. A., and Wiggins, S. M. (2007). Blue and fin whale call source levels and propagation range in the Southern Ocean. *The Journal of the Acoustical Society of America*, **122**, 1208–1215.

Skyrms, B., and Pemantle, R. (2000). A dynamic model of social network formation. *Proceedings of the National Academy of Sciences*, **97**, 9340–9346.

Smith, S. M. (1984). Flock switching in chickadees: why be a winter floater? *American Naturalist*, **123**, 81–98.

Smith, W. J. (1977). *The behavior of communicating*. Harvard University Press, Cambridge, MA.

Smolker, R., and Pepper, J. W. (1999). Whistle convergence among allied male bottlenose dolphins (Delphinidae, *Tursiops* sp.). *Ethology*, **105**, 595–617.

Smolker, R., Richards, A., Connor, R., Mann, J., and Berggren, P. (1997). Sponge carrying by dolphins (Delphinidae, *Tursiops* sp.): a foraging specialization involving tool use? *Ethology*, **103**, 454–465.

Smuts, B. B., Cheney, D. L., Seyfarth, R. M., Wrangham, R. W., and Struhsaker, T. T., eds. (1987). *Primate societies*. University of Chicago Press, Chicago.

Snijders, T. A. B. (2002). Markov chain Monte Carlo estimation of exponential random graph models. *Journal of Social Structure*, **3**, 1–40.

Snijders, T. A. B. (2011). Statistical models for social networks. *Annual Review of Sociology*, **37**, 131–153.

Snowberg, L. K., and Bolnick, D. I. (2008). Assortative mating by diet in a phenotypically unimodal but ecologically variable population of stickleback. *American Naturalist*, **172**, 733–739.

Snyder-Mackler, N., Beehner, J. C., and Bergman, T. J. (2012). Defining higher levels in the multilevel societies of geladas (*Theropithecus gelada*). *International Journal of Primatology*, 33, 1054–1068.

Soczka, L. (1974). Ethologie sociale et sociometrie: analyse de la structure d'un group de singes crabiers (*Macaca fascicularis*) en captivite. *Behaviour*, **50**, 254–269.

Sokal, R. R., and Oden, N. L. (1978). Spatial autocorrelation in biology: 1. Methodology. *Biological Journal of the Linnaean Society*, **10**, 199–228.

Solomon, N. G., and French, J. A. (1996). *Cooperative breeding in mammals*. Cambridge University Press, Cambridge.

Soltis, J., Leong, K., and Savage, A. (2005). African elephant vocal communication I: Antiphonal calling behaviour among affiliated females. *Animal Behaviour*, **70**, 579–587.

Sosnowka-Czajka, E., Skomorucha, I., Herbut, E., and Muchacka, R. (2007). Effect of management system and flock size on the behaviour of broiler chickens. *Annals of Animal Science*, **7**, 329–335.

Speed, C. W., Meekan, M. G., and Bradshaw, C. J. A. (2007). Spot the match—wildlife photo-identification using information theory. *Frontiers in Zoology*, **4**, 1–11.

Sridhar, H., Beauchamp, G., and Shanker, K. (2009). Why do birds participate in mixed-species foraging flocks? A large-scale synthesis. *Animal Behaviour*, **78**, 337–347.

Stamps, J. A., and Groothuis, T. G. G. (2010). Developmental perspectives on personality: implications for ecological and evolutionary studies of individual differencese. *Philosophical Transactions of the Royal Society B: Biological Sciences*, **365**, 4029–4041.

Stamps, J. A., and Krishnan, V. V. (1995). Territory acquisition in lizards: III. Competing for space. *Animal Behaviour*, **49**, 679–693.

Stamps, J. A., and Krishnan, V. V. (1997). Functions of fights in territory establishment. *American Naturalist*, **150**, 393–405.

Stamps, J. A., and Krishnan, V. V. (1998). Territory acquisition in lizards. IV. Obtaining high status and exclusive home ranges. *Animal Behaviour*, **55**, 461–472.

Stanca, L. (2009). Measuring indirect reciprocity: whose back do we scratch? *Journal of Economic Psychology*, **30**, 190–202.

Staniczenko, P. P. A., Kopp, J. C., and Allesina, S. (2013). The ghost of nestedness in ecological networks. *Nature Communications*, **4**, 1391.

Stanley, C. R., and Dunbar, R. I. M. (2013). Consistent social structure and optimal clique size revealed by social network analysis of feral goats, *Capra hircus*. *Animal Behaviour*, **85**, 771–779.

Stanton, M. A., and Mann, J. (2012). Early social networks predict survival in wild bottlenose dolphins. *PLoS ONE*, **7**, e47508.

Steele, J. H. (1985). A comparison of terrestrial and marine ecological systems. *Nature*, **313**, 355–358.

Stegmann, U. (2013). Information and influence in animal communication: a primer. In: Stegmann, U., ed. *Animal communication theory: information and influence*, pp. 1–42. Cambridge University Press, Cambridge.

Sterling, E. J. (1993). Patterns of range use and social organization in aye-ayes (*Daubentonia madagascariensis*) on Nosy Mangabe. In: Kappeler, P. M., and Ganzhorn, J. U., eds. *Lemur social systems and their ecological basis*, pp. 1–10. Plenum, New York.

Stevens, J. R., Cushman, F. A., and Hauser, M. D. (2005). Evolving the psychological mechanisms for cooperation. *Annual Review of Ecology Evolution and Systematics*, **36**, 499–518.

Stevens, J. R., and Hauser, M. D. (2004). Why be nice? Psychological constraints on the evolution of cooperation. *Trends in Cognitive Sciences*, **8**, 60–65.

Stevens, L., Goodnight, C. J., and Kalisz, S. (1995). Multilevel selection in natural populations of *Impatiens capensis*. *American Naturalist*, **145**, 513–526.

Stevick, P. T., Neves, M. C., Johansen, F., Engel, M. H., Allen, J., Marcondes, M. C. C., et al. (2011). A quarter of a world away: female humpback whale moves 10,000 km between breeding areas. *Biology Letters*, **7**, 299–302.

Stopher, K., Walling, C., Morris, A., Guinness, F. E., Clutton-Brock, T. H., Pemberton, J. M. et al. (2012). Shared spatial effects on quantitative genetic parameters: accounting for spatial autocorrelation and home range overlap reduces estimates of heritability in wild red deer. *Evolution*, **66**, 2411–2426.

Stowers, L., Holy, T. E., Meister, M., Dulac, C., and Koentges, G. (2002). Loss of sex discrimination and male–male aggression in mice deficient for TRP2. *Science*, **295**, 1493–1500.

Stumpf, M. P. H., Wiuf, C., and May, R. M. (2005). Subnets of scale-free networks are not scale-free: sampling properties of networks. *Proceedings of the National Academy of Sciences*, **102**, 4221–4224.

Sueur, C., Jacobs, A., Amblard, F., Petit, O., and King, A. J. (2011). How can social network analysis improve the study of primate behavior? *American Journal of Primatology*, **73**, 703–719.

Sueur, C., and Petit, O. (2008). Organization of group members at departure is driven by social structure in *Macaca*. *International Journal of Primatology*, **29**, 1085–1098.

Sueur, C., Petit, O., De Marco, A., Jacobs, A. T., Watanabe, K., and Thierry, B. (2011). A comparative network analysis of social style in macaques. *Animal Behaviour*, **82**, 845–852.

Sueur, C., Petit, O., and Deneubourg, J. L. (2010). Short-term group fission processes in macaques: a social networking approach. *Journal of Experimental Biology*, **213**, 1338–1346.

Sugihara, G., and Ye, H. (2009). Complex systems: cooperative network dynamics. *Nature*, **458**, 979–980.

Sukumar, R. (1992). *The Asian elephant*. Cambridge University Press, Cambridge.

Sumpter, D. J. T. (2006). The principles of collective animal behaviour. *Philosophical Transactions of the Royal Society B: Biological Sciences*, **361**, 5–22.

Sumpter, D. J. T. (2010). *Collective animal behavior*. Princeton University Press, Princeton.

Sumpter, D. J. T., Krause, J., James, R., Couzin, I. D.,and Ward, A. J. W. (2008). Consensus decision-making by fish. *Current Biology*, **18**, 1773–1777.

Sundaresan, S. R., Fischhoff, I. R., and Dushoff, J. (2009). Avoiding spurious findings of nonrandom social structure in association data. *Animal Behaviour*, **77**, 1381–1385.

Sundaresan, S. R., Fischhoff, I. R., Dushoff, J., and Rubenstein, D. I. (2007). Network metrics reveal differences in social organization between two fission–fusion species, Grevy's zebra and onager. *Oecologia*, **151**, 140–149.

Sundaresan, S. R., Fischhoff, I. R. and Rubenstein, D. I. (2007). Male harassment influences female movements and associations in Grevy's zebra (*Equus grevyi*). *Behavioral Ecology*, **18**, 860–865.

Suzuki, R., Buck, J. R., and Tyack, P. L. (2006). Information entropy of humpback whale songs. *The Journal of the Acoustical Society of America*, **119**, 1849–1866.

Takahashi, H., and Furuichi, T. (1998). Comparative study of grooming relationships among wild Japanese macaques in Kinkazan A troop and Yakushima M troop. *Primates*, **39**, 365–374.

Tamashiro, K. L., Sakai, R. R., Shively, C. A., Karatsoreos, I.N., and Reagan, L. P. (2011). Chronic stress, metabolism, and metabolic syndrome. *Stress*, **14**, 468–474.

Tanner, C. J., and Jackson, A. L. (2012). Social structure emerges via the interaction between local ecology and individual behaviour. *Journal of Animal Ecology*, **81**, 260–267.

Tantipathananandh, C., and Berger-Wolf, T. Y. (2011). Finding communities in dynamic social networks. In: *2011 IEEE 11th International Conference On Data Mining*, pp. 1236–1241. IEEE Computer Society, Los Alamitos, CA.

Tarnita, C. E., Antal, T., Ohtsuki, H., and Nowak, M. A. (2009). Evolutionary dynamics in set structured populations. *Proceedings of the National Academy of Sciences*, **106**, 8601–8604.

Tatarenkov, A., Lima, S. M. Q., Taylor, D. S., and Avise, J. C. (2009). Long-term retention of self-fertilization in a fish clade. *Proceedings of the National Academy of Sciences*, **106**, 14456–14459.

Taylor, M. I., and Knight, M. E. (2008). Mating systems in fishes. In: Rocha, M. J., Arukwe, A., and Kapoor, B. G., eds. *Fish reproduction*, pp. 277–309. Science Publishers, Enfield, NH.

Taylor, P. D., Day, T., and Wild, G. (2007). Evolution of cooperation in a finite homogeneous graph. *Nature*, **447**, 469–472.

Templeton, C. N., and Greene, E. (2007). Nuthatches eavesdrop on variations in heterospecific chickadee mobbing alarm calls. *Proceedings of the National Academy of Sciences*, **104**, 5479–5482.

Templeton, C. N., Reed, V. A., Campbell, S. E., and Beecher, M. D. (2012). Spatial movements and social networks in juvenile male song sparrows. *Behavioral Ecology*, **23**, 141–152.

Terry, A. M. R., Peake, T. M., and McGregor, P. K. (2005). The role of vocal individuality in conservation. *Frontiers in Zoology*, **2**, 10.

Teunis, P., Heijne, J. C. M., Sukhrie, F., van Eijkeren, J., Koopmans, M., and Kretzschmar, M. (2013). Infectious disease transmission as a forensic problem: who infected whom? *Journal of the Royal Society Interface*, **10**, 20120955.

Thakar, J., Pathak, A. K., Murphy, L., Albert, R., and Cattadori, I. M. (2012). Network model of immune responses reveals key effectors to single and coinfection kinetics by a respiratory bacterium and a gastrointestinal helminth. *PLoS Computational Biology*, **8**, e100234.

Theraulaz, G., Bonabeau, E., and Deneubourg, J. L. (1998). Response threshold reinforcement and division of labour in insect societies. *Proceedings of the Royal Society B: Biological Sciences*, **265**, 327–332.

Thierry, B., Singh, M., and Kaumanns, W. (2004). *Macaque societies: a model for the study of social organization*. Cambridge University Press, Cambridge.

Thomas, P. O. R., Croft, D. P., Morrell, L. J., Davis, A., Faria, J. J., Dyer, J. R. G., et al. (2008). Does defection during predator inspection affect social structure in wild shoals of guppies? *Animal Behaviour*, **75**, 43–53.

Thompson, A. B., and Hare, J. F. (2010). Neighbourhood watch: multiple alarm callers communicate directional predator movement in Richardson's ground squirrels, *Spermophilus richardsonii*. *Animal Behaviour*, **80**, 269–275.

Thompson, C. W., and Moore, M. C. (1991). Throat colour reliably signals status in male tree lizards, *Urosaurus ornatus*. *Animal Behaviour*, **42**, 745–753.

Thorne, T. E. and Williams, E. S. (2005). Disease and endangered species: the black-footed ferret as a recent example. *Conservation Biology*, **2**, 66–74.

Thornhill, R. (1983). Cryptic female choice and its implications in the scorpionfly. *American Naturalist*, **122**, 765–788.

Thornhill, R., and Alcock, J. (1983). *The evolution of insect mating systems*. Harvard University Press, Cambridge, MA.

Thornton, A., and Malapert, A. (2009). The rise and fall of an arbitrary tradition: an experiment with wild meerkats. *Proceedings of the Royal Society* B: Biological Sciences, **276**, 1269–1276.

Tiddi, B., Aureli, F., Schino, G., and Voelkl, B. (2011). Social relationships between adult females and the alpha male in wild tufted capuchin monkeys. *American Journal of Primatology*, **73**, 812–820.

Tinbergen, N. (1951). *The study of instinct*. Oxford University Press, Oxford.

Tofts, C., and Franks, N. R. (1992). Doing the right thing: ants, honeybees and naked mole-rats. *Trends in Ecology and Evolution*, **7**, 346–349.

Tomley, F. M., and Shirley, M. W. (2009). Livestock infectious diseases and zoonoses. *Philosophical Transactions of the Royal Society B: Biological Sciences*, **364**, 2637–2642.

Tompkins, D. M., Dunn, A. M., Smith, M. J., and Telfer, S. (2011). Wildlife diseases: from individuals to ecosystems. *Journal of Animal Ecology*, **80**, 19–38.

Toth, C. A., Mennill, D. J., and Ratcliffe, L. M. (2012). Evidence for multicontest eavesdropping in chickadees. *Behavioral Ecology*, **23**, 836–842.

Traulsen, A., and Claussen, J. C. (2004). Similarity-based cooperation and spatial segregation. *Physical Review E*, **70**, 046128.

Trianni, V., and Dorigo, M. (2006). Self-organisation and communication in groups of simulated and physical robots. *Biological Cybernetics*, **95**, 213–231.

Trierweiler, C., Mullie, W. C., Drent, R. H., Exo, K. M., Komdeur, J., Bairlein, F., et al. (2013). A Palaearctic migratory raptor species tracks shifting prey availability within its wintering range in the Sahel. *Journal of Animal Ecology*, **82**, 107–120.

Trivers, R. L. (1971). Evolution of reciprocal altruism. *Quarterly Review of Biology*, **46**, 35–57.

Tsuji, K. (1995). Reproductive conflicts and levels of selection in the ant *Pristomyrmex pungens*: contextual analysis and partitioning of covariance. *American Naturalist*, **146**, 586–607.

Turlings, T. C. J., Tumlinson, J. H., and Lewis, W. J. (1990). Exploitation of herbivore-induced plant odors by host-seeking parasitic wasps. *Science*, **250**, 1251–1253.

Tuttle, M. D., and Ryan, M. J. (1982). The role of synchronized calling, ambient light, and ambient noise in anti-bat-predator behavior of a tree frog. *Behavioral Ecology and Sociobiology*, **11**, 125–131.

Tyack, P. L. (1986). Population biology, social behavior, and communication in whales and dolphins. *Trends in Ecology and Evolution*, **1**, 144–150.

Tyack, P. L., and Sayigh, L. S. (1997). Vocal learning in cetaceans. In: Snowdon, C. T., and Hausberger, M., eds. *Social influences on vocal development*. Cambridge University Press, Cambridge.

Ugelvig, L. V., and Cremer, S. (2012). Effects of social immunity and unicoloniality on host–parasite interactions in invasive insect societies. *Functional Ecology*, **26**, 1300–1312.

Ulrich, W., Almeida-Neto, M., and Gotelli, N. J. (2009). A consumer's guide to nestedness analysis. *Oikos*, **118**, 3–17.

Valente, T. (2005). Network models and methods for studying the diffusions of innovations. In: Carrington, P., Scott, J., and Wasserman, S., eds. *Models and methods in social network analysis*, pp. 98–116. Cambridge University Press, Cambridge.

Valone, T. J. (2007). From eavesdropping on performance to copying the behavior of others: a review of public information use. *Behavioral Ecology and Sociobiology*, **62**, 1–14.

Van de Waal, E., Renevey, N., Favre, C. M., and Bshary, R. (2010). Selective attention to philopatric models causes directed social learning in wild vervet monkeys. *Proceedings of the Royal Society B: Biological Sciences*, **277**, 2105–2111.

Van den Brink, R., and Gilles, R. P. (2000). Measuring domination in directed networks. *Social Networks*, **22**, 141–157.

Van Doorn, G. S., and Taborsky, M. (2012). The evolution of generalized reciprocity on social interaction networks. *Evolution*, 66, 651–664.

Van Gils, J. A., Munster, V. J., Radersma, R., Liefhebber, D., Fouchier, R. A. M., and Klaassen, M. (2007). Hampered foraging and migratory performance in swans infected with low-pathogenic avian influenza A virus. *PLoS ONE*, **2**, e184.

VanderWaal, K. L., Atwill, E. R., Hooper, S., Buckle, K., and McCowan, B. (2013). Network structure and prevalence of *Cryptosporidium* in Belding's ground squirrels. *Behavioral Ecology and Sociobiology*, **67**, 1951–1959.

VanderWaal, K. L., Atwill, E. R., Isbell, L. A., and McCowan, B. (2014a). Linking social and pathogen transmission networks using microbial genetics in giraffe (*Giraffa camelopardalis*). *Journal of Animal Ecology*, **83**, 406–414.

VanderWaal, Atwill, Isbell, McCowan, (2014b) VanderWaal, K. L., Atwill, E. R., Isbell, L. A., and McCowan, B. (2014b). Quantifying microbe transmission networks for wild and domestic ungulates in Kenya. *Biological Conservation*, **169**, 136–146.

VanderWaal, K. L., Wang, H., McCowan, B., Fushing, H., and Isbell, L. A. (2014). Multilevel social organization and space use in reticulated giraffe (*Giraffa camelopardalis*). *Behavioral Ecology*, **25**, 17–26.

Vasseur, D. A., and Yodzis, P. (2004). The colour of environmental noise. *Ecology*, **85**, 1146–1152.

Ventura, R., Majolo, B., Koyama, N. F., Hardie, S., and Schino, G. (2006). Reciprocation and interchange in wild Japanese macaques: grooming, cofeeding, and agonistic support. *American Journal of Primatology*, **68**, 1138–1149.

Vicente, J., Delahay, R. J., Walker, N. J., and Cheeseman, C. L. (2007). Social organization and movement influence the incidence of bovine tuberculosis in an undisturbed high-density badger *Meles meles* population. *Journal of Animal Ecology*, **76**, 348–360.

Vicsek, T., Czirók, A., Ben-Jacob, E., Cohen, I., and Shochet, O. (1995). Novel type of phase transition in a system of self-driven particles. *Physical Review Letters*, **75**, 1226–1229.

Vital, C., and Martins, E. P. (2011). Strain differences in zebrafish (*Danio rerio*) social roles and their impact on group task performance. *Journal of Comparative Psychology*, **125**, 278–285.

Voelkl, B. (2010). The 'Hawk–Dove' game and the speed of the evolutionary process in small heterogeneous populations. *Games*, **1**, 103–116.

Voelkl, B., and Kasper, C. (2009). Social structure of primate interaction networks facilitates the emergence of cooperation. *Biology Letters*, **5**, 462–464.

Voelkl, B., Kasper, C., and Schwab, C. (2011). Network measures for dyadic interactions: stability and reliability. *American Journal of Primatology*, **73**, 731–740.

Voelkl, B., and Noë, R. (2008). The influence of social structure on the propagation of social information in artificial primate groups: a graph-based simulation approach. *Journal of Theoretical Biology*, **257**, 77–86.

Voelkl, B., and Noë, R. (2010). Simulation of information propagation in real-life primate networks: longevity, fecundity, fidelity. *Behavioral Ecology and Sociobiology*, **64**, 1449–1459.

Von Keyserlingk, M. A. G., Olenick, D., and Weary, D. M. (2008). Acute behavioral effects of regrouping dairy cows. *Journal of Dairy Science*, **91**, 1011–1016.

Von Rohr, C. R., Koski, S. E., Burkart, J. M., Caws, C., Fraser, O. N., Ziltener, A., et al. (2012). Impartial third-party interventions in captive chimpanzees: a reflection of community concern. *PLoS ONE*, **7**, e32494.

Wagner, D., Tissot, M., and Gordon, D. M. (2001). Task-related environment alters the cuticular hydrocarbon composition of harvester ants. *Journal of Chemical Ecology*, **27**, 1805–1819.

Walker, M. L., and Herndon, J. G. (2008). Menopause in nonhuman primates? *Biology of Reproduction*, **79**, 398–406.

Ward, A. J. W., Botham, M. S., Hoare, D. J., James, R., Broom, M., Godin J-G. J., and Krause, J. (2002). Association patterns and shoal fidelity in the three-spined stickleback. *Proceedings of the Royal Society B: Biological Sciences*, **269**, 2451–2455.

Ward, A. J. W., Hart, P. J. B., and Krause, J. (2004). The effects of habitat- and diet-based cues on association preferences in three-spined sticklebacks. *Behavioral Ecology*, **15**, 925–929.

Ward, A. J. W., Herbert-Read, J. E., Sumpter, D. J. T., and Krause, J. (2011). Fast and accurate decisions through collective vigilance. *Proceedings of the National Academy of Sciences*, **108**, 2312–2315.

Ward, A. J. W., Holbrook, R. I., Krause, J., and Hart, P. (2005). Social recognition in sticklebacks: the role of direct experience and habitat cues. *Behavioral Ecology and Sociobiology*, **57**, 575–583.

Ward, A. J. W., Sumpter, D. J. T., Couzin, I. D., Hart, P. J. B., and Krause, J. (2008). Quorum decision-making facilitates information transfer in fish shoals. *Proceedings of the National Academy of Sciences*, **105**, 6948–6953.

Ward, A. J. W., Webster, M. M., Magurran, A. E., Currie, S., and Krause, J. (2009). Species and population differences in social recognition between fishes: a role for ecology? *Behavioral Ecology*, **20**, 511–516.

Waret-Szkuta, A., Ortiz-Peláez, A., Pfeiffer, D. U., Roger, F., and Guitian, F. J. (2011). Herd contact structure based on shared use of water points and grazing points in the highlands of Ethiopia. *Epidemiology and Infection*, **139**, 875–885.

Wasserman, S., and Faust, K. (1994). *Social network analysis: methods and applications*. Cambridge University Press, Cambridge.

Waters, J. S., and Fewell, J. H. (2012). Information processing in social insect networks. *PLoS ONE*, **7**, e40337.

Waters, J. S., Holbrook, C. T., Fewell, J.H., and Harrison, J.F. (2010). Allometric scaling of metabolism, growth, and activity in whole colonies of the seed-harvester ant *Pogonomyrmex californicus*. *American Naturalist*, **176**, 501–510.

Watt, D. J. (1986). Relationship of plumage variability, size and sex to social dominance in Harris' sparrows. *Animal Behaviour*, **34**, 16–27.

Watt, M. J., and Joss, J. M. P. (2003). Structure and function of visual displays produced by male jacky dragons, *Amphibolurus muricatus*, during social interactions. *Brain Behaviour and Evolution*, **61**, 172–183.

Watts, D. J., and Strogatz, S. H. (1998). Collective dynamics of 'small-world' networks. *Nature*, **393**, 440–442.

Watwood, S. L., Miller, P. J. O., Johnson, M., Madsen, P. T., and Tyack, P. L. (2006). Deep-diving foraging behaviour of sperm whales (*Physeter macrocephalus*). *Journal of Animal Ecology*, **75**, 814–825.

Watwood, S. L., Tyack, P. L., and Wells, R. S. (2004). Whistle sharing in paired male bottlenose dolphins, *Tursiops truncatus*. *Behavioral Ecology and Sociobiology*, **55**, 531–543.

Wearmouth, V. J., and Sims, D. W. (2008). Sexual segregation in marine fish, reptiles, birds and mammals: behaviour patterns, mechanisms and conservation implications. *Advances in Marine Biology*, **54**, 107–170.

Weber, N., Carter, S. P., Dall, S. R. X., Delahay, R. J., McDonald, J. L., Bearhop, S., et al. (2013). Badger social networks correlate with tuberculosis infection. *Current Biology*, **23**, 915–916.

Webster, J. P. (2007). The impact of *Toxoplasma gondii* on animal behaviour: playing cat and mouse. *Schizophrenia Bulletin*, **33**, 752–756.

Webster, M. M., Atton, N., Hoppitt, W. J. E., and Laland, K. N. (2013). Environmental complexity influences association network structure and network-based diffusion of foraging information in fish shoals. *American Naturalist*, **181**, 235–244.

Webster, M. M., Goldsmith, J., Ward, A. J. W., and Hart, P. J. B. (2007). Habitat-specific chemical cues influence association preferences and shoal cohesion in fish. *Behavioral Ecology and Sociobiology*, **62**, 273–280.

Webster, M. M., and Laland, K. N. (2009). Evaluation of a non-invasive tagging system for laboratory studies using three-spined sticklebacks *Gasterosteus aculeatus*. *Journal of Fish Biology*, **75**, 1868–1873.

Webster, M. M., Ward, A. J. W., and Hart, P. J. B. (2009). Individual boldness affects interspecific interactions in sticklebacks. *Behavioral Ecology and Sociobiology*, **63**, 511–520.

Wedekind, C. (1992). Detailed information about parasites revealed by sexual ornamentation. *Proceedings of the Royal Society B: Biological Sciences*, **247**, 169–174.

Wedekind, C., and Milinski, M. (2000). Cooperation through image scoring in humans. *Science*, **288**, 850–852.

Weimerskirch, H., Bonadonna, F., Bailleul, F., Mabille, G., Dell'Omo, G., and Lipp, H. P. (2002). GPS tracking of foraging albatrosses. *Science*, **295**, 1259.

Weinig, C., Johnston, J. A., Willis, C. G., and Maloof, J. N. (2007). Antagonistic multilevel selection on size and architecture in variable density settings. *Evolution*, **61**, 58–67.

Weinrich, M. T. (1991). Stable social associations among humpback whales (*Megaptera novaeangliae*) in the southern Gulf of Maine. *Canadian Journal of Zoology*, **69**, 3012–3019.

Weinrich, M. T., and Kuhlberg, A. E. (1991). Short-term association patterns of humpback whale (*Megaptera novaeangliae*) groups on their feeding grounds in the southern Gulf of Maine. *Canadian Journal of Zoology*, **69**, 3005–3011.

Wells, R. S. (1991). The role of long-term study in understanding the social structure of a bottlenose dolphin community. In: Pryor, K., and Norris, K. S., eds. *Dolphin societies: discoveries and puzzles*, pp. 199–225. University of California Press, Berkeley.

Wessnitzer, J., Adamatzky, A., and Melhuish, C. (2001). Towards self-organising structure formations: a decentralized approach. *Advances in Artificial Life*, **2159**, 573–581.

West, S. A., Griffin, A., and Gardner, A. (2006). Social semantics: altruism, cooperation, mutualism, strong reciprocity and group selection. *Journal of Evolutionary Biology*, **20**, 415–432.

West, S. A., Griffin, A. S., and Gardner, A. (2007). Evolutionary explanations for cooperation. *Current Biology*, **17**, R661–R672.

West, S. A., Pen, I., and Griffin, A. S. (2002). Cooperation and competition between relatives. *Science*, **296**, 72–75.

West-Eberhard, M. J. (1989). Phenotypic plasticity and the origins of diversity. *Annual Review of Ecology and Systematics*, **20**, 249–278.

Wey, T. W., and Blumstein, D. T. (2010). Social cohesion in yellow-bellied marmots is established through age and kin structuring. *Animal Behaviour*, **79**, 1343–1352.

Wey, T. W., and Blumstein, D. T. (2012). Social attributes and associated performance measures in marmots: bigger male bullies and weakly affiliating females have higher annual reproductive success. *Behavioral Ecology and Sociobiology*, **66**, 1075–1085.

Wey, T., Blumstein, D. T., Shen, W., and Jordán, F. (2008). Social network analysis of animal behaviour: a promising tool for the study of sociality. *Animal Behaviour*, **75**, 333–344.

Wheeler, B. C., Scarry, C. J., and Koenig, A. (2013). Rates of agonism among female primates: a cross-taxon perspective. *Behavioral Ecology*, **24**, 1369–1380.

While, G. M., Uller, T., and Wapstra, E. (2009a). Family conflict and the evolution of sociality in reptiles. *Behavioral Ecology*, **20**, 245–250.

While, G. M., Uller, T., and Wapstra, E. (2009b). Within-population variation in social strategies characterize the social and mating system of an Australian lizard, *Egernia whitii*. *Austral Ecology*, **34**, 938–949.

Whitehead, H. (1996). Babysitting, dive synchrony, and indications of alloparental care in sperm whales. *Behavioral Ecology and Sociobiology*, **38**, 237–244.

Whitehead, H. (1997). Analysing animal social structure. *Animal Behaviour*, **53**, 1053–1067.

Whitehead, H. (1998). Cultural selection and genetic diversity in matrilineal whales. *Science*, **282**, 1708–1711.

Whitehead, H. (2003). *Sperm whales: social evolution in the ocean*. University of Chicago Press, Chicago.

Whitehead, H. (2007). Learning, climate and the evolution of cultural capacity. *Journal of Theoretical Biology*, **245**, 341–350.

Whitehead, H. (2008a). *Analyzing animal societies*. University of Chicago Press, Chicago.

Whitehead, H. (2008b). Precision and power in the analysis of social structure using associations. *Animal Behaviour*, **75**, 1093–1099.

Whitehead, H., Antunes, R., Gero, S., Wong, S., Engelhaupt, D., and Rendell, L. (2012). Multilevel societies of female sperm whales (*Physeter macrocephalus*) in the Atlantic and Pacific: why are they so different? *International Journal of Primatology*, **33**, 1142–1164.

Whitehead, H., Bejder, L., and Ottensmeyer, C. A. (2005). Testing association patterns: issues arising and extensions. *Animal Behaviour*, **69**, e1–e6.

Whitehead, H., and Carlson, C. (1988). Social behaviour of feeding finback whales off Newfoundland: comparisons with the sympatric humpback whale. *Canadian Journal of Zoology*, **66**, 217–221.

Whitehead, H., and Dufault, S. (1999). Techniques for analyzing vertebrate social structure using identified individuals: review and recommendations. *Advances in the Study of Behavior*, **28**, 33–74.

Whitehead, H., and Lusseau, D. (2012). Animal social networks as substrate for cultural behavioural diversity. *Journal of Theoretical Biology*, **294**, 19–28.

Whitehead, H., Waters, S., and Lyrholm, T. (1991). Social organization of female sperm whales and their offspring: constant companions and casual acquaintances. *Behavioral Ecology and Sociobiology*, **29**, 385–389.

Whitehead, H., and Weilgart, L. (1991). Patterns of visually observable behavior and vocalizations in groups of female sperm whales. *Behaviour*, **118**, 275–296.

Whiten, A., and Byrne, R. W. (1997). *Machiavellian intelligence II: extensions and evaluations*. Cambridge University Press, Cambridge.

Whiten, A., Goodall, J., McGrew, W. C., Nishida, T., Reynolds, V., Sugiyama, Y., et al. (1999). Cultures in chimpanzees. *Nature*, **399**, 682–685.

Wilbur, A. K., Engel, G. A., Rompis, A., Putra, I. A., Lee, B. P. H., Aggimarangsee, N., et al. (2012). From the mouths of monkeys: detection of *Mycobacterium tuberculosis* complex, DNA from buccal swabs of synanthropic macaques. *American Journal of Primatology*, **74**, 676–686.

Wilbur, H. M. (1980). Complex life cycles. *Annual Review of Ecology and Systematics*, **11**, 67–93.

Wiley, D., Ware, C., Bocconcelli, A., Cholewiak, D., Friedlaender, A., Thompson, M., et al. (2011). Underwater components of humpback whale bubble-net feeding behaviour. *Behaviour*, **148**, 575–602.

Wilkinson, A., Kuenstner, K., Mueller, J., and Huber, L. (2010). Social learning in a non-social reptile (*Geochelone carbonaria*). *Biology Letters*, **6**, 614–616.

Wilkinson, D. (2012). *Stochastic modelling for systems biology*, 2nd edn. Chapman and Hall/CRC Mathematical and Computational Biology Series, Boca Raton, FL.

Wilkinson, G. S. (1984). Reciprocal food sharing in the vampire bat. *Nature*, **308**, 181–184.

Williams, R., and Lusseau, D. (2006). A killer whale social network is vulnerable to targeted removals. *Biology Letters*, **2**, 497–500.

Williams, T. M. (1999). The evolution of cost effective swimming in marine mammals: limits to energetic optimization. *Philosophical Transactions of the Royal Society B: Biological Sciences*, **354**, 193–201.

Wilson, A. D. M., and Krause, J. (2012a) Metamorphosis and animal personality: a neglected opportunity. *Trends in Ecology and Evolution*, **27**, 529–531.

Wilson, A. D. M., and Krause, J. (2012b) Personality and metamorphosis: is behavioral variation consistent across ontogenetic niche shifts? *Behavioral Ecology*, **23**, 1316–1323.

Wilson, A. D. M., Krause, S., Dingemanse, N. J., and Krause, J. (2013). Network position: a key component in the characterization of social personality types. *Behavioral Ecology and Sociobiology*, **67**, 163–173.

Wilson, A. D. M., Krause, S., James, R., Borner, K., Clement, R., Ramnarine, I. W., et al. (2014). Dynamic social networks in guppies (*Poecilia reticulata*). *Behavioral Ecology and Sociobiology*, **68**, 915–925.

Wilson, A. D. M., and McLaughlin, R. L. (2007). Behavioural syndromes in brook charr, *Salvelinus fontinalis*: prey-search in the field corresponds with space use in novel laboratory situations. *Animal Behaviour*, **74**, 689–698.

Wilson, D. S. (1975). A theory of group selection. *Proceedings of the National Academy of Sciences*, **72**, 143–146.

Wilson, D. S. (2008). Social semantics: toward a genuine pluralism in the study of social behaviour. *Journal of Evolutionary Biology*, **21**, 368–373.

Wilson, E. O. (1975). *Sociobiology: the new synthesis*. Harvard University Press, Cambridge, MA.

Wilson, E. O. (2005). Kin selection as the key to altruism: its rise and fall. *Social Research: An International Quarterly*, **72**, 1–8.

Wilson, E. O., and Holldobler, B. (1988). Dense heterarchies and mass communication as the basis of organization in ant colonies. *Trends in Ecology and Evolution*, **3**, 65–68.

Wirtu, G., Pope, C. E., Vaccaro, J., Sarrat, E., Cole, A., Godke, R. A., et al. (2004). Dominance hierarchy in a herd of female eland antelope (*Taurotragus oryx*) in captivity. *Zoo Biology*, **23**, 323–333.

Wiszniewski, J., Allen, S. J., and Möller, L. M. (2009). Social cohesion in a hierarchically structured embayment population of Indo-Pacific bottlenose dolphins. *Animal Behaviour*, **77**, 1449–1457.

Wiszniewski, J., Lusseau, D., and Möller, L. M. (2010). Female bisexual kinship ties maintain social cohesion in a dolphin network. *Animal Behaviour*, **80**, 895–904.

Wiszniewski, J., Brown, C., and Möller, L. M. (2012). Complex patterns of male alliance formation in a dolphin social network. *Journal of Mammalogy*, **93**, 239–250.

Witte, K., and Nöbel, S. (2011). Learning and mate choice. In: Brown, C., Laland K., and Krause, J., eds. *Fish cognition and behavior*, pp. 70–95. Wiley-Blackwell, Oxford.

Wittemyer, G., Douglas-Hamilton, I., and Getz, W. M. (2005). The socioecology of elephants: analysis of the processes creating multitiered social structures. *Animal Behaviour*, **69**, 1357–1371.

Wittig, R. M., Crockford, C., Lehmann, J., Whitten, P. L., Seyfarth, R. M., and Cheney, D. L. (2008). Focused grooming networks and stress alleviation in wild female baboons. *Hormones and Behavior*, **54**, 170–177.

Wohlfeil, C. K., Leu, S., Godfrey, S., and Bull, C. M. (2013). Testing the robustness of transmission network models to predict ectoparasite loads. One lizard, two ticks and four years. *International Journal for Parasitology*, **2**, 271–277.

Wolf, J. B., Brodie, E. D., III, and Moore, A. (1999). Interacting phenotypes and the evolutionary process. II. Selection resulting from social interactions. *American Naturalist*, **153**, 254–266.

Wolf, J. B. W., Mawdsley, D., Trillmich, F., and James, R. (2007). Social structure in a colonial mammal: unravelling hidden structural layers and their foundations by network analysis. *Animal Behaviour*, **74**, 1293–1302.

Wolf, J. B. W., Traulsen, A., and James, R. (2011). Exploring the link between genetic relatedness r and social contact structure k in animal social networks. *The American Naturalist*, **177**, 135–142.

Wolf, J. B. W., and Trillmich, F. (2008). Kin in space: social viscosity in a spatially and genetically substructured network. *Proceedings of the Royal Society B: Biological Sciences*, **275**, 2063–2069.

Wolf, M., van Doorn, G. S., Leimar, O., and Weissing, F. J. (2007). Life-history trade-offs favour the evolution of animal personalities. *Nature*, **447**, 581–584.

Wolf, M., and Weissing, F. J. (2012). Animal personalities: consequences for ecology and evolution. *Trends in Ecology and Evolution*, **27**, 452–461.

Wood, A. J. (2010). Strategy selection under predation; evolutionary analysis of the emergence of cohesive aggregations. *Journal of Theoretical Biology*, **264**, 1102–1110.

Woolhouse, M. E. J., Dye, C., Etard, J. F., Smith, T., Charlwood, J. D., Garnett, G. P., et al. (1997). Heterogeneities in the transmission of infectious agents: implications for the design of control programs. *Proceedings of the National Academy of Sciences*, **94**, 338–342.

Wursig, B., and Wursig, M. (1977). The photographic determination of group size, composition, and stability of coastal porpoises (*Tursiops truncatus*). *Science*, **198**, 755–756.

Yasukawa, K., and Bick, E. I. (1983). Dominance hierarchies in dark-eyed juncos (*Junco hyemalis*): a test of a game-theory model. *Animal Behaviour*, **31**, 439–448.

Yeager, C. P. (1990). Proboscis monkey (*Nasalis larvatus*) social organization: group structure. *American Journal of Primatology*, **20**, 95–106.

Yeung, K., Dombek, K., Lo, K., Mittler, J., Zhu, J., Schadt, E., et al. (2011). Construction of regulatory networks using expression time-series data of a genotyped population. *Proceedings of the National Academy of Sciences*, **108**, 19436–19441.

Yurk, H., Barrett-Lennard, L., Ford, J. K. B., and Matkin, C. O. (2002). Cultural transmission within maternal lineages: vocal clans in resident killer whales in southern Alaska. *Animal Behaviour*, **63**, 1103–1119.

Zahavi, A. (1979). Why shouting? *American Naturalist*, **113**, 155–156.

Zentall, T. R., Sutton, J. E., and Sherburne, L. M. (1996). True imitative learning in pigeons. *Psychological Science*, **7**, 343–346.

Zhang, P., B. Li, B. G., Qi, X. G., MacIntosh, A. J., and Watanabe, K. (2012). A proximity-based social network of a group of Sichuan snub-nosed monkeys. *International Journal of Primatology*, **33**, 1081–1095.

Zhou, W.-X., Sornette, D., Hill, R. A., and Dunbar, R. I. M. (2005). Discrete hierarchical organization of social group sizes. *Proceedings of the Royal Society B: Biological Sciences*, **272**, 439–444.

Zimmer, W. M. X., Johnson, M. P., Madsen, P. T., and Tyack, P. L. (2005). Echolocation clicks of free-ranging Cuvier's beaked whales (*Ziphius cavirostris*). *The Journal of the Acoustical Society of America*, **117**, 3919–3927.

Zohdy, S., Kemp, A. D., Durden, L. A., Wright, P. C., and Jernvall, J. (2012). Mapping the social network: tracking lice in a wild primate (*Microcebus rufus*) population to infer social contacts and vector potential. *BMC Ecology*, **12**, 4.

Index

A

Acanthiza spp. 179
Accipiter cooperi 90
activity 22, 37, 53, 55–56, 137, 164, 205
actor 5, 13, 24, 62
adjacency matrix 5–7, 63, 64, 68–71
Aetobatus narinari 151, 152, 154
affiliations 74, 80, 82, 83, 190
African
  buffalo 105
  elephant 76, 148, 172, 192–193
Agelaius phoeniceus 90
aggregation 92, 150, 193, 199–200
aggression 22, 53, 55–56, 59, 89, 90, 105, 111–113, 116–121, 125, 126, 130, 152, 158, 180, 181, 187
alarm call 88, 90–91, 92
Alcelaphus buselaphus 185
algorithm
  clustering 185, 190, 195
  community finding 189, 193
  genetic 80
  Metropolis Hastings 45
  ranking 63, 65–66
alliance 62, 118, 140, 147, 178
allogrooming, *see* grooming
altruism 154, 155
Amazon molly 150, 158
anglerfish 150
Apis meliferra 161–168
Apodemus flavicollis 97, 124
aquaculture 151, 156, 158, 207
Aquarius remigis 55
arc, *see* edge
arms race 86
Assamese macaque 137
association
  data 6, 8, 126, 128, 132, 137, 186, 193
  index 6, 130, 140, 142, 185, 187, 193–194
  pattern 57, 123, 124, 144, 151, 156, 173, 198
assortative mixing 35
asssortment 7, 8, 9, 13–23, 24, 32–36, 57, 76, 118, 123, 129, 151, 153, 211, 212
Astatotilapia burtoni 89
Ateles geoffroyi 120, 130
Atlantic cod 150
audience 54, 89
authority score 7, 71
aye-aye 125

B

baboon 20, 56, 88, 91, 119–120, 125–127, 130, 133, 136–137
badger, European 96, 103, 104, 106, 109
baleen whale 140
bats 20, 85, 87, 118, 120, 124, 175
Bayesian approach 38–52
Bechstein's bat 124
behaviour
  aggressive 22, 53, 55–56, 59, 89, 90, 105, 111–113, 116–121, 125, 126, 130, 152, 158, 180, 181, 187
  cooperative 3, 9, 13–23, 129, 152, 155, 211
behavioural syndrome 22, 53, 170
Belding's ground squirrel 124
betweenness 7, 54, 56, 65, 71, 77, 97, 98, 106, 107, 114, 117, 130, 140, 178, 186, 191, 195
biologging 151, 213
bipartite networks 32, 34, 181
black-capped chickadee 88
blacktip reef shark 152, 154
blue tit 39
body size 18, 26, 77, 151, 184, 203
boldness 18, 53, 56–57, 186
Bolitotherus cornutus 37, 55
bond models 71
bootstrap methods 133
Bos taurus 79, 111, 116, 120
bottlenose dolphin 77, 92, 118, 140, 143–144, 146, 147, 149
boundary effects 100
brokerage 54, 131, 145, 148
brushtail possum 97, 98, 107, 114, 124

C

Calidris alba 173
canine distemper virus 97
Capreolus capreolus 85
Carcharhinus melanopterus 152, 154
Carpodacus mexicanus 27, 56, 178, 212
Catharus fuscescens 88
catshark 152, 155
census 77, 189
  dyad 7
  triad 7, 62, 63, 65
centrality 6, 7, 36, 37, 55, 56–57, 62, 69, 71, 101, 106–107, 113, 114, 117, 119, 120, 121, 128, 129, 130, 131, 134, 135, 138, 144, 177, 178, 186, 189, 192, 194
  alpha 71
  betweenness centrality 106, 114, 178, 186
  Bonacich 71
  degree 7, 106–107, 114, 120, 186
  eigenvector 7, 71, 128, 130, 134, 144, 186, 189, 192
  group closeness centrality 135, 186
  information centrality 177
  Katz centrality 7
  node 6, 62, 71, 119
Cephalophus spp. 184
Cervus canadensis 97
Cervus elaphus 96
Cetacea 38, 126, 139–149, 151
chacma baboon 133
chicken 25, 111, 116
chimpanzee 92, 137
Chiroxiphia linearis 56, 58, 177
Chlorocebus pygerythrus 38, 91
cichlid 89, 92, 150, 152, 155, 158
clan 142, 147
cleaner wrasse 155

cliques, *see* clustering coefficient; community
cluster analysis 128, 130
clustering coefficient 7, 35, 55, 65, 107, 120, 129, 132, 134, 135, 168
  weighted 135
cohesion, *see* group cohesion
collective
  behaviour 10, 73–83, 119, 151, 156, 158, 170, 180–181, 211
  cognition 155, 156
colony 36, 57, 160–170
  interaction network 160–170
  organisation 57, 160–170
  size 163, 164, 167, 170
communication
  networks 10, 74, 84–94, 124, 142, 147–148, 181, 201, 212
  vocal 76, 142, 143, 147
community
  structure 8, 27–28, 35, 36, 62, 93, 115, 124, 130, 131–132, 135, 140, 141–142, 144, 145, 147, 151, 153, 154, 189–191, 193, 194, 195, 203, 214
competition 17, 53, 121, 140, 184, 185, 193, 197
  sexual 24–38, 144, 180, 181
component 6
connections 5
connectivity 16, 22, 56, 58, 59, 116, 145, 148, 153, 161–168, 178, 186, 191, 192, 193, 202
Connochaetes taurinus 185
contest matrix 63–71
cooperation 3, 9, 13–23, 129, 152, 155, 211
Cooper's hawk 90
correlation coefficient 34
cow 79, 111, 116, 120
crayfish 89
Crenicichla frenata 155
crested macaque 130
crickets 88
crocodilians 199, 203, 207
Crocuta crocuta 123
culture 38, 144, 146, 155
cuticular hydrocarbon 162
Cyanistes caeruleus 39

D

Damaliscus lunatus 185
damselfish 59, 150, 158
damselfly 26
Danio rerio 152, 158
data
  collection 4, 83, 96, 98, 103–104, 105, 107, 109, 138, 173, 213
  missing 80, 81, 100, 106, 140, 186
Daubentonia madagascariensis 125
David deer 195
deer mouse 97
degree
  correlation 34
  distribution 7, 16, 106, 107, 108, 114, 131, 153, 162, 163, 168
  weighted 7, 163
dendrogram 128
density
  network 7, 131, 132, 134, 158, 168, 178, 190
  population 11, 26, 116, 161
design, *see* experimental design
Desmodus rotundus 20
devil facial tumour disease 97, 101, 103, 105, 124
diameter (of a network) 190
diet preference 36, 180
diffusion (in networks) 10, 38–52, 146, 178, 211
dik-dik 184
disease transmission 95–110, 113–121, 124
dispersal 17, 24, 53, 57, 58, 182
distribution
  contact 97, 106
  degree 7, 16, 106, 107, 108, 114, 131, 153, 162, 163, 168
  exponential 41
  gamma 45
  MAN 7, 8
  negative binomial distribution 97, 99
  non-random 17–18
  posterior 41, 45, 46, 48, 50
  power law 76
  prior 41, 44, 51
  resource 24, 116, 168–169, 184
  spatial 14, 17, 35, 93
  uniform 51
division of labour 160–170
dolphin 77, 92, 118, 140, 143–144, 146, 147, 149
domestic animals 109, 111–121, 194
dominance 55, 56, 57, 59, 61–72, 80, 83, 89, 92, 107, 116–117, 121, 152, 155, 164, 169, 172, 181, 189, 193, 201
duiker 184
dyad 5, 7, 8, 54, 61, 62, 63, 65, 70, 74, 84, 119, 121, 126, 133, 142, 148, 151, 172, 182, 185, 200, 203
dynamics 5, 9, 10, 11, 13, 15, 16, 17, 20, 23, 25, 53, 54, 55, 58, 59, 60, 62, 63, 71, 72, 73, 74, 75, 76, 77, 79, 80, 81, 82, 93, 95, 99, 101, 104, 105, 107, 108, 109, 111, 116, 117, 118, 119, 124, 139, 140, 142, 144, 146, 147, 149, 150, 156, 162, 165, 166, 168, 170, 178, 182, 184, 185, 187, 188, 189, 190, 191, 193, 195, 211, 212, 213

E

eagle ray 151, 152, 154
Eastern
  chipmunk 88
  gorilla 125
  kangaroo 124
  kingbird 90
eavesdropping 86, 87–89, 93
ectothermy 199
edge
  directed 5, 71, 78, 116
  undirected 6, 7, 74, 78, 116, 132, 187
  weighted 5, 6, 7, 34, 63–64, 74, 100–101, 106, 133, 134, 135, 140–142, 163, 175, 182, 186, 188
Egernia
  stokesii 198
  striolata 199, 200, 205
eigenvector 7, 56, 71, 128, 130, 134, 144, 182, 186
Eimeria vermiformis 108
Elaphurus davidianu 195
elasmobranch 151, 152, 154
elephant 145, 192–193, 194, 195
  African 76, 148, 172, 185, 192–193
  Asian 92, 192–193
elk 97
elo rating 62–67, 71
encounternet 60, 103, 174, 176, 177
epidemics 35, 95, 96, 107
epidemiology 10, 96, 101, 105, 109, 110
Equus
  africanus 188
  burchelli 188, 190, 192, 193, 194
  grevyi 185, 188, 189, 190, 191
  hemionus khur 188
  quagga 185
Erdős-Rényi random graph 8
Esox lucius 150
Eurosta solidaginis 88
experimental design 149, 180
exponential random graph models 8, 72, 214

F
familiarity 60, 81, 94, 154–155
fecundity 25, 55
female choice 25
fiddler crab 88
field vole 88
fish 19, 56–57, 150–159
fission-fusion systems 57, 60, 75, 76, 77, 118, 124, 138, 140, 147, 153, 156, 157, 181, 188–189
fitness 13, 14, 15, 22, 24–27, 32, 54–56, 58–60, 77, 79, 80, 94, 124, 177, 178, 179, 182, 189, 192, 197
flamingo 92
flow 7, 65, 71
forked fungus beetle 37
friendship networks 127
fringe-lipped bat 87
Fugu rubripes 158

G
Gadus morhua 150
Galápagos sea lion 36, 123
Gallus gallus 25, 111, 116
gambit of the group 76, 77, 143, 148, 174, 175, 176, 181, 186, 203
Gambusia affinis 57
Gambusia holbrooki 154, 213
game theory 184
Gasterosteus aculeatus 56, 150, 152, 153, 155, 158
Gazella granti 185
Gazella rufifrons 185
gelada baboon 125, 130
geodesic distance 6, 71
Gerris odontogaster 26
gidgee skink 199, 200, 205
giraffe 185, 193–194, 195
Giraffa camelopardalis 185, 193–194, 195
Gorilla
  beringei 125
  gorilla 125
GPS 96, 104, 174, 176, 177, 186, 194, 195, 198, 205, 206, 213
Grant's gazelle 185
graphs
  random 8, 72, 132, 214
  regular 15
graph theory 14, 15, 15, 17, 22, 64
great tit 39, 56, 87, 137, 172, 174, 178
Grevy's zebra 185, 188, 189, 190, 191
grooming 92, 105, 113, 118, 119, 120, 121, 123, 125–138, 187, 192, 202
ground squirrel 85, 91, 114, 115, 124
group cohesion 57, 73, 75
group size 11, 26, 82, 107, 114, 116, 132, 134, 143, 145, 150, 172, 182, 200
growth 58, 158, 203, 212
Gryllodes supplicans 88
guppy 13, 16, 57, 74, 75, 82, 150, 153–154, 156, 157
Gyrodactylus 107

H
harbour seal 88
half-weight index 140, 142, 188
hantavirus 97, 103, 105
hartebeest 185
harvester ant 162, 163, 164
Heligmosomoides polygyrus 108
helminth 96, 97, 103
Hemidactylus turcicus 88
hermaphrodite 29
heterarchy 161, 165
hidden Markov field 8
hierarchy 36, 55, 59, 61–73, 80, 83, 89, 92, 93, 107, 116, 117, 135, 155, 169, 172, 181, 187, 188, 192, 193, 195
HIV 95, 100
homeostasis 162, 166
homophily 7
Homo sapiens 7, 8, 11, 16, 17, 20, 21, 23, 32, 35, 73, 78, 79, 80, 93, 95, 96, 100, 101, 109, 113, 125, 145, 146, 147, 153, 164, 182, 203, 214
honeybee 161–168
horse 116, 188, 192
host-parasite interaction 95, 96, 98, 99, 107, 109, 169
house
  finch 27, 56, 178, 212
  gecko 88
  mouse 111
hub score 7, 70, 71
human 7, 8, 11, 16, 17, 20, 21, 23, 32, 35, 73, 78, 79, 80, 93, 95, 96, 100, 101, 109, 113, 125, 145, 146, 147, 153, 164, 182, 203, 214
humpback whale 38, 39, 140, 146, 147
hyena 123
hypothesis testing 171, 187
hyrax 194

I
identification of individuals 27, 36, 57, 60, 103, 111, 114, 143, 151–152, 174, 179, 204, 213
immune response 108, 109, 119
in-degree 7, 99, 106, 114, 117, 168, 187
infection 35, 36, 95–110, 113–115, 168, 202
information transmission 38–52, 84–94, 164
insects 160–169
inspecting predators 13, 16, 18, 20, 54, 57, 151, 153, 154
interaction
  directed 5, 71, 78, 116
  pair-wise 5, 7, 8, 54, 61, 62, 63, 65, 71, 74, 84, 119, 121, 126, 133, 142, 148, 151, 172, 182, 185, 200, 203
  undirected 6, 7, 74, 78, 116, 132, 187
  weighted 5, 6, 7, 34, 63–64, 74, 100–101, 106, 133, 134, 135, 140–142, 163, 175, 182, 186, 188
internet 71, 96, 176
Ischnura elegans 26

J
Japanese macaque 97, 106, 130, 135

K
kangaroo 124
killer whale 88, 141, 142, 144, 145, 147, 148
kinship 13, 59, 118, 119, 155, 193, 194, 211
klip-springer 184
Kobus ellipsiprymnus 185
Kryptolebias marmoratus 150, 158

L
Labroides dimidiatus 155
leadership 53, 77, 79, 152, 156, 181, 186, 196
leaf roosting bat 118, 124
learning
  asocial 38–52
  social 38, 40, 129, 131, 144, 146, 155, 158, 197, 207
least weasel 88
lek 28, 29, 30, 31, 92, 178, 185
lemon shark 152
Lemur catta 39
Leucaspius delineatus 152, 154
life history 58, 59, 60, 110, 150, 170, 171, 172, 179
likelihood function 40, 43, 44
link 5, 16, 28, 91, 92, 99, 119, 124, 128, 129, 135, 142, 144, 154, 193. *see also* edge
lion 97, 103, 123
Liopholis slateri 205
local structure 26

logistic regression 72
longitudinal network 118
long-tailed manakin 56, 58, 177
Loxondonta africana 76, 148, 172, 185, 192–193

M
Macaca
assamensis 137
fuscata 97, 106, 130, 135
mulatta 74, 92, 116, 127
nemestrina 57, 92, 118, 129
nigra 130
tonkeana 74, 130
Macropus giganteus 124
Madame Berthe's mouse lemur 125
Madoqua spp. 184
mandrill 132
mangrove killifish 150, 158
Manta birostris 152
manta ray 152
Mantel test 8
marking techniques 152
Markov chain 40, 41, 45, 49
mark-recapture 103, 107
Marmota flaviventris 59, 92, 124
marsh tit 39
mate choice 26, 89, 108, 151, 156, 178, 179, 180
mating
behaviour 9, 25, 34, 35, 169, 185
success 24, 30, 32, 35, 37
system 9, 28, 35–36, 150, 171, 178, 179
matriline 118, 126, 131, 135, 141, 142, 146, 147
matrix
adjacency 5–7, 63, 64, 68–71
maximum likelihood 40, 43
medaka 158
meerkats 13, 38, 101–102, 103, 105, 107, 123
Megaptera novaeangliae 38, 39, 140, 146, 147
Meles meles 96, 103, 104, 106, 109
Microcebus berthae 125
Microtus agrestis 88
Microtus rossiae meridionalis 88
migration 73, 131
missing data 80, 81, 100, 106, 140, 186
modularity 8, 21, 112, 114, 115, 116, 118, 120, 121, 131, 132, 135, 146, 188
monkeys 125–138
monogamy 35, 150, 184
Monte Carlo 40, 41, 45, 46, 49, 78
mosquitofish 57, 152, 153, 213
motif 7, 63, 164
multiple networks 119
Mus musculus 111
Mustela nivelis 88
mutualism 13
Mycobacterium bovis 96, 97, 98, 114
Myotis bechsteinii 124
Myotis septentrionalis 124
Mysticeti 140

N
natal philopatry 146
navigation 80, 179
nearest neighbour 15, 27, 39, 74, 118, 156, 157, 187, 192
Negaprion brevirostris 152
nematode 97, 108, 202
neo-cortex 132
Neolamprologus pulcher 92, 152
Neotragus spp. 184
network
based diffusion analysis (NBDA) 39
bipartite 32, 34, 181
boundary 100
community 8, 27–28, 35, 36, 62, 93, 115, 124, 130, 131–132, 135, 140, 141–142, 144, 145, 147, 151, 153, 154, 189–191, 193, 194, 195, 203, 214
comparison 118
component 6
composition 57
cumulative 77
density 7, 131, 132, 134, 158, 168, 178, 190
dynamics 72, 107
ego-centric 157
measures 5–8, 28, 36–37, 99, 105–106, 107, 111, 113, 114, 115, 116, 117, 121, 129, 130, 132, 133, 134, 135, 137, 144, 145, 175, 176
motifs 7, 63, 164
neighbour 20
pattern 7, 72, 111, 116, 117, 155
position 36, 37, 53–60, 77, 112, 123, 178, 180, 201, 212
process 60
random 8, 132
scale free 8, 106
small-world 164
sparse 33, 67, 71, 132, 133, 169, 177, 197, 212
static 5, 15, 105, 169, 189, 190
structure 5–8
temporal 5, 213
time-ordered 105, 170
weighted 5, 6, 7, 34, 63–64, 74, 100–101, 106, 133, 134, 135, 140–142, 163, 175, 182, 186, 188
network measures
global 5–8
local 5–8
Nipah virus 109
node
attribute 5–8
degree 32, 34, 191
label 8
measures 7–8
size 5
norovirus 109
northern long-eared bat 124
null
hypothesis 8, 214
models 8, 128

O
oarfish 150
Odontoceti 140
olfaction 81, 162, 197, 200
onager 188, 190–191
ontogeny 54, 58, 59
operational sex ratio 35
Orcinus orca 88, 141, 142, 144, 145, 147, 148
Order of Acquisition Diffusion Analysis (OADA), 39
Oreotragus oreotragus 184
Oryctolagus cuniculus 97
Oryzias latipes 158
outcome matrix 63–68
out-degree 7, 106, 114, 117, 158, 168, 187
ovenbird 88

P
pair-wise interactions 5, 7, 8, 54, 61, 62, 63, 65, 71, 74, 84, 119, 121, 126, 133, 142, 148, 151, 172, 182, 185, 200, 203
Pan troglodytes 92, 137
Panthera leo 97, 103, 123
Papio
anubis 20, 120
cynocephalus 133
hamadryas ursinus 88
parasite transmission 95, 96, 101, 107, 152, 158, 198, 199, 201, 202
parasitoid 35, 93, 94
parental care 86, 124, 146, 150, 199
Parus major 39, 56, 87, 137, 172, 174, 178
passive integrated transponder 137, 152, 158, 174, 175, 207

paternity 25, 92
path geodesic 6, 71
path length 6, 128, 131, 132, 144, 164, 188
permutation test 8, 102, 141. *see also* randomization
Peromyscus maniculatus 97
personality 10, 18, 22, 53–60, 108, 112, 130, 152, 153, 170, 178, 182, 186, 191, 196, 198, 200, 201, 212
phenotypic assortment 27
pheromone 88, 162, 166, 169
Phoca vitulina 88
Phoenicopterus spp 92
Photocorynus spiniceps 150
phylogenetic comparisons 126, 131, 156
pig 96, 111, 116
pigeon 79, 80, 83, 156, 181
pig-tailed macaque 57, 92, 118, 129
pike cichlid 155
Pipra filicauda 56, 92
plains zebra 188, 190, 192, 193, 194
plants 29, 32, 37, 88, 93–94
plasticity 22, 161
platyfish 158
Poecile atricapillus 88
Poecile palustris 39
Poecilia
  formosa 150, 158
  reticulata 13, 16, 57, 74, 75, 82, 150, 153–154, 156, 157
Poisson random graph 8
Polybia occidentalis 167
pollen 29, 32, 166
polyandry 25, 30, 34, 150, 184
polygamy 35
Pomacentridae 150
population structure 29, 30, 146, 151
power 7
power law distributions 76
predation 13, 18, 53, 73, 77, 82, 93, 140, 145, 150, 184, 188, 189, 194, 199, 205
primates 125–138
priors 44–45, 51
Procambarus clarkii 89
Procavia capensis 194
Procyon lotor 123
promiscuity 30, 35
protozoa 108
proximity logger 96, 103, 152, 153, 158, 159, 174
pygmy bluetongue lizard 198, 199, 202, 204, 206

Q
Q, *see* modularity
quadratic assignment procedure 8

R
rabbit, European 97
racoon 123
radio
  frequency identification tags 137, 213
  tracking 103, 107, 176, 177, 178, 207
randomization 8, 15, 29, 65, 66, 67, 68, 76, 99, 114, 132, 163, 170, 173, 187, 191, 198, 199, 202, 214
random network 8, 132
ranking algorithm 63–68
rat 22, 111
Rattus norvegicus 22, 111
reach 7, 56, 129, 132, 186, 195
reachability 116, 117
reality mining 23, 125, 213
receiver 5, 13, 74, 86–93, 106, 127, 148, 152, 158, 167, 176, 181
reciprocity 13–23
recognition individual 20, 23, 59, 60, 80, 124, 154, 195, 213
redbreasted nuthatch 88
red deer 96
red junglefowl 25
red-winged blackbird 90
Regalecus glesne 150
reproductive
  state 184, 188, 194, 195
  success, *see* fitness
resampling 100
Reversible Jump Markov Chain Monte Carlo (RJMCMC) 45
rhesus monkey 74, 92, 116, 127
Rhincodon typus 151
Rhinopithecus roxellana 130
rhynchocephalian 103, 198, 199, 202, 205
ring-tailed lemur 39
roe deer 85
Ropalidia marginata 57
Rutilus rutilus 78

S
Salmo salar 117, 145, 152, 158, 202
salmon 117, 145, 152, 158, 202
Salmonella 202
sample size 96–99, 103, 107, 132, 174, 175, 177, 179, 203
sanderling 173
Sarcophilus harrisii 97, 101, 103, 105, 124
scale-free network 8, 106
Scyliorhinus canicula 152, 155
sea-lion 36, 123
seasonality 102, 105, 109, 178
Seiurus aurocapilla 88
selection
  gradient 25, 31
  pressure 26, 86, 87, 88, 89, 178
  sexual 24–37, 123, 150, 151, 158, 180, 211
self organisation 61
sender 84–94
sex
  ratio 35, 112
sexual
  harassment 185, 192
  selection 24–37, 123, 150, 151, 158, 180, 211
sexually transmitted diseases 35, 36, 101, 103, 180
sheep 104, 111, 114
shoaling 18, 57, 73, 74, 78, 82, 150, 151–155
shortest path 71, 106, 186, 189, 190
shyness 153
signalling 10, 84–94, 119, 130, 200, 201, 212
site fidelity 154
Sitta canadensis 88
Slater's skink 205
sleepy lizard 198, 199, 201, 205, 206, 207
small-world network 164
snub-nose monkey 130, 132
social
  insects 160–169
  learning 38, 40, 129, 131, 144, 146, 155, 158, 197, 207
  niche 112, 118, 136, 137, 138, 179
  organization 63, 74, 124, 126, 134, 150, 160, 168, 195, 197, 198–202
  recognition 23, 60, 154
sociogram 6, 64, 126, 127, 128, 151, 211
Solidago altissima 88
sparseness 33, 67, 71, 132, 133, 169, 177, 197, 212
sperm competition 24–37, 180
Spermophilus beldingi 114, 124
Spermophilus richardsonii 91
Sphenodon punctatus 103, 198, 199, 202, 205
spider monkey 120, 130
Spix's disc-winged bat 118, 124
squamates 197, 203, 207
starling 39, 78, 181
sticklebacks 38, 39, 56, 150, 152, 153, 155, 158

Sturnus vulgaris 39, 78, 181
submission 61, 125
subordinance 62, 63, 71
sub-structures 78, 153, 195
sunbleak 152, 154
suni 184
superspreader 7, 97, 98, 99, 101, 103, 106, 109, 168
Suricata suricatta 13, 38, 101–102, 103, 105, 107, 123
Sus scrofa 96, 111, 116
Syncerus caffer 105

T
tagging 153
Tamias striatus 88
Tasmanian devil 97, 101, 103, 105, 124
TB 97–98, 101–105, 107, 109, 113
telemetry 152–153, 158, 174, 176–177, 178, 207
teleost fishes 150–159
temporal models 5, 213
territory 56, 67, 100, 104, 134, 197, 200, 201
Tetraodon nigroviridis 158
Theropithecus gelada 125, 130
Thomson's gazelle 185
thornbill 179
three-spine stickleback 38, 39, 56, 150, 152, 153, 155, 158
Thyroptera tricolor 124, 179
tick 96, 97, 202
tie strength 144
ties weak 27, 154, 166
Tiliqua adelaidensis 198, 199, 204
Tiliqua rugosa 198, 199, 201, 205, 206, 207
Time of Acquisition Diffusion Analysis (TADA) 39
tit-for-tat 20–21
toothed whale 140
topi 185
topology 5, 8, 16, 17, 32, 34, 35, 36, 74, 78, 108, 112, 144, 213
Toxoplasma gondii 108
Trachops cirrhosus 87
trail 6, 71, 88
transitivity 117, 121
transmission
  cultural 146
  network 95–109, 114
tree skink 199
triad 7, 61–65
Trichosurus vulpecula 97, 114, 124
tuatara 103, 198, 199, 202, 205
Tursiops truncatus 16, 92, 140, 144, 146, 147
Tyrannus tyrannus 90

U
Uca tangeri 88
undirected graphs 6, 7, 74, 78, 116, 132, 187
ungulates 10, 22, 113, 117, 123, 126, 184–196, 213

V
vampire bat 20
veery 88
vertex, *see* node
vervet monkey 38, 91

W
walk 6, 45, 71, 73, 164
wasp 57, 93, 163, 164, 167
waterbuck 185
water strider 26, 55
weighted network 5, 6, 7, 34, 63–64, 74, 100–101, 106, 133, 134, 138, 140–142, 163, 175, 182, 186, 188
welfare 58, 94, 111–121, 129, 151, 152, 156, 158, 212
whale shark 151, 152
wild asses 188, 190, 195
wildebeest 185
wire-tailed manakin 56, 92

X
Xiphophorus maculatus 158

Y
yellow-bellied marmot 59, 92, 124
yellow-necked mouse 97, 124

Z
Zalophus wollebaeki 36, 123
zebra 185–196
zebra fish 152, 158

Printed and bound by CPI Group (UK) Ltd, Croydon, CR0 4YY